Learning Materials in Biosciences

Learning Materials in Biosciences textbooks compactly and concisely discuss a specific biological, biomedical, biochemical, bioengineering or cell biologic topic. The textbooks in this series are based on lectures for upper-level undergraduates, master's and graduate students, presented and written by authoritative figures in the field at leading universities around the globe.

The titles are organized to guide the reader to a deeper understanding of the concepts covered.

Each textbook provides readers with fundamental insights into the subject and prepares them to independently pursue further thinking and research on the topic. Colored figures, step-by-step protocols and take-home messages offer an accessible approach to learning and understanding.

In addition to being designed to benefit students, Learning Materials textbooks represent a valuable tool for lecturers and teachers, helping them to prepare their own respective coursework.

More information about this series at ► http://www.springer.com/series/15430

Gabriela Rodrigues • Bernard A. J. Roelen

Editors

Concepts and Applications of Stem Cell Biology

A Guide for Students

Editors
Gabriela Rodrigues
Centro de Ecologia, Evolução e Alterações
Ambientais, Departamento de Biologia
Animal, Faculdade de Ciências da
Universidade de Lisboa
Lisbon, Portugal

Bernard A. J. Roelen
Embryology, Anatomy and Physiology,
Department Clinical Sciences, Faculty
of Veterinary Medicine
Utrecht Universtity
Utrecht, The Netherlands

ISSN 2509-6125   ISSN 2509-6133 (electronic)
Learning Materials in Biosciences
ISBN 978-3-030-43938-5   ISBN 978-3-030-43939-2 (eBook)
https://doi.org/10.1007/978-3-030-43939-2

© Springer Nature Switzerland AG 2020, corrected publication 2020
This work is subject to copyright. All rights are reserved by the Publisher, whether the whole or part of the material is concerned, specifically the rights of translation, reprinting, reuse of illustrations, recitation, broadcasting, reproduction on microfilms or in any other physical way, and transmission or information storage and retrieval, electronic adaptation, computer software, or by similar or dissimilar methodology now known or hereafter developed.
The use of general descriptive names, registered names, trademarks, service marks, etc. in this publication does not imply, even in the absence of a specific statement, that such names are exempt from the relevant protective laws and regulations and therefore free for general use.
The publisher, the authors and the editors are safe to assume that the advice and information in this book are believed to be true and accurate at the date of publication. Neither the publisher nor the authors or the editors give a warranty, expressed or implied, with respect to the material contained herein or for any errors or omissions that may have been made. The publisher remains neutral with regard to jurisdictional claims in published maps and institutional affiliations.

This Springer imprint is published by the registered company Springer Nature Switzerland AG
The registered company address is: Gewerbestrasse 11, 6330 Cham, Switzerland

To all our dear students, challenging them to reach new horizons in stem cell biology.

Preface

More than a decade has gone by since the Master's Programme in Evolutionary and Developmental Biology (MSc EDB) started in the Department of Animal Biology, Faculdade de Ciências da Universidade de Lisboa (Faculty of Sciences of the University of Lisbon (FCUL)) Portugal. In 2007, several colleagues within the Department joined efforts to put together a programme that would provide comprehensive courses to its students on the topic "how organisms are built, maintained and changed" by covering areas of Biology ranging from Cell and Developmental Biology and Genetics to Evolution. It was, and still is, our objective that students perceive the developmental and evolutionary processes that have shaped organismal complexity. The driving force of the MSc EDB is to inspire students and ignite their curiosity and enthusiasm regarding the biological processes that underlie the building of an organism, as well as the integration of these processes in an evolutionary perspective. This Master's Programme puts its major emphasis on understanding and connecting concepts, providing students a strong practical training and encouraging them to think and act independently, as well as to be creative in their laboratory work. Critical analysis of papers and presenting seminars covering a variety of themes are another important aspect of the study programme. The students are also challenged to expand their interests to societal aspects, such as ethics, or to biomedical issues, such as cancer and regenerative medicine. The enthusiastic coordination team of the MSc EDB covers expertise in evolution, evolutionary developmental biology and development, and one of the coordinators, Sólveig Thorsteinsdóttir, authors a chapter in this book (see ▶ Chap. 9).

The Stem Cell Biology and Technology (SCBT) course is one of the courses taught during the first year of the programme, leading to the preparation of this book authored and edited by Gabriela Rodrigues, who is also responsible for the above-mentioned course. It is expected to be an "immersive" experience where students enjoy learning and discussing stem cell biology, as well as many of the impacts of the use of stem cells in biomedical research and in society. During the second year, students enrol in their master's thesis learning to be part of a team and a lab, either indoors or outdoors, and learn to navigate the excitement and frustrations of a research project.

Several collaborators of the MSc EDB Programme have come from abroad. Bernard Roelen (from Utrecht University, Netherlands) has been a long-time colleague and friend of Gabriela Rodrigues. He kindly offered to participate in the SCBT course almost since the beginning of the programme. Susana Chuva de Sousa Lopes (from Leiden University Medical Centre, Leiden, Netherlands) is a well-known scientist in the field of primordial germ cell biology and pluripotent stem cells who annually provides an invaluable contribution to the course, including cells, reagents and protocols. Bernard Roelen and Gabriela Rodrigues are coeditors of this book as a natural outcome of their ongoing common work in the programme. Moreover, Gabriela and both Bernard and Susana, with their students, have authored several chapters of this book (see ▶ Chaps. 1, 2, 3 and 6).

Other collaborations from colleagues in Portugal, particularly from the Gulbenkian Institute of Science, the Institute of Molecular Medicine (University of Lisboa)

Preface

and the Champalimaud Foundation (Lisbon), and from several institutes and universities in Porto, Braga and Faro deserve to be mentioned and acknowledged. For the last 13 years, many researchers and professors from these institutions have been lecturers of the course. Some collaborators very generously brought practical exercises to the class lab: António Salgado (University of Minho) and Isabel Alcobia and Bruno Bernardes de Jesus (Institute of Molecular Medicine, University of Lisbon) have been contributing cells, tissues and practical protocols for the SCBT course. Bruno, who is currently in the University of Aveiro, has also authored ▶ Chap. 5. A more general list of contributors is listed in our website: ▶ http://bed.campus.ciencias.ulisboa.pt/. These collaborations, together with the ones coming from research teams working in the FCUL campus, as well as the lab managing (Marta Palma) and technical support (especially Luís Marques, FCUL Microscopy Facility) teams, have allowed the students to access high-quality teaching material and logistics.

These committed and enthusiastic groups of collaborators have challenged our students with their conceptual knowledge and at the same time urging them to critically discuss what they learned. This way of teaching has been a hallmark of the success of this course. The professional outcomes of our MSc EDB students have been rather successful, both in Portugal and abroad, either working in a lab, in a company or in an academic career. Internationalization has been a main aim of the master's and Erasmus programme that has been assisting students to travel and to gain international experience, and usually, they succeed very well in doing so.

To organize, author and edit a book of this learning series based on the classes of the SCBT course was a challenging adventure. This book involves everybody who has been contributing to this course in its 13 years of duration. Both teachers and students are part of this ongoing and inspirational project, and, because of that, they are the building blocks of this book. Gabriela Rodrigues, Bernard Roelen, Susana Chuva de Sousa Lopes, Bruno Jesus, Sólveig Thorsteinsdóttir, Diana Nascimento, Evguenia Bekman and Joana Miranda, authors of the chapters in this book, teach in the SCBT course, either directly or indirectly contributing to other courses that build the background of the students. Due to unfortunate different reasons, many of our regular collaborators (Isabel Alcobia, Claudia Lobato da Silva, António Salgado and Jorge Marques da Silva, among others) could not take part in the book but still have had important inputs in the course for many years. Even more important than the professors are the students of the SCBT course, some of them brilliant and dedicated former students are authors in this book: Maria Fernandes (▶ Chap. 3), Beatriz Gonçalves (▶ Chap. 7), André Dias and Rita Aires (▶ Chap. 8), André Gonçalves (▶ Chap. 9), Vasco Sampaio-Pinto (▶ Chap. 10), Carolina Nunes (▶ Chap. 12) and Sérgio Camões (▶ Chap. 13). Some of these talented students have themselves already taught younger colleagues in class, taking part in this never-ending cycle of transmitting fresh information produced by their own work coming out of their research projects. A special word needs to be dedicated to Ana Bernardo, who has not only been a student of the MSc EDB but later went to do her PhD in Leiden in Susana Chuva de Sousa Lopes' lab. She hand-drew an image inspired in each one of the subjects covered in this book that nicely decorates the front page of each chapter. The image illustrating the introductory chapter (▶ Chap. 1) represents the Biology

Building in the FCUL where we teach the course, a beautiful environment that houses so may students living intensely their academic life every day.

We expect this book to be useful to anyone who needs to learn about stem cell biology in general and about how cells, from the beginning of the morphogenesis of the embryo, give rise to a whole organism. We hope that this book will also be a helpful tool for those who want to know further about stem cell applications and for biomedical research. The simple approach used in this book makes it amenable to the academic universe worldwide. We hope to have modestly contributed with our share to reveal the mesmerizing field of stem cell biology!

Gabriela Rodrigues
Lisbon, Portugal

Bernard A. J. Roelen
Utrecht, The Netherlands

Acknowledgements

We, the editors, would like to thank all the students who participated in the Stem Cell Biology and Technology course as part of the Master's Programme in Evolutionary and Developmental Biology at the University of Lisbon. Their enthusiasm and questions have been inspirational. Also, we are extremely thankful to all teachers who helped to shape and contributed to the course. Thanks are also due to the past and present members of the Rodrigues and Thorsteinsdóttir Lab in Lisbon for their help with the course and the past and present members of the Roelen lab in Utrecht and of the Chuva de Sousa Lopes lab in Leiden, Netherlands, for their contribution, both intellectually and materially. We want to leave here a special word of appreciation to the work of our manager, Marta Palma, and of our FCUL Microscopy Unit technician, Luís Marques. Their constant "backstage" work is surely a part of the success of this course.

And without the course, there would not have been a book. We thank all the authors of the chapters for their contribution and for being patient with us. Special thanks to Ana Bernardo for providing the opening illustrations of the chapters.

Contents

1. **Introduction: A Decade Teaching Stem Cell Biology** 1
   Gabriela Rodrigues

2. **Preimplantation Development: From Germ Cells to Blastocyst** 11
   Bernard A. J. Roelen

3. **Origins of Pluripotency: From Stem Cells to Germ Cells** 29
   Maria Gomes Fernandes and Susana M. Chuva de Sousa Lopes

4. **Human Induced Pluripotent Stem (hiPS) Cells: Generation and Applications** 57
   Christian Freund

5. **Cellular Reprogramming and Aging** 73
   Sandrina Nóbrega-Pereira and Bruno Bernardes de Jesus

6. **Cloning** 93
   Bernard A. J. Roelen

7. **Stem Cells in Plant Development** 115
   Beatriz Gonçalves

8. **Axial Stem Cells and the Formation of the Vertebrate Body** 131
   André Dias and Rita Aires

9. **Skeletal Muscle Development: From Stem Cells to Body Movement** 159
   Marianne Deries, André B. Gonçalves, and Sólveig Thorsteinsdóttir

10. **Cardiac Regeneration and Repair: From Mechanisms to Therapeutic Strategies** 187
    Vasco Sampaio-Pinto, Ana C. Silva, Perpétua Pinto-do-Ó, and Diana S. Nascimento

11. **Reproducing Human Brain Development In Vitro: Generating Cerebellar Neurons for Modelling Cerebellar Ataxias** 213
    Evguenia Bekman, Teresa P. Silva, João P. Cotovio, and Rita Mendes de Almeida

12. **Neurotoxicology and Disease Modelling** 229
    Carolina Nunes and Marie-Gabrielle Zurich

13 **Mesenchymal Stem Cells for Cutaneous Wound Healing** 247
    Sérgio P. Camões, Jorge M. Santos, Félix Carvalho, and Joana P. Miranda

Correction to: Cellular Reprogramming and Aging C1

Contributors

Rita Aires   Instituto Gulbenkian de Ciência, Oeiras, Portugal

DFG-Center for Regenerative Therapies Dresden, Center for Molecular and Cellular Bioengineering (CMCB), Technische Universität Dresden Dresden, Germany
rita.aires@tu-dresden.de

Evguenia Bekman   iBB – Institute of Bioengineering and Biosciences, Instituto Superior Técnico, Universidade de Lisboa, Lisboa, Portugal

The Discoveries Centre for Regenerative and Precision Medicine, Lisbon Campus, Porto Salvo, Portugal
evguenia.bekman@tecnico.ulisboa.pt

Sérgio P. Camões   Research Institute for Medicines (iMed.ULisboa), Faculty of Pharmacy, Universidade de Lisboa, Lisbon, Portugal
sergiocamoes@campus.ul.pt

Félix Carvalho   UCIBIO, REQUIMTE, Laboratory of Toxicology, Department of Biological Sciences, Faculty of Pharmacy, Universidade do Porto, Oporto, Portugal
felixdc@ff.up.pt

Susana M. Chuva de Sousa Lopes   Department of Anatomy and Embryology Leiden University Medical Center, Leiden, The Netherlands

Department for Reproductive Medicine Ghent University Hospital, Ghent, Belgium
lopes@lumc.nl

João P. Cotovio   iBB – Institute of Bioengineering and Biosciences, Instituto Superior Técnico, Universidade de Lisboa, Lisboa, Portugal
joaocotovio@tecnico.ulisboa.pt

Rita Mendes de Almeida   iMed.ULisboa – Research Institute for Medicines, Faculty of Pharmacy, Universidade de Lisboa, Lisboa, Portugal
rmalmeida@farm-id.pt

Bruno Bernardes de Jesus   Department of Medical Sciences and Institute of Biomedicine – iBiMED, University of Aveiro, Aveiro, Portugal
brunob.jesus@ua.pt

Marianne Deries   Centro de Ecologia, Evolução e Alterações Ambientais, Departamento de Biologia Animal, Faculdade de Ciências da Universidade de Lisboa, Lisboa, Portugal
mederies@fc.ul.pt

André Dias   Instituto Gulbenkian de Ciência, Oeiras, Portugal
amdias@igc.gulbenkian.pt

Maria Gomes Fernandes Central Laboratory Animal Facility Leiden University Medical Center, Leiden, The Netherlands
maria@meatable.com; marymgf@gmail.com

Christian Freund, PhD Department of Anatomy & Embryology, LUMC hiPSC Hotel, Leiden University Medical Center, Leiden, The Netherlands
c.m.a.h.freund@lumc.nl

André B. Gonçalves Centro de Ecologia, Evolução e Alterações Ambientais, Departamento de Biologia Animal, Faculdade de Ciências da Universidade de Lisboa, Lisboa, Portugal
andre.b.goncalves88@hotmail.com

Beatriz Gonçalves John Innes Centre, Norwich Research Park, Colney Lane, Norwich, UK
beatriz.pinto-goncalves@jic.ac.uk

Joana P. Miranda Research Institute for Medicines (iMed.ULisboa), Faculty of Pharmacy Universidade de Lisboa, Lisbon, Portugal
jmiranda@ff.ulisboa.pt

Diana S. Nascimento i3S – Instituto de Investigação e Inovação em Saúde, Universidade do Porto, Porto, Portugal

INEB – Instituto Nacional de Engenharia Biomédica, Universidade do Porto, Porto, Portugal

ICBAS – Instituto de Ciências Biomédicas de Abel Salazar, Universidade do Porto, Porto, Portugal
dsn@ineb.up.pt

Sandrina Nóbrega-Pereira Instituto de Medicina Molecular João Lobo Antunes, Faculdade de Medicina, Universidade de Lisboa, Lisbon, Portugal

Department of Medical Sciences and Institute of Biomedicine – iBiMED, University of Aveiro, Aveiro, Portugal
sandrina.pereira@ua.pt

Carolina Nunes Department of Biomedical Sciences, University of Lausanne, Lausanne, Switzerland

Swiss Centre for Applied Human Toxicology (SCAHT), Lausanne, Switzerland
carolinarnunes@gmail.com

Perpétua Pinto-do-Ó i3S – Instituto de Investigação e Inovação em Saúde, Universidade do Porto, Porto, Portugal
INEB – Instituto Nacional de Engenharia Biomédica, Universidade do Porto, Porto, Portugal

ICBAS – Instituto de Ciências Biomédicas de Abel Salazar, Universidade do Porto, Porto, Portugal
perpetua@ineb.up.pt

Contributors

Gabriela Rodrigues   Centro de Ecologia, Evolução e Alterações Ambientais, Departamento de Biologia Animal, Faculdade de Ciências da Universidade de Lisboa, Lisbon, Portugal
mgrodrigues@fc.ul.pt

Bernard A. J. Roelen   Embryology, Anatomy and Physiology, Department Clinical Sciences, Faculty of Veterinary Medicine, Utrecht University, Utrecht, The Netherlands
b.a.j.roelen@uu.nl

Vasco Sampaio-Pinto   i3S – Instituto de Investigação e Inovação em Saúde, Universidade do Porto, Porto, Portugal

INEB – Instituto Nacional de Engenharia Biomédica, Universidade do Porto, Porto, Portugal

ICBAS – Instituto de Ciências Biomédicas de Abel Salazar, Universidade do Porto, Porto, Portugal

Department of Cardiology, CARIM School for Cardiovascular Diseases, Faculty of Health, Medicine and Life Sciences, Maastricht University, Maastricht, The Netherlands

Department of Molecular Genetics, Faculty of Sciences and Engineering, Maastricht University, Maastricht, The Netherlands
vascomspinto@gmail.com

Jorge M. Santos   Instituto de Tecnologia Química e Biológica (ITQB), Universidade Nova de Lisboa, Oeiras, Portugal
miguel.santos@itqb.unl.pt

Ana C. Silva   i3S – Instituto de Investigação e Inovação em Saúde, Universidade do Porto, Porto, Portugal

INEB – Instituto Nacional de Engenharia Biomédica, Universidade do Porto, Porto, Portugal

ICBAS – Instituto de Ciências Biomédicas de Abel Salazar, Universidade do Porto, Porto, Portugal

Gladstone Institutes, University of California San Francisco, San Francisco, CA, USA
ana.silva@gladstone.ucsf.edu

Teresa P. Silva   iBB – Institute of Bioengineering and Biosciences, Instituto Superior Técnico, Universidade de Lisboa, Lisboa, Portugal
teresamsilva@medicina.ulisboa.pt

Sólveig Thorsteinsdóttir   Centro de Ecologia, Evolução e Alterações Ambientais, Departamento de Biologia Animal, Faculdade de Ciências da Universidade de Lisboa, Lisboa, Portugal
solveig@fc.ul.pt

Marie-Gabrielle Zurich   Department of Biomedical Sciences, University of Lausanne, Lausanne, Switzerland

Swiss Centre for Applied Human Toxicology (SCAHT), Lausanne, Switzerland
mzurich@unil.ch

Abbreviations

2i	Two small-molecule inhibitors: CHIR99021 and PD0325901	*BAM*	*BARELY ANY MERISTEM*
3R	Replacement, reduction and refinement	BARHL1	BarH like homeobox 1
		bFGF	Basic fibroblast growth factor
4i	Four small-molecule inhibitors: CHIR99021, PD0325901, SB203580 and SP600125	bHLH	Basic helix–loop–helix
		BM	Bone marrow
		BMMNCs	Bone marrow mononuclear cells
A	Anterior	BMP	Bone morphogenetic protein
Abcg-2	ATP-binding cassette superfamily G member 2	BMPRI	Bone morphogenetic protein receptor type 1
ActA	ActivinA	BOEC	Blood outgrowth endothelial cell
ActRI	Activin receptor type 1		
AF	Amniotic fluid	CALB	Calbindin
AFP	α-Fetoprotein	Cas	Castor
AKT	Protein kinase B	CAS	CRISPR associated
ALCAPA	Anomalous left coronary artery from the pulmonary artery	CBF1	C-repeat/DRE binding factor 1
		CCL	C-C motif chemokine ligand
all	Allantois	CD	Cluster of differentiation
ALS	Amyotrophic lateral sclerosis	CDKs	Cyclin-dependent kinases
ALT	Alternative lengthening of telomeres	CENP-B	Centromere protein B
		CG	Cytosine guanine
am	Amnion	Check1	Checkpoint kinase 1
AMD	Age-related macular degeneration	CK	Cytokeratin
		c-Kit	Proto-oncogene receptor tyrosine kinase
ARF	*AUXIN RESPONSE FACTOR*[1]		
ARR	*ARABIDOPSIS RESPONSE REGULATOR*	CLE	Caudal lateral epiblast
		CLE	*CLAVATA3/EMBRYO SURROUNDING REGION-RELATED*
AT	Adipose tissue		
ATOH1	Atonal homolog 1		
AVE	Anterior visceral endoderm		
		CLV	*CLAVATA*
		CM	Conditioned medium
		CMs	Cardiomyocytes
		CNH	Chordo-neural hinge
		CNS	Central nervous system
		CPCs	Cardiac progenitor cells

[1] Note: Plant genes are written in *ITALICIZED UPPERCASE*, both in long form and short form. Alleles and mutants are written in *italicized lowercase*. When referring to the protein or gene product, the gene name is written in ROMAN UPPERCASE.

Abbreviations

CRISPR	Clustered regularly interspaced short palindromic repeats	Epo	Erythropoietin
		ERK	Extracellular signal-regulated kinase
CRN	*CORYNE*	ES	Embryonic stem
CSCs	Cardiac stem cells	ESCs	Embryonic stem cells
CVD	Cardiovascular diseases	EVs	Extracellular vesicles
CVH	Chicken vasa homologue	ExE	Extraembryonic ectoderm
CX3CR1	CX3C chemokine receptor 1		
CXCL	C-X-C motif chemokine	FBS	Fetal bovine serum
CXCR	C-X-C chemokine receptor	FCS	Fetal calf serum
CZ	Central zone	FCUL	Faculdade de Ciências da Universidade de Lisboa/ Faculty of Sciences of the University of Lisbon
Dan	Distal antenna		
DDR	DNA-damage response		
Dll1	Delta-like ligand 1	FDA	Food and Drug Administration
DMD	Duchenne muscular dystrophy	FGF	Fibroblast growth factor
DMEM	Dulbecco modified Eagles minimal essential medium	FGFR	Fibroblast growth factor receptor
DMSO	Dimethylsulfoxide	FZ	Frizzled
DNT	Developmental neurotoxicity testing	GABA	Gamma-aminobutyric acid
		GBX2	Gastrulation brain homeobox 2
D-V	Dorso-Ventral		
DVL	Dishevelled	G-CSF	Granulocyte colony-stimulating factor
E	Embryonic day	GFAP	Glial fibrillary acidic protein
EB	Embryoid body	GM-CSF	Granulocyte-macrophage colony-stimulating factor
E-cad	E-cadherin		
ECM	Extracellular matrix	GP130	Glycoprotein 130
EDTA	Ethylenediamine tetraacetic acid	GSK	Glycogen synthase kinase
		GSK-3	Glycogen synthase kinase-3
EGCs	Embryonic germ cells	GSK3β	Glycogen synthase kinase 3 beta
EGF	Epidermal growth factor		
EGL	External germinal layer		
EMT	Epithelial to mesenchymal transition	H3K27me3	Histone 3 lysine 27 trimethylation
EN2	Engrailed 2	Hb	Hunchback
END-2 cell	Visceral endoderm-like cell	HBEGF	Heparin-binding endothelial growth factor
Epi	Epiblast		
EpiSCs	Epiblast stem cells	hCG	Human chorionic gonadotrophin

hES cell	Human embryonic stem cell	lb	Limb bud
hESCs	Human embryonic stem cells	LD_{50}	Median lethal dose
HGF	Hepatocyte growth factor	LIF	Leukaemia inhibitory factor
HH	Hamburger–Hamilton stage		
HHT	Hereditary hemorrhagic telangiectasia	LIFR	Leukaemia inhibitory factor receptor
HIF	Hypoxia-inducible factor	LNA	Locked nucleic acid
hiPS cell	Human induced pluripotent stem cell	lncRNAs	Long non-coding RNAs
		LPMPs	Lateral and paraxial mesoderm progenitors
hiPSCs	Human-induced pluripotent stem cells	LPS	Lipopolysaccharide
HOX	Homeobox	LRP	Low-density lipoprotein receptor-related protein
HoxPGs	Hox paralogous groups		
hPSCs	Human pluripotent stem cells	LRR	Leucine rich repeat
HSA	Heat-stable antigen	LRR-LRK	LRR-receptor-like kinases
HSC	Hematopoietic stem cells		
ht	Heart	MAP	Mitogen-activated protein
		MAP2	Microtubule-associated protein 2
I/R	Ischemia/reperfusion		
ICM	Inner cell mass	MAPK	Mitogen-activated protein kinase
IGF	Insulin-like growth factor		
IL	Interleukin	MDC1A	Merosin-deficient congenital muscular dystrophy 1A
INM	Interkinetic nuclear migration		
iPS cell	Induced pluripotent stem cell	ME	Mesendodermal
iPS	Induced pluripotent stem	MEFs	Mouse embryonic fibroblasts
iPSCs	Induced pluripotent stem cells		
IsO	Isthmic organizer	mES cell	Mouse embryonic stem cell
IVF	In vitro fertilization		
		MET	Mesenchymal-to-epithelial transition
JAK	Janus kinase		
JNKi	Jun N-terminal kinase inhibitor	MHB	Midbrain-hindbrain boundary
		MHCII	Major histocompatibility complex class II
KGF	Keratinocyte growth factor		
Klf4	Krüppel-like factor 4	MI	Myocardial infarction
Kp	Kruppel	miPS cell	Mouse induced pluripotent stem cell
kPa	Kilopascal		
		miRs/miRNAs	MicroRNAs
LAD	Left anterior descending	MMPs	Matrix metalloproteinases
LATS	Large tumor suppressor		
		MN	Motor neuron

Abbreviations

MP	*MONOPTEROS*	PAI	Plasminogen activator inhibitor
MRF	Myogenic regulatory factor	PAR3	Partitioning defective protein 3
MSc EDB	Master's Programme in Evolutionary and Developmental Biology	PAX2	Paired box gene 2
		PBMC	Peripheral blood mono-nuclear cell
MSCs	Mesenchymal stem/stromal cells	PBS	Phosphate-buffered saline
mTOR	Target of rapamycin	PCNA	Proliferating cell nuclear antigen
MuSCs	Muscle stem cells		
		PDGF	Platelet-derived growth factor
NCAD	N-Cadherin		
ncRNAs	Non-coding RNAs	PDGFR	Platelet-derived growth factor receptor
NDM	Neural differentiation medium	PDGFR-α	Platelet-derived growth factor receptor alpha
NEM	Neural expansion medium	Pdm	POU domain protein
NF-κB	Nuclear factor kappa B	PECAM	Platelet–endothelial cell adhesion molecule
NGF	Nerve growth factor		
NICD	Notch intracellular domain	PG	Prostaglandin
		PGC	Primordial germ cell
NK	Natural killer	PGCLCs	Primordial germ cells-like cells
NMP	Neuro-mesodermal progenitor		
		PI3K	Phosphatidylinositol 3-kinase
NPCs	Neural progenitor cells	PICMI	Post-ICM intermediate
NRC	National Research Council	*PIN1*	*PIN-FORMED1*
		piRISC	piRNA-induced silencing complexes
NRGN	Neurogranin		
NSB	Node-streak border	PIWIL	P-element-induced wimpy testis-like
NSCs	Neural stem cells		
NT	Neurotoxicity testing	PKB	Protein kinase B
		PLC	Phosphoinositide phospholipase C
OC	Organizing Centre		
Oct-4	Octamer-binding transcription factor 4	*PLL1*	*POLTERGEIST-LIKE*
		PLT	*PLETHORA*
OTX2	Orthodenticle homeobox 2	*POL*	*POLTERGEIST*
P	Phosphorylation	PRC	Polycomb repressive complexes
P	Posterior	PrE	Primitive endoderm
P	Postnatal day	PS	Primitive streak
p38i	p38 inhibitor	PSCs	Pluripotent stem cells

PTF1a	Pancreas-specific transcription factor 1a	*SHR*	*SHORTROOT*
PVALB	Parvalbumin	shRNAs	Short hairpin RNAs
PZ	Peripheral zone	siRNAs	Small interference RNAs
		sm	Somites
QC	Quiescent Centre	SMA	Smooth muscle actin
		SMAD	Mothers against decapentaplegic homolog
R	Receptor	Sox	(Sex-determining region Y)-box
RA	Retinoic acid	SOX1	SRY (Sex determining region Y)-box 1
RAF	Rapidly accelerated fibrosarcoma		
RAM	Root apical meristem	Sox2	Sex-determining region Y-box 2
RAS	Rat sarcoma oncogene	STAT	Signal transducer and activator of transcription
Rb	Retinoblastoma		
R-C	Rostro-caudal	*STM*	*SHOOT MERISTEMLESS*
RE cell	Renal epithelial cell	SVZ	Subventricular zone
RII	Receptor type 2		
RL	Rhombic lip	T	*Brachyury*
RM	Rib meristem	TAC	Transverse aortic constriction
ROCKi	RHO-associated protein kinase 1 inhibitor	TCF	T-cell factor
		TE	Trophectoderm
ROS	Reactive oxygen species	TERRA	Telomeric repeat-containing RNA
RPE	Retinal pigment epithelial		
RPE cell	Retinal pigment epithelial cell	TERT	Telomerase reverse transcriptase
RPK2	*RECEPTOR-LIKE PROTEIN KINASE*	TF	Transcription factor
		TGF	Transforming growth factor
		TGFβ	Transforming growth factor beta
SAM	Shoot apical meristem	TLR	Toll-like receptor
SASP	Senescence-associated secretory phenotype	TNF	Tumour necrosis factor
		TNFR	Tumour necrosis factor receptor
Sca-1	Stem cell antigen 1	TrE	Transposable elements
SCBT	Stem cell biology and technology	TSC	Trophoblast stem cell
SCF	Stem cell factor	TSG	Tumour necrosis factor-stimulated gene
SCNT	Somatic cell nuclear transfer		
SCR	*SCARECROW*		
SDF	Stromal cell-derived factor	UC	Umbilical cord
SDF1	Stromal cell-derived factor 1	UCB	Umbilical cord blood
SeV	Sendai virus		
Shh	Sonic hedgehog	VE	Visceral endoderm
SHH	Sonic hedgehog	VEGF	Vascular endothelial growth factor
SHP	SH2 domain-containing tyrosine phosphatase	VZ	Ventricular zone

Abbreviations

W	Week	WT-1	Wilms' tumour 1
Wnt	Wingless/Int-1	*WUS*	*WUSCHEL*
WNT	Wingless/integrated		
Wnt	Wingless-type family member	XEN	Extraembryonic endoderm
WOX	*WUSCHEL-RELATED HOMEOBOX*	YAP	Yes-associated protein

List of Tables

Table 6.1	List of mammalian species cloned by somatic cell nuclear transfer that led to live births.	103
Table 9.1	MuSC terminology.	167
Table 13.1	Summary of the studies using mesenchymal stem cell-derived conditioned media for the treatment of cutaneous wounds including study design, mechanisms and outcomes.	257
Table 13.2	Summary of studies suggesting the use of mesenchymal stem cell-derived exosomes for the treatment of cutaneous wounds.	259

Introduction: A Decade Teaching Stem Cell Biology

Gabriela Rodrigues

Contents

1.1 Introduction – 2

1.2 The Embryo and the Stem Cell Concept – 2

1.3 Pluripotent Stem Cells – 4
1.3.1 Practical Exercise: Evaluation of ESCs Differentiation – 5

1.4 Adult Stem Cells – 6

1.5 Stem Cells and the Changing Paradigm of Cell Biology – 7

References – 8

© Springer Nature Switzerland AG 2020
G. Rodrigues, B. A. J. Roelen (eds.), *Concepts and Applications of Stem Cell Biology*,
Learning Materials in Biosciences, https://doi.org/10.1007/978-3-030-43939-2_1

What Will You Learn in This Chapter?

This chapter is aimed at introducing the structure of this book, which is inspired by the Stem Cell Biology and Technology course running in the Faculty of Sciences of the University of Lisbon. Early mouse development is revisited as a basis for introducing the concept of pluripotent stem cells. A simple practical protocol is described in which the pluripotent differentiation potential of mouse embryonic stem cells is highlighted. Finally, students and readers are invited to navigate through some groundbreaking subjects that will be detailed in the following chapters, such as epigenetics, reprogramming, cloning, regenerative medicine and other innovative ideas spinning off the stem cell biology field.

1.1 Introduction

For more than a decade, the Stem Cell Biology and Technology (SCBT) course has been part of the Master's Programme in Evolutionary and Developmental Biology (MSc EDB) at the Faculdade de Ciências da Universidade de Lisboa/Faculty of Sciences of the University of Lisbon (FCUL), represented in the front page image of this chapter [1–3], and it was the driving force for composing this book. In the SCBT course, students learn the basics of stem cell biology, linking these concepts with the knowledge they already have on the developing embryo and navigating through subjects such as epigenetics, differentiation, plasticity and cell fate decisions. Other areas rooted in the stem cell field, such as regeneration, reprogramming, cloning and ageing, are discussed as well. Finally, applied and translational areas are introduced to the students, highlighting the biotechnological and biomedical use of stem cells at the forefront of modern stem cell biology.

The structure of this book follows the organisation of the course, in the sense that the first chapters are based on more fundamental science and, progressively, subsequent chapters describe more translational applications of stem cell biology. In harmony with the mammalian embryo developmental timeline, the book begins with preimplantation embryology (► Chap. 2), followed by the specification of germ cells and the generation of pluripotency (► Chap. 3). Because these early stages of development are particularly dynamic and several epigenetic landmarks are established in the embryo at that time, chapters approaching induced pluripotent stem cells (► Chap. 4), ageing (► Chap. 5) and cloning (► Chap. 6) follow. Then, ► Chaps. 7 (plants) and ► 8 (mouse) explain how stem cells contribute to the establishment and the morphogenesis of the body plan. In the subsequent four chapters, stem cells in specific tissues (skeletal muscle in ► Chap. 9, cardiac muscle in ► Chap. 10 and neuronal tissues in ► Chaps. 11 and 12) are described. The use of stem cells as disease models is discussed in ► Chaps. 11 and 12. The final chapter approaches translational aspects of stem cells, where mesenchymal stem cells and their secretome are presented as a potential biomedical tool to ameliorate wound healing (► Chap. 13).

1.2 The Embryo and the Stem Cell Concept

Fertilisation triggers the development of a new organism and, since that moment, a series of events occur in the embryo that will give rise to an organism with anatomy, physiology and behaviour that enables it to live and interact with its environment.

This complex process is not achieved in one single step, but instead through a series of events that progressively specify different types of cells, tissues and organ systems that eventually end up building a coherent organism.

The blastomeres resulting from the first rounds of cleavage in the early mammalian embryo are totipotent since these cells can give rise to the embryo itself and to extra-embryonic membranes, including the foetal contribution to the placenta. After several divisions, the first polarised cells appear in the embryo leading to lineage segregation. The blastocyst consists of two types of cells: the trophoblast cells, needed for implantation in placental mammals and which will contribute the foetal part of the placenta, and the inner cell mass (ICM) that holds the pluripotent cells that will generate the whole embryo and some extra-embryonic membranes as well. These pluripotent cells will differentiate to all the populations of cells that derive from the germ layers of the embryo: ectoderm, mesoderm and endoderm, in addition to the germ cells. The ICM cells can be retrieved from the blastocyst and cultivated to generate embryonic stem cell (ESC) lines (see ► Chap. 2 for more information; see ◘ Fig. 1.1a). Stem cells with a more limited developmental potential are continuously generated throughout embryonic and foetal development. In fact, these stem cells are of paramount importance for a properly organised and oriented organogenesis to take place in the right place at the right time. The ability of nascent stem cells to pattern the embryo is illustrated in plants (► Chap. 7), the vertebrate body (► Chap. 8), skeletal muscle (► Chap. 9), the heart (► Chap. 10) and the brain (► Chap. 11) throughout this book.

After birth, some cells remain in the adult body in specific organ micro-environments, which are still multipotent. These cells can give rise to a multi-lineage progeny but limited to the germ layer of origin.

This progression of differentiation has been beautifully represented by Waddington in the 1950s in an image where marbles (totipotent cells) roll downhill on an 'epigenetic landscape' and become progressively more differentiated, finally stably resting at the bottom of a valley. Each marble would then lie still in its own valley, unable to climb back the hill or jump to side valleys, therefore representing a terminally differentiated cell [4]. This representation of cell differentiation occurring in a unidirectional flow has been the dogma for many years but has now been replaced by a more plastic and accurate interpretation, greatly due to the contribution of the stem cell biology field.

The epigenetic mechanisms by which cells acquire their cell fate – or, by analogy, roll down Waddington's hills – fall mainly under two molecular mechanisms operating on their genetic material: DNA methylation and histone modification. In ► Chap. 2, students can learn more and understand thoroughly the processes that underlie fertilisation and lineage specification in the early embryo and in ► Chap. 6 the epigenetic mechanisms that condition cell lineage restriction are described, as well as the epigenetic regulation variability that occurs during cloning. The reversion of the differentiated state to a pluripotent state either by somatic cell nuclear transfer (► Chap. 6) or by direct reprogramming generating the so-called induced pluripotent stem (iPS) cells (► Chap. 4) not only defies classical biology concepts, but also opens avenues to the use of patient-specific cells in personalised medicine and custom-tailored drug development (see ► Chaps. 11 and 12). However, the cells that are used for cloning or reprogramming are usually aged somatic cells, and epigenetic constraints can hamper the reprogramming efficiency or compromise the differentiation capacity of these cells (see the discussion in ► Chap. 5).

Similarly to what happens during development in the embryo and later in postnatal life, stem cells grown in a Petri dish can exhibit several levels of developmental potency. However, one compulsory condition to be defined as a 'stem cell' is the ability to self-renew, i.e. upon dividing, the stem cell has to give rise to a cell identical to itself, thereby maintaining the original pool of stem cells. In addition, stem cells can give rise to cells with a more specialised fate. Therefore, although it is not a simple task, a putative definition of 'stem cell' can be stated as the following: a stem cell is a cell that has to be able to self-renew and give rise to at least one differentiated cell type.

1.3 Pluripotent Stem Cells

The history of stem cell biology can be traced back to the late nineteenth century when experiments with rabbit embryos were performed, as well as with mouse embryonal carcinoma cells [5, 6]. In the 1960s, Till & McCulloch, working on bone marrow transplantation in mice, paved the way for the conceptualisation of the idea of 'stemness' [7]. However, the first successful isolation of a mouse embryonic stem cell line occurred in 1981 [8, 9]. Only much later – due to ethical restrictions and technical difficulties – in 1998, Thomson and his team were able to derive a human embryonic stem cell line [10]. Ever since then, the field of stem cell biology has boosted exponentially and diversified in several branches of research.

During embryonic development, the inner cell mass of the mouse blastocyst eventually separates in two layers: the epiblast, which will give rise to the embryo, and the hypoblast, an extraembryonic sheet of cells (▶ Chap. 2). Similar to the ESCs, the cells of the epiblast are also pluripotent (they are the so-called epiblast stem cells – EpiSC), although epigenetically they are more limited in terms of pluripotency than ESCs. Therefore, EpiSCs are generally termed 'primed pluripotent stem cells' in comparison to ESCs which are characterised as more 'naïve' pluripotent cells. One of the differences between these two types of pluripotent stem cells is that while germ cells can be derived from naïve ESCs, it is quite difficult to differentiate them from primed EpiSCs (see ▶ Chap. 3).

Additionally, another line of pluripotent stem cells has been generated: these are the embryonal carcinoma cell lines and they derive from tumours called teratomas or teratocarcinomas, which generally arise in the gonads [5].

To evaluate if the cell lines that are generated are indeed pluripotent, they have to comply with several conditions tested in the laboratory. One of these experiments consists of injecting these cells into a mouse blastocyst. This so-called chimeric blastocyst can be implanted into the womb of a foster mother and the mice that are born can be tested for the presence of the injected cells in various organs, so as to ascertain that the tested cells contributed to the derivatives of the three germ layers and the germ cells. Alternatively, and especially in the case of human ESCs, with which the generation of chimeras is ethically and legally restrained, a different experiment can be undertaken. In this case, the pluripotent cells line can be injected into an immunocompromised mouse, and an ectopic teratoma is expected to form in the mouse. This teratoma can then be analysed for the presence of tissues representative of the three germ layers derivatives [5].

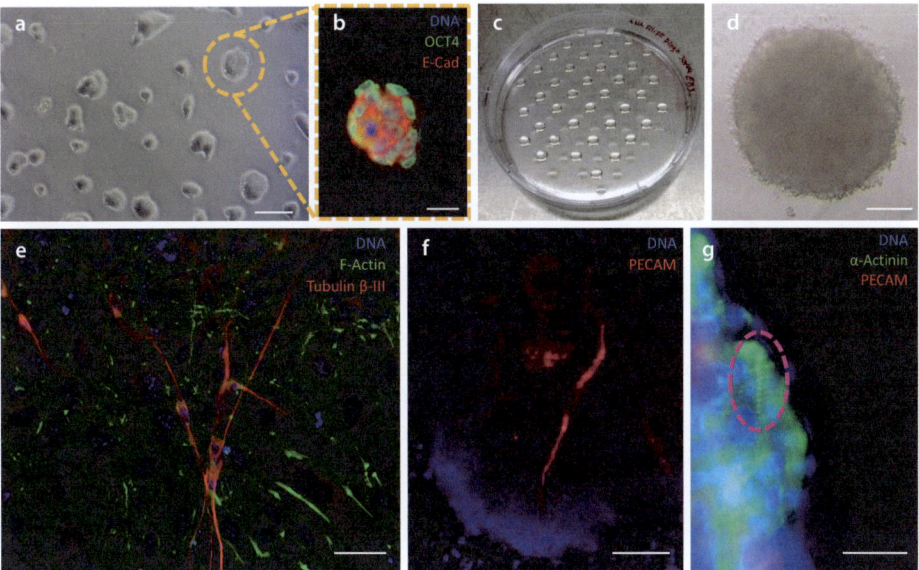

Fig. 1.1 Results of the practical exercise on the evaluation of ESCs differentiation potential: **a** Mouse ESCs growing on a gelatine-coated Petri dish for 2 days in a culture medium specific for the maintenance of pluripotency: small and round clumps of ESCs can be observed. **b** Immunocytochemical staining of ESCs obtained in **a** (blue: DNA, green: OCT4, red: E-cadherin); the stained ESCs shown is an example of the general staining obtained in the cultured cells (yellow circle in **a**); all the cells in the clumps display OCT4 expression, a marker of pluripotency; in addition, E-cadherin, an epithelial membranar adherent junctions' marker, is a hallmark of ESCs pluripotent cells as well. **c** 20 µl drops hanging upside-down on the lid of a 60 mm diameter Petri dish containing ESCs resuspended in a culture medium that no longer maintains pluripotency. **d** EB obtained after 4 days of incubating the cells in the hanging drops. **e–g** Immunocytochemical staining obtained from several experiments using cells grown as explants from the EBs obtained in D after an additional 2 days period upon plating the EBs in gelatine-coated Petri dishes growing in the same culture medium; in **e** (DNA in blue, F-actin in green and tubulin β3 in red), it is possible to observe cells' actin cytoskeleton in all cells (green) and tubulin β3 staining reveals that some cells differentiated towards the neural fate (red cells); in **f** (DNA in blue, PECAM in red), some 'endothelial-like cells' (red) are highlighted and these cells were able to form very rudimentary blood vessels; in **g** a few cells express a protein (α-Actinin, in green) present in the z-lines of cardiomyocytes' sarcomeres; the striped pattern of α-Actinin staining (see magenta circle) strongly suggests that this cell differentiated to the cardiomyocyte fate (DNA in blue and PECAM in red, the red marker does not show any specific staining). Scale bars: 50 µm in **a, d, e** and **f**; 20 µm in **b** and **g**. EB Embryoid Body, E-cad E-cadherin, ESC Embryonic Stem Cell, PECAM platelet–endothelial cell adhesion molecule, SCBT Stem Cell Biology and Technology. (All the images were obtained from experiments performed in different years by the students in the practical classes of the SCBT course)

1.3.1 Practical Exercise: Evaluation of ESCs Differentiation

The ability of pluripotent stem cells to differentiate into several types of cells can be tested in a number of ways. In the practical classes of the SCBT course, a simple protocol is undertaken by the students to learn about mouse embryonic stem cell biology. In a first step, students learn the basics of cell culture and practice several standard procedures with regular cell lines in the cell culture facility. Once trained, the students perform a simple protocol to maintain ESCs and to differentiate them in embryoid bodies (EBs). EBs are forced aggregations of dissociated ESCs in suspension and they are intermediate structures, no longer pluripotent, which can be channelled to differentiate to several cell lineages. In this exercise, students are expected to acquire skills in ESCs culture, to analyse stem cell behaviour and also to train the capacity to troubleshoot problems and critically interpret the obtained results.

Briefly, a line of mouse ESCs is cultivated in a humidified cell culture incubator (37 °C, 5% CO_2) on gelatine-coated Petri dishes for 2 days growing in stem cell culture medium whose formula was developed to maintain ESCs' pluripotency (◘ Fig. 1.1a). A subset of cells is transferred to a suspension in a different culture medium (that no longer maintains pluripotency) for the preparation of EBs in hanging drops (◘ Fig. 1.1c). Drops of 20 μl of the cell suspension are deposited in Petri dish lids, turned upside-down and further cultured for 4 days to force aggregation and, therefore, the formation of EBs (◘ Fig. 1.1d). The EBs are harvested and plated on gelatine-coated Petri dishes for an additional period of 2 days in the same culture medium. At the end of the incubation time, ESCs and EB cells are fixed and prepared for immunocytochemistry, using markers of pluripotency and differentiation. Students will evaluate the maintenance of the pluripotent state in the ESCs and the differentiation of the plated EB cells towards some cell fates under study. In this way, students will also have the opportunity to gain expertise in immunocytochemistry and imaging techniques. Because the culture medium used for the formation and cultivation of EBs is not specific for the differentiation of any particular type of cell, students will have to carefully analyse the results using a panel of markers and try to understand if any specific type of cell has appeared in the cell populations growing out of the plated EBs. Indeed, while ESCs consistently express pluripotent markers, such as OCT4 and NANOG (◘ Fig. 1.1b; see ► Chap. 2), cells growing in the EBs assumed neuronal (◘ Fig. 1.1e), endothelial (◘ Fig. 1.1f), or cardiac (◘ Fig. 1.1g) cell fates in different sets of experiments.

1.4 Adult Stem Cells

In the adult organism reservoirs of stem cells are present in almost every organ, generally in a quiescent state. These cells are usually multipotent, which means that their proliferative and multi-lineage differentiation capacities are limited compared to those of ESCs. Some adult stem cells, like the epithelial cells that constantly renew the epidermis or the gut lining in our bodies, are very active. On the contrary, some others, such as neural stem cells, are not easily activated. When skeletal muscle is injured, satellite cells, the stem cells of skeletal muscle, can become activated, fuse with the damaged muscle fibre and repair the local defect. Muscle satellite stem cells, and others, such as neural stem cells are approached in several chapters in this book (► Chaps. 9 and 11).

One of the best studied types of adult stem cells is the mesenchymal stem cells [11]. These cells have the potential to give rise to several kinds of cells, including condrogenic, osteogenic and adipogenic lineages. Mesenchymal stem cells can be harvested from the bone marrow, from fat tissue – even from liposuction material if collected under good medical practice – and from the umbilical cord matrix – the Wharton jelly. These cells have been largely studied and used due to their plasticity, safety and immunomodulatory properties. In addition, the secretome produced by mesenchymal stem cells has been investigated in view of many applications (see ▶ Chap. 13).

1.5 Stem Cells and the Changing Paradigm of Cell Biology

Roughly a century has passed since the first experiments leading to the definition of the stem cell concept. From then up to our days, an enormous amount of research has been conducted, ideas have emerged and dogmas were broken. After the demonstration that cells could revert the epigenetic memory imposed during cell differentiation, through cloning experiments (▶ Chap. 6) and direct reprogramming using iPS technology (▶ Chaps. 4 and 5) [12], Waddington's landscape had necessarily to be redrawn. Indeed, an updated epigenetic landscape has been represented with marbles jumping up, down and sideways on the hills [13–15].

More than being a major breakthrough in cell biology and a new vision of cell differentiation and plasticity, the reprogramming of somatic cells has opened promising avenues for the use of human iPS cells in many biomedical applications [14]. Given the possibility of using patient-specific cells from skin biopsies, for instance, to differentiate tissues or even to generate organoids in vitro, diseases can now be studied under a completely new perspective. Organoids are 'miniature organs' totally produced in vitro from stem cells. Organoids can reproduce part of the architectural composition of the native organ, preserving some of its functionality as well. Therefore, organoids have been used as tools to study disease progression, serving also as platforms to test drug toxicity and efficacy. The 'minibrains' [16] are an extraordinary example of a model that has been used to understand the progression of a pathological condition, in this particular case the infection of brain neuro-progenitors with the Zika virus, otherwise impossible to study directly in human brains [17]. Another very elegant application of organoids has been the use of cystic fibrosis patient-derived gut organoids to test drugs that are potentially particularly suited for each one of the patients [18].

The field of stem cell biology has greatly benefited from tissue engineering technology [5]. Expertise on natural and synthetic biomaterials, along with the fast development of 3D printing equipment, has allowed this field to thrive enormously. The project of producing an artificial human heart is no longer an idea only possible coming out of a science fiction movie [19]. Bioreactor technology can now be used to expand cells efficiently in such a way that it is possible to harvest enough cells for efficient transplantation procedures. The scaling up of cells is not only important for the urgent need of cells for transplantation but has also fueled cutting-edge 'outside the box' ideas, such as the production of in vitro meat [20]. The possibility of producing a hamburger in a Petri dish was a very important proof-of-concept and it will, or not, be scaled-up for common use in the food industry, depending on commercial constraints and societal decisions.

The stem cell field, much like Waddington's marbles, has rolled down and climbed up many hills, in a way that was difficult to imagine few years ago. Our expectation is that the students of the SCBT course and the readers of this book learn more about the fascinating and ever-growing world of stem cell biology and open their spirits to what is yet to come in this field in the near future.

> **Take-Home Message**
>
> - Pluripotent cells can be retrieved from the ICM of mouse or human blastocysts and be used to derive ESC lines.
> - Stem cells can self-renew and give rise to at least one differentiated cell type.
> - Patient-specific pluripotent stem cells can be generated by cloning (somatic cell nuclear transfer) or direct reprogramming originating iPS cells.
> - Adult stem cells are multipotent and, therefore, more restricted in their differentiation capacity compared to ESCs.
> - The traditional epigenetic landscape imagined by Waddington in the 1950s, with a unidirectional flow of cell differentiation, has been reinterpreted in the light of new data as a more dynamic and multidirectional 'cellular flowchart'.
> - The expertise to grow organoids in vitro, together with considerable improvements in the technology to produce iPS cells, has propelled drug toxicity and efficacy testing in personalised medicine.
> - Teaching stem cell biology is fascinating and research in the field of stem cell biology will certainly lead to exciting new frontiers.

Acknowledgements I would like to express my deep gratitude to all the students of the SCBT course for their hard work, enthusiasm and friendly interaction. I also want to thank Bernard Roelen for his inspirational lectures and discussions in the course, and Susana Chuva de Sousa Lopes for her continued support with cells, antibodies, and for sharing technical skills. Finally, I want to address a word of recognition to my colleague Sólveig Thorsteinsdóttir for her constant support, to Marta Palma and Luís Marques for their long-lasting cheerful help in the cell culture and the microscopy facilities, and to Ana Bernardo for drawing beautiful illustrations for this book.

References

1. https://ciencias.ulisboa.pt/en.
2. http://bed.campus.ciencias.ulisboa.pt/.
3. Thorsteinsdóttir S, Rodrigues G, Crespo EG. History and present state of teaching and research in Developmental Biology in Portugal – a personal account. Int J Dev Biol. 2009;53:1235–43.
4. Waddington CH. The strategy of genes: a discussion of some aspects of theoretical biology. London: George Allen and Unwin; 1957.
5. Mummery C, van de Stolpe A, Roelen BAJ, Clevers H. Stem cells: scientific facts and fiction. Amsterdam: Academic Press/Elsevier; 2014.
6. Solter D. From teratocarcinomas to embryonic stem cells and beyond: a history of embryonic stem cell research. Nat Rev Genet. 2006;7:319–27.
7. McCulloch EA, Till JE. Perspectives on the properties of stem cells. Nat Med. 2005;11(10):1026–8.
8. Evans MJ, Kaufman MH. Establishment in culture of pluripotential cells from mouse embryos. Nature. 1981;292(5819):154–6.
9. Martin GR. Isolation of a pluripotent cell line from early mouse embryos cultured in medium conditioned by teratocarcinoma stem cells. Proc Natl Acad Sci U S A. 1981;78(12):7634–8.
10. Thomson JA, Itskovitz-Eldor J, Shapiro SS, Waknitz MA, Swiergiel JJ, Marshall VS, et al. Embryonic stem cell lines derived from human blastocysts. Science. 1998;282(5391):1145–7.
11. Caplan AI, Correa D. The MSC: an injury drugstore. Cell Stem Cell. 2011;9(1):11–5.
12. Takahashi K, Tanabe K, Ohnuki M, Narita M, Ichisaka T, Tomoda K, et al. Induction of pluripotent stem cells from adult human fibroblasts by defined factors. Cell. 2007;131(5):861–72.

13. Hochedlinger K, Plath K. Epigenetic reprogramming and induced pluripotency. Development. 2009;136:509–23.
14. Takahashi K, Yamanaka S. Induced pluripotent stem cells in medicine and biology. Development. 2013;140:2457–61.
15. Takahashi K, Yamanaka S. A developmental framework for induced pluripotency. Development. 2015;142:3274–85.
16. Lancaster MA, Renner M, Martin C-A, et al. Cerebral organoids model human brain development and microcephaly. Nature. 2013;501(7467):373–9.
17. Qian X, Nguyen HN, Jacob F, Song H, Ming GL. Using brain organoids to understand Zika virus-induced microcephaly. Development. 2017;144(6):952–7.
18. Dekkers JF, Wiegerinck CL, de Jonge HR, et al. A functional CFTR assay using primary cystic fibrosis intestinal organoids. Nat Med. 2013;19(7):939–45.
19. Noor N, Shapira A, Edri R, Gal I, Wertheim L, Dvir T. 3D printing of personalized thick and perfusable cardiac patches and hearts. Adv Sci. 2019;6:1900344.
20. Post MJ. Cultured meat from stem cells: challenges and prospects. Meat Sci. 2012;92(3):297–301.

Preimplantation Development: From Germ Cells to Blastocyst

Bernard A. J. Roelen

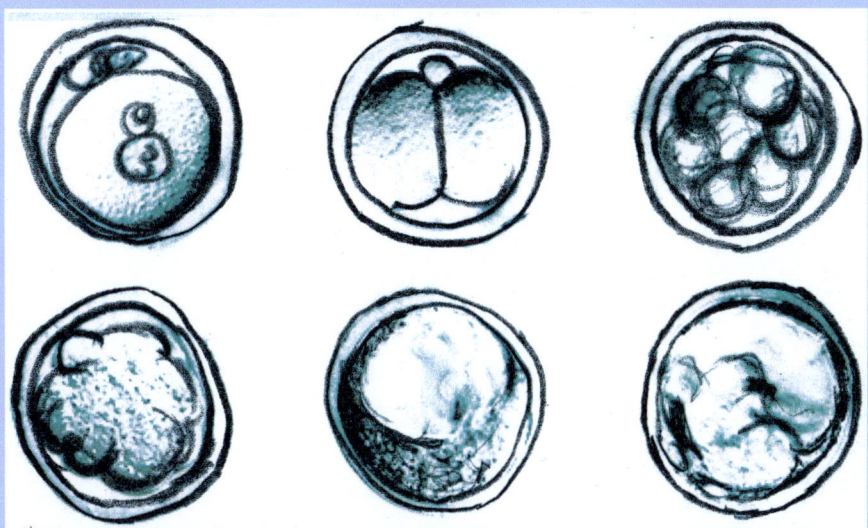

Contents

2.1 The Beginning of Life – 13

2.2 Development of the Germ Cells – 13
2.2.1 Development of the Oocyte – 13
2.2.2 Development of the Sperm Cell – 14

2.3 Cleavage Divisions – 15

© Springer Nature Switzerland AG 2020
G. Rodrigues, B. A. J. Roelen (eds.), *Concepts and Applications of Stem Cell Biology*,
Learning Materials in Biosciences, https://doi.org/10.1007/978-3-030-43939-2_2

2.4	**Differentiation – 18**	
2.4.1	First Lineage Segregation: Formation of Trophectoderm and Inner Cell Mass – 18	
2.4.2	Second Lineage Segregation: Formation of Epiblast and Hypoblast – 19	
2.5	**Pluripotent Stem Cells – 21**	
2.5.1	Testing Pluripotency: Chimera Formation – 21	
2.6	**Implantation – 23**	
	References – 25	

2 Preimplantation Development: From Germ Cells to Blastocyst

What Will You Learn in This Chapter?
In mammals, after germ cells have arrived in the gonads, they have to undergo several steps to become cells that are competent of forming a totipotent zygote after fusion. The zygote undergoes a series of cleavage divisions which will give rise to a morula stage embryo. At around this stage, the first lineage segregation event takes place leading to the formation of an outer trophectoderm layer and a pluripotent inner cell mass that will give rise to the fetus. In a second lineage segregation event, NANOG-expressing cells form the pluripotent epiblast while GATA4/6-expressing cells will give rise to the yolk sac. Around this time, the blastocyst stage embryo arrives at the uterus for implantation. Blastocyst stage embryos of mouse and human have an invasive type of implantation, while embryos of other mammalian species can have a more superficial type of implantation.

2.1 The Beginning of Life

A mammalian organism starts as a fusion of two cells: the sperm cell and the oocyte, or egg. The sperm and egg are both haploid cells, containing half the number of chromosomes of the adult animal. The resulting cell is termed zygote, and as this cell inherits the chromosomes from the sperm cell and from the egg, the zygote is diploid. The number of chromosomes of diploid cells is constant within a species but differs largely among species. Derby's woolly opossum (*Caluromys derbianus*) for instance, a marsupial from Central and South America, has a diploid number of 14 chromosomes, humans have 46 chromosomes, and dogs (*Canis lupus familiaris*) for instance have 78 chromosomes.

At the time of ovulation, the oocyte of most mammalian species is arrested at the metaphase stage of the second meiotic division. After fusion of the sperm plasma membrane with that of the oocyte, inositol 1,4,5-triphosphate (IP3) and the IP3 receptor bind which leads to calcium waves that cause the oocyte to resume meiosis and formation of the maternal pronucleus.

The zygote is a totipotent cell, meaning that, in the case of placental mammals, it will give rise to all embryonic and extraembryonic tissues. As the embryo develops, however, the developmental potential of the cells decreases. While the zygote is spherical with no discernable axis, the embryo and fetus will adapt a bilateral symmetry with anterior–posterior, dorsal–ventral, and left–right axes.

2.2 Development of the Germ Cells

The germ cell lineage is one of the first differentiated cells that is set aside from the developing fetus. In mammals, the primordial germ cells migrate from the posterior region of the embryo through the hindgut to reach the developing gonads via the dorsal mesentery (see also ▶ Chap. 3).

2.2.1 Development of the Oocyte

A finite number of oocytes colonize the gonads during fetal development. Under influence of the somatic cells from the ovary, the oocytes initiate meiosis and become arrested at the prophase of the first meiotic division. This arrest is maintained until

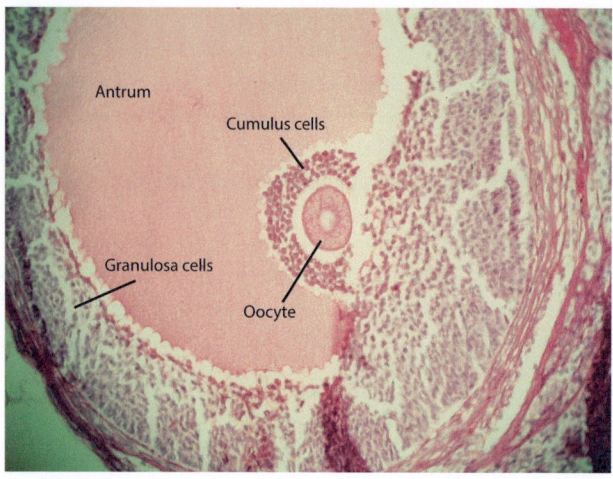

Fig. 2.1 Antral follicle. In this cross section of a bovine antral follicle counterstained with hematoxylin, the granulosa cells, cumulus cells, and enclosed oocyte are visible

shortly before ovulation, which in the case of women means that some oocytes remain at this stage for decades. Together with somatic cells, the oocytes form primordial follicles and around birth the stock of these follicles is at its peak and from then on will only decrease in number. The somatic cells surrounding the oocyte are called granulosa cells and are first flattened. As the oocyte grows, the granulosa cells become cuboidal and increase in number, and the follicle transforms into a secondary follicle. Subsequently, the formation of a zona pellucida is induced. This protein shell protects the oocyte and the developing embryo until implantation into the uterus. The origin of the zona pellucida, either granulosa cells or the oocyte, has been debated but at least in the mouse, the oocyte appears to be the site of zona pellucida biosynthesis [1]. A fluid-filled cavity called antrum is formed, separating the granulosa cell population in the granulosa cells that line the antrum and the cumulus cells that remain closely associated with the oocyte (Fig. 2.1). Physical interaction between oocytes and cumulus cells is maintained by transzonal cytoplasmic projections of the cumulus cells, ending in gap junctions. During the estrus cycle, follicles are further activated, among others by follicle-stimulating hormone (FSH) to become antral follicles of which a restricted number will eventually ovulate [2].

The principle of selection for those follicles that ovulate and those that will go into regression is largely unknown.

In order to acquire developmental competence, the oocyte synthesizes large quantities of mRNA and proteins. Synthesis of mRNA has to be completed before meiotic resumption since from that moment on the cells are transcriptionally quiescent until activation of the embryonic genome. For long-term storage, the mRNAs in the oocyte are deadenylated and are selectively polyadenylated during maturation and the first cleavage stages [3, 4].

2.2.2 Development of the Sperm Cell

In contrast to the oocytes, sperm cells are continuously produced during male adult life spermatogonial stem cells that are located in the seminiferous tubules in the testes. Similar to for oocytes, somatic Sertoli cells are also important for the correct

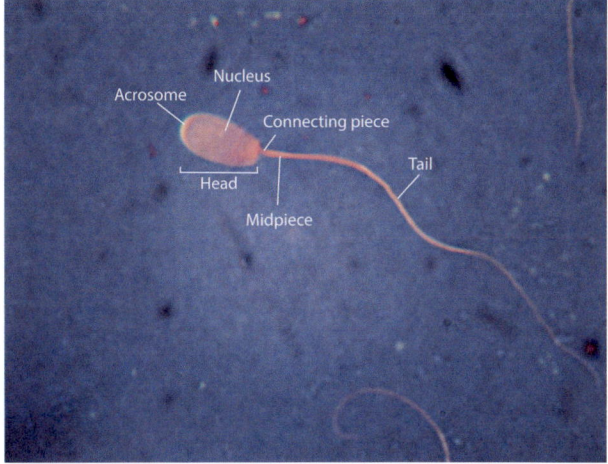

Fig. 2.2 Sperm cell. Microscopic image of an ejaculated bovine sperm cell with various structures indicated

formation of functional sperm cells. Spermatogenesis is a complex but highly organized process with different developmental steps having a fixed duration with every area of the seminiferous epithelium following similar sequences of events [5]. An important step in the formation of functional sperm cells is the compaction of the DNA to a transportable format. Therefore, a genome-wide removal of the canonical histones takes place and chromatin stability and compaction are facilitated by protamines [6]. While the oocyte is a large cell that only passively moves through the oviduct, the mature sperm cell is small and highly motile. Morphologically, the mature sperm cell consists of a head with an acrosome vesicle separated from the compact DNA by membranes, a midpiece containing mitochondria and a flagellum that is needed for motility (Fig. 2.2).

The acrosome is located on the anterior part of the sperm cell. It is basically a vesicle containing proteolytic enzymes that are released after binding to the zona pellucida [7]. The zona pellucida degrades locally under influence of the enzymes, and the sperm cell can penetrate the zona to make contact with the oocyte's plasmalemma (Fig. 2.3). Fusion of the two cells results in the formation of male and female pronuclei. The protamines in the paternal DNA are rapidly exchanged by histones prior to fusion of the maternal and paternal pronuclei [6] (Fig. 2.4).

2.3 Cleavage Divisions

Following syngamy, the totipotent diploid cell will start mitotic divisions without cell growth. As a result, the blastomeres become smaller in size with each division while the total volume of the embryo remains the same. At the first cleavage division, individual blastomeres can yield a viable healthy fetus after transplantation to a surrogate mother, demonstrating that the cells are still totipotent. At the zygote stage, the embryo is transcriptionally silent and is depending on the maternal transcripts formed in the oocyte. Shortly after formation of the diploid nucleus, the novel zygotic genome can be transcribed. The timing of zygotic genome activation varies among the mammalian species. In the mouse, the first transcripts from the zygotic genome have been detected as early as 7 hours after pronucleus formation. In embryos of

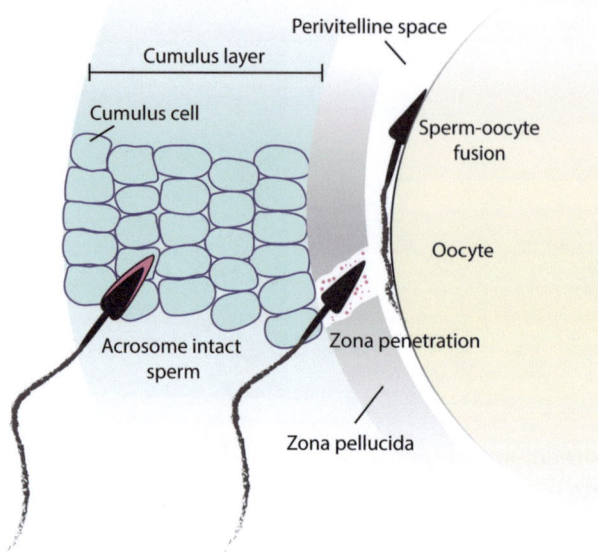

Fig. 2.3 Fertilization. The acrosome intact sperm reaches the cumulus cell layer and migrates through these cells. Upon binding to the zona pellucida, the membrane of the acrosome fuses releasing proteolytic enzymes that locally degrade the zona matrix. The hypermotile sperm can penetrate the zona and the sperm plasma membrane can adhere to and fuse with the oolemma

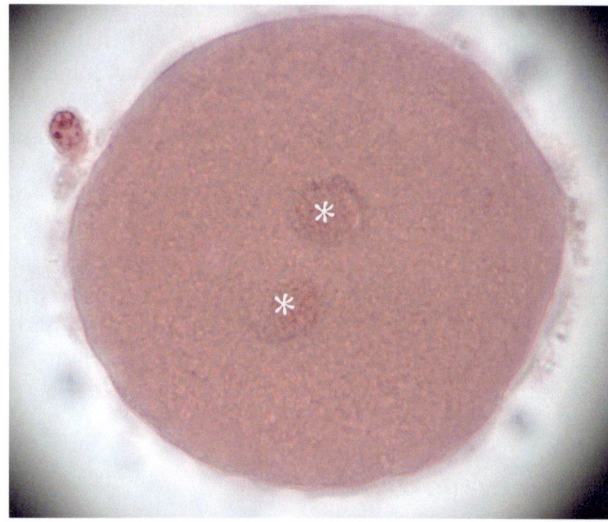

Fig. 2.4 Zygote. Microscopic image of a bovine zygote shortly after fertilization. The zygote has been stained with aceto-orcein, and pronuclei are indicated by white asterisks

other well-studied mammals, zygotic genome activation starts later. Human embryos are thought to undergo zygotic genome activation between the 4- and 8-cell stages, while the majority of new bovine transcripts were detected between the 8- and 16-cell stages [8] (Fig. 2.5).

The developmental capacity of the individual blastomeres diminishes after the second and third cleavage division, depending on the animal species.

A decreased cell volume is partly responsible for this, indicated by the observation that split morula or blastocyst stage embryos can give rise to two healthy individuals as is the case in spontaneous monozygotic twinning. From the morula, Latin for

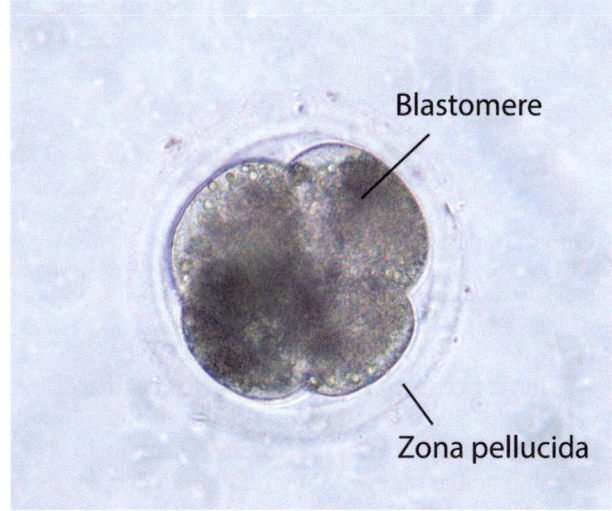

Fig. 2.5 Cleavage stage embryo. Light microscopic image of a porcine 4-cell stage embryo. At this stage, the embryo is still enclosed within the zona pellucida and individual blastomeres are totipotent

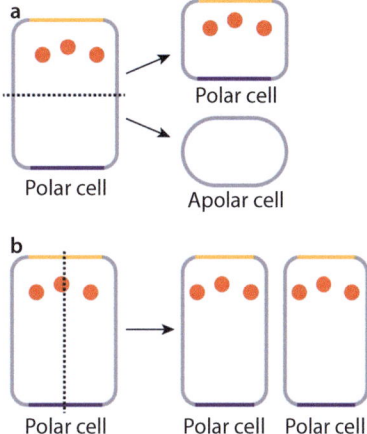

Fig. 2.6 Polarity model of fate specification. The polarity model suggests that individual blastomeres are already polarized, indicated by yellow (apical) and purple (basal) lines. Differentiation into inner cell mass (apolar cell) and trophectoderm (polar cell) requires a differentiative division **a**. A conservative division **b** generates to polar, outside, cells. (Redrawn from Ref. [13])

'mulberry' reflecting the shape, stage onwards the developmental capacity of cells diminishes accompanied by the first morphological differences between the blastomeres. Cells that constitute the outside of the embryo adopt a polarized state with microvilli indicating the apical part, while inside blastomeres remain apolar (Fig. 2.6). It appears that cell number and position are not solely responsible for the differences in cell behavior but that distribution of cell organelles together with asymmetric and symmetric division also gives rise to different cells [9]. Between the 8- and 16-cell stages, the embryo undergoes a process known as compaction, resulting in an embryo in which individual cells are no longer recognizable. This is particularly evident for embryos with relatively translucent cytoplasm, such as mouse and human embryos. Individual blastomeres of porcine and bovine embryos for instance are morphologically less distinct even before compaction due to a high content of lipid droplets. Compaction is achieved by spreading of the contacts between the cells and reducing the contact area with the surrounding and is suggested to be important for cellular diversification [9]. Compaction is a process resulting from the ratio of two

Fig. 2.7 Compaction. Compaction of a morula stage embryo results from changes of the balance between tensions at the cell–cell contact (purple arrows) and surface tension at the cell–medium interface (bordeaux arrows). A decrease in the cell–cell contact tension promotes contact spreading leading to compaction. (Redrawn from Ref. [10])

basic forces: cell–cell interfacial tension at the cell contacts, and surface tension, which is the tension between the outside of the cells and their surrounding and which gives cells their stiffness. The mechanics of this process are not very different from those that define the spreading of a liquid droplet on a surface [10] (Fig. 2.7).

2.4 Differentiation

Until these stages, no embryonic axes such as dorsal–ventral or anterior–posterior have been set up in the embryo and all cells can still contribute to all lineages.

2.4.1 First Lineage Segregation: Formation of Trophectoderm and Inner Cell Mass

Eventually, two different cell populations arise in the embryo: the outer cells epithelialize and become the trophectoderm (TE) that in eutherians will give rise to the embryonic portion of the placenta while the inner cells form the inner cell mass (ICM), precursor of the epiblast and hypoblast. It is the inner cell mass from which embryonic stem (ES) cells are derived. An important molecule for the differentiation of TE is the caudal-type homeodomain transcription factor CDX2. Mouse embryos that are genetically deficient in *Cdx2* do not form trophectoderm and fail to implant [11]. Conversely, *Oct3/4* expression is needed for the formation of the inner cell mass, and CDX2 and OCT4 reciprocally repress their transcription [12]. In the mouse, the Hippo signaling pathway has been identified as responsible for the differential activation of CDX2. In inside cells, the Hippo (in mammals Mst1/2) pathway is active leading to activation of the large tumor suppressor kinase (LATS) and subsequent phosphorylation of Yes-associated protein (YAP). Upon phosphorylation, YAP is rapidly sequestered in the cytoplasm and thereby inactivated. In outside cells, Hippo signaling is weaker enabling YAP to translocate to the nucleus where it partners with the transcription factor TEAD4 to promote *Cdx2* expression [13] (Fig. 2.8). Whether these pathways and molecules also function for trophoblast specification in other mammals remains to be established, but several differences have been noticed between species.

While in the mouse, OCT4 and CDX2 expression is mutually exclusive, in pig and cattle embryos, for instance, OCT4 remains expressed in the trophectoderm until

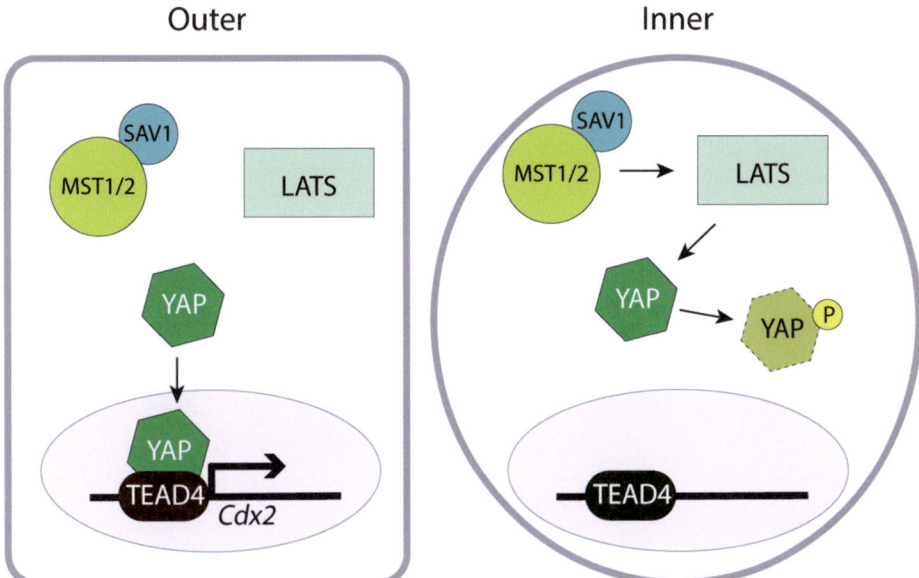

Fig. 2.8 Lineage segregation by Hippo signaling. In outside cells (left), Hippo signaling is absent, facilitating nuclear accumulation of Yes-associated protein (YAP), where it together with the transcription factor TEAD4 activates the *Cdx2* enhancer leading to *Cdx2* transcription. When Hippo signaling is active, in inner cells (right), the protein kinases MST1/2 (Hippo) and their coactivator Salvador (SAV1) increase YAP phosphorylation via large tumor suppressor kinase (LATS). Phosphorylated YAP is degraded via ubiquitination and can therefore not stimulate *Cdx2* expression

after inner cell mass formation [14, 15]. This could indicate a difference in the mechanisms of lineage segregation or in the timing of lineage commitment. In the tammar wallaby, a marsupial, although POU5F1 (OCT4) is expressed in the pluriblast (equivalent to the inner cell mass) cell, so is CDX2. In this species, the hippo signaling pathway components YAP and its close relative WWTR1 are also differentially expressed in the pluriblast and trophoblast [16].

Formation of TE and inner cell mass is accompanied by formation of a fluid-filled cavity with the inner cell mass adjacent to it. The embryo is now at the blastocyst stage (Fig. 2.9). The outer TE is epithelialized and supports blastocoelic expansion mediated by water influx from the oviductal fluid, or culture medium in case of in vitro fertilization (IVF). Active sodium ion transport through transmembrane pumps facilitates the water influx; Na^+/H^+ exchangers at the apical membranes, and Na^+/K^+-ATPases on the basal membranes. In addition, aquaporins mediate water transport across the trophectoderm [17].

2.4.2 Second Lineage Segregation: Formation of Epiblast and Hypoblast

Although the inner cell mass appears to be a homogenous pluripotent cell population, it further differentiates into two lineages, epiblast and hypoblast. The epiblast gives rise to all fetal and adult tissues, and in the mouse also extraembryonic mesoderm and part of the embryonic gut tube [18]. The hypoblast, in mouse embryos

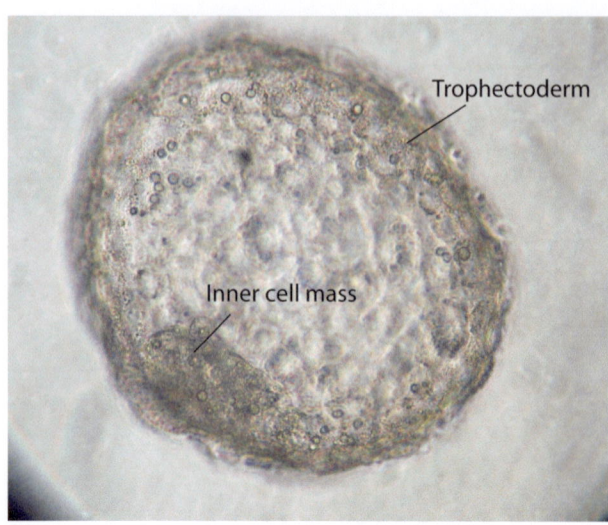

Fig. 2.9 Blastocyst. Light microscopic image of an 8-day-old bovine blastocyst. The inner cell mas and the outer trophectoderm layer are indicated. This embryo has already escaped (hatched) from its surrounding protein shell, the zona pellucida

often referred to as primitive endoderm, is the source of all extraembryonic endodermal tissue of the yolk sac [19]. Exactly when lineage segregation occurs and at what stage the cells have become restricted to a lineage is unknown but varies among species. Two transcription factors are central in the specifying of epiblast or hypoblast fate: NANOG and GATA family members. Mouse embryos that are genetically deficient for *Nanog* do not form epiblast but take up an endoderm-like fate. Conversely, *Gata6* knockout embryos initiate hypoblast formation but fail to form visceral endoderm, a derivative of primitive endoderm, after implantation. In early embryonic day (E)3.5, mouse blastocysts, *Nanog,* and *Gata6* are mutually exclusively expressed in what has been described as a salt-and-pepper-like pattern, indicating that the inner cell mass is a mosaic of NANOG- and GATA6-positive cells [20]. By E4.5, just before implantation, the GATA6-positive primitive endoderm precursors have sorted as a morphologically distinct monolayer on the surface of the ICM, facing the blastocoelic cavity. Pluripotent stem cells have also been derived from epiblast cells, so-called Epi stem cells (EpiScs), and their pluripotency is at a primed state of pluripotency (see for more details ▶ Chap. 3 and ◘ Fig. 3.1 in that Chapter).

In mouse embryos, fibroblast growth factor (FGF) and mitogen-activated protein (MAP) kinase are crucial for the NANOG-epiblast and GATA6-primitive endoderm cell fate. Culture of embryos from E2.5-E4.5 in the presence of an FGF receptor inhibitor and a MAPK/ERK kinase (MEK) inhibitor resulted in embryos with all ICM cells positive for NANOG, while the numbers of ICM cells had not changed. By contrast, in embryos cultured in the presence of FGF4, all ICM cells have become GATA6-positive primitive endoderm at E4.5. In mouse embryos, although acquisition of epiblast or primitive endoderm fates already starts at around E3.0, cell fate is not irreversibly determined until around E4.0 [21]. Endogenous FGF4 secreted by the presumptive epiblast cells and signaling through FGF receptor (FGFR)1, expressed in all ICM cells, seems critical for primitive endoderm formation, with FGFR2, although specifically expressed in presumptive primitive endoderm cells, being less decisive [22, 23].

What is evident, however, is that although the NANOG and GATA transcription factors are key molecules for the determination of, respectively, epiblast and primitive endoderm fate in various examined mammals, the mechanisms by which these factors are activated are different between species [14, 24]. When bovine embryos are cultured in the presence of a MEK inhibitor, the percentage of cells in the inner cell mass that express NANOG is increased, similar to what has been described for mouse embryos. Importantly, however, although the percentage of GATA6 expressing cells decreases, there are still fair numbers of GATA6-positive cells in the inner cell mass of day 8 bovine blastocysts after MAP kinase inhibition. Culture with FGF4 on the other hand leads to inner cell masses completely composed of GATA6-positive cells. Surprisingly, inhibition of FGFR1 activity with the small molecule PD173074 does not affect the percentages of NANOG- and GATA6-positive cells [25]. Most likely other FGF receptors transduce the FGF signal in presumptive endoderm cells. Combined, these data indicate that in cattle embryos while MAP kinase activity inhibits NANOG expression, thereby releasing the inhibition of GATA6 expression, there must be other as yet unknown signals that activate GATA6 expression [25].

When MAP kinase activity is inhibited during the culture of human embryos, this surprisingly has no effect on the percentages of NANOG- and GATA-positive cells of days 6 and 7 blastocysts [25, 26]. Similarly, inhibition of FGF receptor activity in human embryos using a small molecule did not lead to reduced numbers of GATA4-positive cells in day 7 blastocysts [26]. Whether activation of FGF signaling can lead to primitive endoderm formation in human embryos has so far not been examined, but segregation of epiblast and primitive endoderm is clearly not dependent on FGF/MAP kinase signaling.

2.5 Pluripotent Stem Cells

Cell from the ICM of mouse preimplantation embryos can be cultured in vitro such that they commence to self-renew and maintain pluripotent. When cultured in the presence of leukemia inhibitory factor (LIF) and bone morphogenetic protein (BMP), these embryonic stem (ES) cells grow in dome-shaped colonies and can be passaged by trypsinization. Pluripotent stem cell lines can also be derived from the epiblast layer of early postimplantation embryos, but these epiblast stem cells (EpiSCs) cells do not respond to LIF but rely on FGF and Activin/NODAL signaling instead [27, 28]. Also these cells grow in flat colonies and do not survive single cell-passaging. Mouse ES cells are in a naïve or ground state of pluripotency, similar to cells of the ICM [29]. EpiSCs and epiblast cells of the postimplantation embryo on the other hand are in a primed state of pluripotency, ready for differentiation. Human ES cells, although derived from blastocyst-stage embryos, are similar to mouse EpiScs, in terms of culture requirements (see for more details ▶ Chap. 3 and ◘ Fig. 3.1 in that Chapter).

The pluripotency of mouse stem cells can be tested via teratoma formation and chimaera formation. Pluripotent cells injected into the testis or kidney capsule of immunocompromised mice will form teratomas containing a multilinage variety of tissues.

2.5.1 Testing Pluripotency: Chimera Formation

Several experimental procedures exist by which pluripotency of cells can be demonstrated. The most straightforward is the culture of cells in the presence of activators or inhibitors of signaling pathways and analysis of the cells that have differentiated. When cells from the three germ layers (endoderm, ectoderm, and mesoderm) are formed from a clonal cell population, this gives a good indication of cellular pluripotency, although it is not possible to demonstrate that all cell types can be formed and functional. A more conclusive demonstration is the formation of teratomas after subcapsular kidney or intratesticular transplantation. Analysis of the teratomas should reveal that they are formed from the transplanted cells and contain derivatives of all three germ layers. As an advantage to the in vitro differentiation analysis, in teratomas relatively organized tissue rudiments can be formed such as renal tissue and hair follicles. Functionality is, however, difficult to demonstrate in the disorganized cell mass that constitutes the teratoma.

The most rigorous evidence of cellular pluripotency is the formation of germline transmitting chimeric animals. In modern-day science, a chimera is a single individual from different embryonic origins. A chimera can be formed spontaneously, when two embryos from two different fertilized eggs fuse before implantation but whether this occurs in laboratory animals (mice and rats) is not known. Chimeras have been made experimentally, predominantly in mice, and in order to identify the contribution of the embryos, different breeds with different coat colors can be used. Interspecies chimeras using embryos of two different species have only been successfully generated with embryos from closely related species such as house mouse *Mus musculus* and Ryukyu mouse (*Mus caroli*) [30] or sheep (*Ovis aries*) with goat (*Capra hircus*) [31].

Chimerism can also occur spontaneously in humans and in the case of same-sex embryos without genetic analysis these will almost never be recognized as such and almost impossible to discriminate from single embryo individuals. Chimeras from different-sex embryos may develop anomalies of sexual anatomy or function which can lead to the discovery of the chimerism [32].

When stem cells are truly pluripotent, they can contribute to the formation of all cells and tissues of the embryo and the fetus. Therefore, chimeras can be formed by aggregation with morula stage embryos or injection of pluripotent cells into the blastocoelic cavity followed by transfer into pseudopregnant recipients. Mouse ES and iPS cells are indeed capable of significantly contributing to the formation of chimeric mice, including the germline, by both morula aggregation [33] and blastocyst injection [34]. When developmentally compromised tetraploid embryos are being used as host for either aggregation or blastocyst injection, the tetraploid cells can only contribute to extraembryonic tissue and as a consequence offspring can be generated that is entirely stem cell derived [35]. This tetraploid system is now considered the gold standard for pluripotency.

Obviously, the pluripotency level of human ES cells cannot be tested by aggregation with, or injection into, human embryos. As a less stringent criterion for pluripotency, human ES cells have been injected into mice for teratoma formation and analysis, and based on formation of representative cell types from all three primary germ layers, the cells are considered pluripotent [36]. The similarities between culture requirements and morphology suggest, however, that human ES cells represent the primed state of pluripotency similar to mouse EpiSCs. Interestingly, mouse EpiSCs, although considered pluripotent, are barely able to contribute to developing mouse chimeras [27, 28] raising the question about the developmental plasticity of human

ES cells. To answer this question, several laboratories have attempted to generate interspecies chimeras by injecting mouse blastocysts with human pluripotent cells. Ethical concerns make it difficult to perform such experiments and many countries ban the implantation of such embryos into mouse uteri and the in vitro culture beyond 14 days or gastrulation. Injection of human ES or iPS cell into mouse blastocysts followed by short-term in vitro culture indicated incorporation into the ICM but no survival of the human cells after several days of culture [37]. Injection of human pluripotent stem cells that had been cultured under 'naïve' conditions did reveal several human cells in midgestation mouse embryos after blastocyst injection. The functionality of the human cell was, however, not further tested [38]. Interestingly, human ES cells did contribute to embryonic and extraembryonic tissues of mouse E10.5 embryos when expression of anti-apoptotic genes was induced and cells were injected into 4-cell stage embryos [39]. Since human pluripotent stem cells are relatively similar to mouse Epi SCs derived from the postimplantation epiblast, it has been similarly investigated whether these cells could participate in mouse development after transplantation to gastrula-stage embryos. Indeed transplanted human pluripotent stem cells could contribute to developing mouse embryos and differentiated into endoderm, ectoderm, and mesoderm. These data indicate that human pluripotent cells (ES or iPS) are bonafide pluripotent and can undergo location-appropriate differentiation [40].

2.6 Implantation

Shortly after entering the uterine cavity, the blastocyst sheds from the zonapellucida which has protected the embryo during its journey through the oviduct. The trophectoderm secretes factors, in the case of a human embryo chorionic gonadotrophin (hCG), that stimulate the continuous release of progesterone by the remnant of the ovulated follicle, the corpus luteum, to prepare the uterus for reception of the embryo (◘ Fig. 2.10). Uterine implantation can be invasive, such as in rodents, carnivores and primates, or noninvasive as seen in ruminants, pigs, and horses [41]. The orientation of implantation also varies among the animal species. In the case of the invasive implanters, mouse embryos attach to the endometrium with the mural trophectoderm, that is the trophectoderm opposite to the inner cell mass.

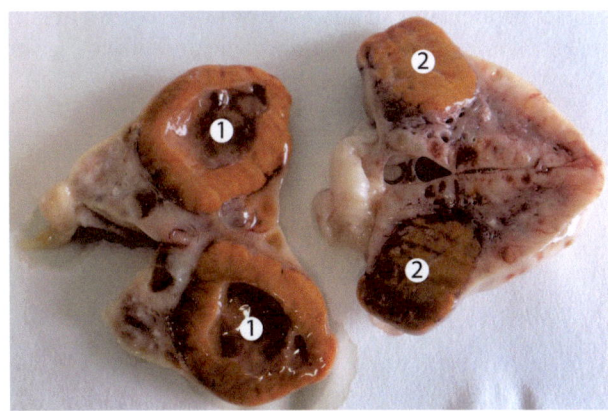

◘ **Fig. 2.10** Corpus luteum. Shown are two bovine ovaries that have been cut in half. The left ovary shows a corpus luteum (1) of which the original antral cavity of the follicle is still present. The corpus luteum (2) in the ovary on the right is older but still active. Note the size of the structures in relation to the size of the total ovaries, and the yellowish-orange color that led to their name (corpus luteum = yellow body)

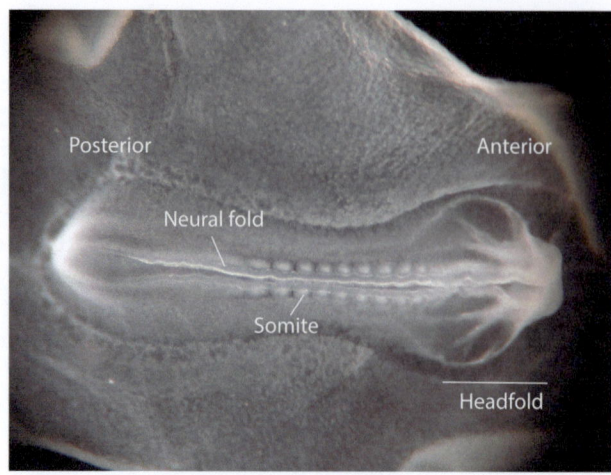

Fig. 2.11 Horse embryo. Light microscopic image of an 18-day-old horse embryo. In various mammals, including horses, the embryo does not implant immediately after arrival in the uterus. In this embryo, neural tube and brain development has already started, and somite pairs are being formed

Human embryos on the other hand have an opposite orientation at implantation. These embryos anchor to the endometrial surface via the polar trophectoderm, the cells that are closest to the inner cell mass. The endometrial stromal fibroblasts transform into secretory decidual cells, a change that in most species, except for the human, is triggered by the embryo [42]. An intimate dialog has to be established between the embryo and the uterine tissue to maintain pregnancy [43]. In human pregnancy, the presence of developmentally impaired embryos leads to reduced interleukin production by the decidualized endometrial cells and this possibly represents a natural selection mechanism [44]. The high incidence of chromosomal abnormalities in embryos obtained by in vitro fertilization may explain the relatively low success rate of human embryo implantation [45]. Whether in vivo-conceived human embryos also exhibit chromosome instability remains unknown, but for different farm animals, a higher percentage of apoptotic cells was detected in in vitro-derived embryos compared with in vivo produced embryos [46].

In animals with a noninvasive type of implantation, the uterine histotroph is an essential nutrient source. In ungulate farm animals, the trophectoderm elongates dramatically presumably to best facilitate nutrient uptake.

Indeed, the trophectoderm of, for instance, porcine embryos grows out to adapt an elongated filamentous shape of over 100 mm in length [47]. In these types of animals, formation of extraembryonic mesoderm, gastrulation, and axis formation occurs before definitive implantation (**Fig. 2.11**).

Upon implantation, the epiblast transitions from a naïve pluripotent state to a polarized state with lumenogenesis leading to the formation of the (pro)amniotic cavity. In the mouse, exit from pluripotency is not required for polarization. Exit from naïve pluripotency is accomplished by downregulation of *Nanog* expression, but this is insufficient to induce lumen formation. Instead, after pluripotency exit *Oct4* together with *Otx2* direct lumen formation through *Rab11* and *Podxl* [48].

Take-Home Message

This chapter covers the following topics:
- Mammalian oocytes, together with somatic cells form follicles in ovaries.
- A finite number of oocytes is present in the ovaries.
- Fusion of an oocyte with a sperm cells gives rise to a totipotent zygote.
- A zygote becomes a multicellular embryo by cleavage divisions.
- The first lineage segregation sets apart the outer trophectoderm cells, needed for implantation into the womb, and the inner cell mass, that will give rise to the fetus.
- In a second lineage segregation event, NANOG-expressing cells in the inner cell mass form the pluripotent epiblast while GATA-expressing cells form the hypoblast that will give rise to the yolk sac.
- In mouse, but not in human embryos, MAP-kinase signaling inhibits epiblast but stimulates hypoblast formation.
- Human embryonic stem cells resemble mouse epiblast stem cells.
- Pluripotency can be tested by formation of chimaera or formation of teratomas when injected into mice.
- Mouse and human embryos implant at the blastocyst stage while embryos of other mammalian species implant at stages when gastrulation has already started.

References

1. Epifano O, Liang LF, Familari M, Moos MC Jr, Dean J. Coordinate expression of the three zona pellucida genes during mouse oogenesis. Development. 1995;121(7):1947–56.
2. Aerts JM, Bols PE. Ovarian follicular dynamics. A review with emphasis on the bovine species. Part II: antral development, exogenous influence and future prospects. Reprod Domest Anim. 2010;45(1):180–7.
3. Reyes JM, Ross PJ. Cytoplasmic polyadenylation in mammalian oocyte maturation. Wiley Interdiscip Rev RNA. 2016;7(1):71–89.
4. Norbury CJ. Cytoplasmic RNA: a case of the tail wagging the dog. Nat Rev Mol Cell Biol. 2013;14(10):643–53.
5. de Rooij DG. The nature and dynamics of spermatogonial stem cells. Development. 2017;144(17):3022–30.
6. Yang P, Wu W, Macfarlan TS. Maternal histone variants and their chaperones promote paternal genome activation and boost somatic cell reprogramming. BioEssays. 2015;37(1):52–9.
7. Flechon JE. The acrosome of eutherian mammals. Cell Tissue Res. 2016;363(1):147–57.
8. Svoboda P. Mammalian zygotic genome activation. Semin Cell Dev Biol. 2018;84:118–26.
9. Saini D, Yamanaka Y. Cell polarity-dependent regulation of cell allocation and the first lineage specification in the preimplantation mouse embryo. Curr Top Dev Biol. 2018;128:11–35.
10. Turlier H, Maitre JL. Mechanics of tissue compaction. Semin Cell Dev Biol. 2015;47–48:110–7.
11. Strumpf D, Mao CA, Yamanaka Y, Ralston A, Chawengsaksophak K, Beck F, et al. Cdx2 is required for correct cell fate specification and differentiation of trophectoderm in the mouse blastocyst. Development. 2005;132(9):2093–102.
12. Niwa H, Toyooka Y, Shimosato D, Strumpf D, Takahashi K, Yagi R, et al. Interaction between Oct3/4 and Cdx2 determines trophectoderm differentiation. Cell. 2005;123(5):917–29.
13. Sasaki H. Mechanisms of trophectoderm fate specification in preimplantation mouse development. Develop Growth Differ. 2010;52(3):263–73.
14. Kuijk EW, Du Puy L, Van Tol HT, Oei CH, Haagsman HP, Colenbrander B, et al. Differences in early lineage segregation between mammals. Dev Dyn. 2008;237(4):918–27.

15. Berg DK, Smith CS, Pearton DJ, Wells DN, Broadhurst R, Donnison M, et al. Trophectoderm lineage determination in cattle. Dev Cell. 2011;20(2):244–55.
16. Frankenberg S, Shaw G, Freyer C, Pask AJ, Renfree MB. Early cell lineage specification in a marsupial: a case for diverse mechanisms among mammals. Development. 2013;140(5):965–75.
17. Marikawa Y, Alarcon VB. Creation of trophectoderm, the first epithelium, in mouse preimplantation development. Results Probl Cell Differ. 2012;55:165–84.
18. Kwon GS, Viotti M, Hadjantonakis AK. The endoderm of the mouse embryo arises by dynamic widespread intercalation of embryonic and extraembryonic lineages. Dev Cell. 2008;15(4):509–20.
19. Frankenberg SR, de Barros FR, Rossant J, Renfree MB. The mammalian blastocyst. Wiley Interdiscip Rev Dev Biol. 2016;5(2):210–32.
20. Chazaud C, Yamanaka Y, Pawson T, Rossant J. Early lineage segregation between epiblast and primitive endoderm in mouse blastocysts through the Grb2-MAPK pathway. Dev Cell. 2006;10(5):615–24.
21. Yamanaka Y, Lanner F, Rossant J. FGF signal-dependent segregation of primitive endoderm and epiblast in the mouse blastocyst. Development. 2010;137(5):715–24.
22. Kang M, Garg V, Hadjantonakis AK. Lineage establishment and progression within the inner cell mass of the mouse blastocyst requires FGFR1 and FGFR2. Dev Cell. 2017;41(5):496–510.e5.
23. Molotkov A, Mazot P, Brewer JR, Cinalli RM, Soriano P. Distinct requirements for FGFR1 and FGFR2 in primitive endoderm development and exit from pluripotency. Dev Cell. 2017;41(5):511–26.e4.
24. du Puy L, Lopes SM, Haagsman HP, Roelen BA. Analysis of co-expression of OCT4, NANOG and SOX2 in pluripotent cells of the porcine embryo, in vivo and in vitro. Theriogenology. 2011;75(3):513–26.
25. Kuijk EW, van Tol LT, Van de Velde H, Wubbolts R, Welling M, Geijsen N, et al. The roles of FGF and MAP kinase signaling in the segregation of the epiblast and hypoblast cell lineages in bovine and human embryos. Development. 2012;139(5):871–82.
26. Roode M, Blair K, Snell P, Elder K, Marchant S, Smith A, et al. Human hypoblast formation is not dependent on FGF signalling. Dev Biol. 2012;361(2):358–63.
27. Brons IG, Smithers LE, Trotter MW, Rugg-Gunn P, Sun B, Chuva de Sousa Lopes SM, et al. Derivation of pluripotent epiblast stem cells from mammalian embryos. Nature. 2007;448(7150):191–5.
28. Tesar PJ, Chenoweth JG, Brook FA, Davies TJ, Evans EP, Mack DL, et al. New cell lines from mouse epiblast share defining features with human embryonic stem cells. Nature. 2007;448(7150):196–9.
29. Nichols J, Smith A. Naive and primed pluripotent states. Cell Stem Cell. 2009;4(6):487–92.
30. Rossant J, Frels WI. Interspecific chimeras in mammals: successful production of live chimeras between Mus musculus and Mus caroli. Science. 1980;208(4442):419–21.
31. Fehilly CB, Willadsen SM, Tucker EM. Interspecific chimaerism between sheep and goat. Nature. 1984;307(5952):634–6.
32. Boklage CE. Embryogenesis of chimeras, twins and anterior midline asymmetries. Hum Reprod. 2006;21(3):579–91.
33. Koller BH, Hagemann LJ, Doetschman T, Hagaman JR, Huang S, Williams PJ, et al. Germ-line transmission of a planned alteration made in a hypoxanthine phosphoribosyl transferase gene by homologous recombination in embryonic stem cells. Proc Natl Acad Sci U S A. 1989;86(22):8927–31.
34. Wood SA, Allen ND, Rossant J, Auerbach A, Nagy A. Non-injection methods for the production of embryonic stem cell-embryo chimaeras. Nature. 1993;365(6441):87–9.
35. Nagy A, Rossant J, Nagy R, Abramow-Newerly W, Roder JC. Derivation of completely cell culture-derived mice from early-passage embryonic stem cells. Proc Natl Acad Sci USA. 1993;90(18):8424–8.
36. Thomson JA, Itskovitz-Eldor J, Shapiro SS, Waknitz MA, Swiergiel JJ, Marshall VS, et al. Embryonic stem cell lines derived from human blastocysts. Science. 1998;282(5391):1145–7.
37. Masaki H, Kato-Itoh M, Umino A, Sato H, Hamanaka S, Kobayashi T, et al. Interspecific in vitro assay for the chimera-forming ability of human pluripotent stem cells. Development. 2015;142(18):3222–30.
38. Gafni O, Weinberger L, Mansour AA, Manor YS, Chomsky E, Ben-Yosef D, et al. Derivation of novel human ground state naive pluripotent stem cells. Nature. 2013;504(7479):282–6.

39. Wang X, Li T, Cui T, Yu D, Liu C, Jiang L, et al. Human embryonic stem cells contribute to embryonic and extraembryonic lineages in mouse embryos upon inhibition of apoptosis. Cell Res. 2018;28(1):126–9.
40. Mascetti VL, Pedersen RA. Human-mouse chimerism validates human stem cell pluripotency. Cell Stem Cell. 2016;18(1):67–72.
41. Bazer FW, Spencer TE, Johnson GA, Burghardt RC, Wu G. Comparative aspects of implantation. Reproduction. 2009;138(2):195–209.
42. Gellersen B, Brosens IA, Brosens JJ. Decidualization of the human endometrium: mechanisms, functions, and clinical perspectives. Semin Reprod Med. 2007;25(6):445–53.
43. Salker M, Teklenburg G, Molokhia M, Lavery S, Trew G, Aojanepong T, et al. Natural selection of human embryos: impaired decidualization of endometrium disables embryo-maternal interactions and causes recurrent pregnancy loss. PLoS One. 2010;5(4):e10287.
44. Teklenburg G, Salker M, Molokhia M, Lavery S, Trew G, Aojanepong T, et al. Natural selection of human embryos: decidualizing endometrial stromal cells serve as sensors of embryo quality upon implantation. PLoS One. 2010;5(4):e10258.
45. Vanneste E, Voet T, Le Caignec C, Ampe M, Konings P, Melotte C, et al. Chromosome instability is common in human cleavage-stage embryos. Nat Med. 2009;15(5):577–83.
46. Pomar FJ, Teerds KJ, Kidson A, Colenbrander B, Tharasanit T, Aguilar B, et al. Differences in the incidence of apoptosis between in vivo and in vitro produced blastocysts of farm animal species: a comparative study. Theriogenology. 2005;63(8):2254–68.
47. Geisert RD, Brookbank JW, Roberts RM, Bazer FW. Establishment of pregnancy in the pig: II. Cellular remodeling of the porcine blastocyst during elongation on day 12 of pregnancy. Biol Reprod. 1982;27(4):941–55.
48. Shahbazi MN, Scialdone A, Skorupska N, Weberling A, Recher G, Zhu M, et al. Pluripotent state transitions coordinate morphogenesis in mouse and human embryos. Nature. 2017;552(7684):239–43.

Origins of Pluripotency: From Stem Cells to Germ Cells

Maria Gomes Fernandes and Susana M. Chuva de Sousa Lopes

Contents

3.1   Early Mouse Embryonic Development – 31

3.2   Mouse Embryonic Stem Cells – 32
3.2.1   States of Pluripotency In Vitro – 34

3.3   Human Embryonic Stem Cells – 38
3.3.1   Primed Versus Naïve Pluripotency in hESCs – 39
3.3.2   Epigenetics in Naïve Versus Primed Pluripotency in Humans – 39

© Springer Nature Switzerland AG 2020
G. Rodrigues, B. A. J. Roelen (eds.), *Concepts and Applications of Stem Cell Biology*,
Learning Materials in Biosciences, https://doi.org/10.1007/978-3-030-43939-2_3

3.4	hESCs as a Model to Study Embryonic Development – 40
3.5	Primordial Germ Cells Have an Underlying Pluripotent State – 41
3.5.1	Origin and Specification of PGCs in Humans and Other Animals – 41
3.5.2	Molecular Mechanisms Regulating Specification of PGCs in Humans and Other Animals – 43
3.5.3	Migration of PGCs in Humans and Other Animals – 43
3.5.4	Arrival and Colonization of the Gonad – 44
3.5.5	Protecting PGC Genome Integrity – 44
3.5.6	Protocols for In Vitro Germ Cell Development – 46

References – 47

3

Origins of Pluripotency: From Stem Cells to Germ Cells

What Will You Learn in This Chapter
In this chapter, the basic concept of pluripotency is explored. You will learn about the origins of pluripotent stem cells, both in mice and in human. The embryonic origins of stem cells will be discussed. The differences and similarities between naïve and primed embryonic stem cells will be described, and you will learn about the gene regulatory networks and signalling pathways that regulate each of them. This will relate to different culture conditions and the use of small molecules to keep pluripotency in vitro. You will also learn about the origin and development of primordial germ cells, the signalling pathways involved and how studies regarding pluripotency in vitro and in vivo helped to establish protocols for the generation of primordial germ cell-like cells in vitro.

Learning Objectives
After completing this chapter, students should be able to:
1. Distinguish different types of pluripotent stem cells and the signalling needed to maintain them in vitro
2. Pinpoint the differences between mouse embryonic stem cells and human embryonic stem cells
3. Understand the origins and development of primordial germ cells and the main differences and similarities between animal models
4. Understand the main signalling pathways and markers associated with specific stages of germ cell development
5. Understand the need for developmental biology studies together with stem cell biology studies for the development of in vitro protocols for the derivation of primordial germ cells and vice versa

Important Concepts Discussed in This Chapter
- Naïve and primed pluripotency
- The ground state of pluripotent stem cells
- Gene regulatory networks
- Epigenetic state of naïve and primed pluripotency
- Signalling pathways and relationship with pluripotency
- Primordial germ cells

3.1 Early Mouse Embryonic Development

Fusion of a mature oocyte with a sperm cell results in the formation of the zygote at embryonic day (E)0, and the life cycle of an organism (and of its germ cells) starts once again. The zygote has the ability to generate all the cells of the embryo, including the extraembryonic cells, essential to support proper embryonic development. The zygote is therefore referred to as totipotent. After a series of cleavage divisions, a morula (4–16 cells) is generated (E2.0–E2.5, in mice). In mice, the cells of the uncompacted morula can still give rise to all extraembryonic and embryonic tissues, and are still considered totipotent.

At the 32-cell stage, the morula undergoes a process called compaction followed, at E3.0, by cavitation – the formation of a central cell-free cavity. In mice, the E3.5 blastocyst has a distinctive inner cell mass (ICM) and an outer layer of trophectoderm cells (TE) [1, 2] (◘ Fig. 3.1a). Expression of the transcription factor *Cdx2* directs TE

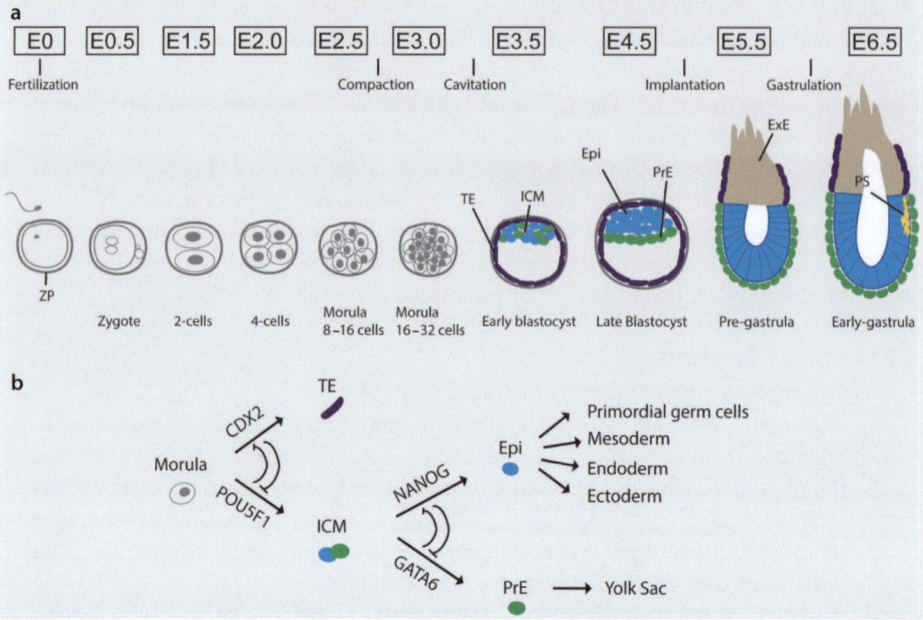

Fig. 3.1 Mouse early embryonic development and lineage choices. **a** Overview of mouse early embryonic development from the moment of fertilization, at embryonic day (E)0, until the onset of gastrulation with the appearance of the primitive streak (PS) at E6.5. **b** Main lineage choices and transcription factors that regulate those choices in the pre-implantation mouse embryo. *CDX2* drives trophectoderm (TE) differentiation while *POU5F1* keeps pluripotency in the inner cell mass (ICM). Later, *GATA6* drives primitive endoderm (PrE) fate while *NANOG*-positive cells give rise to the epiblast (Epi). The epiblast will form the primordial germ cells and the germ layers of the embryo (mesoderm, endoderm, ectoderm). Abbreviations: TE trophectoderm, ICM inner cell mass, Epi epiblast, PrE primitive endoderm, ExE extraembryonic ectoderm, PS primitive streak, E embryonic day

differentiation, while ICM fate is directed by *Pou5f1* (or *Oct4*) expression [3, 4]. At this stage, the ICM can only give rise to the embryonic germ layers (ectoderm, endoderm and mesoderm), extraembryonic mesoderm and germline, and consists of pluripotent cells [5]. In mice, E3.75 ICM is composed of two distinct populations of cells: *Gata6+Nanog-* cells and *Gata6-Nanog+* cells. *Gata6+* cells are the primitive endoderm (PrE) progenitors that, at around E4.0, become lineage restricted and are located between the blastocoel cavity and the *Gata6-Nanog+* epiblast (Epi) [6, 7] (Fig. 3.1b). Thereafter, in mice at E5.0, implantation with the formation of the egg cylinder begins, and at E5.5, the anterior–posterior axis of the embryo is defined. Gastrulation, during which the three embryonic germ layers form, begins around E6.0 (Fig. 3.1a). The Epi loses its functional pluripotency around E8.0 [8–10], becoming the multipotent ectoderm layer that can only give rise to ectoderm derivatives [11].

3.2 Mouse Embryonic Stem Cells

Mouse embryonic stem cells (mESCs) were first established from the ICM of E3.5 blastocysts in 1981 [12, 13] (Fig. 3.2). These cells have an unlimited capacity for self-renewal during in vitro culture, can be expanded clonally and retain pluripo-

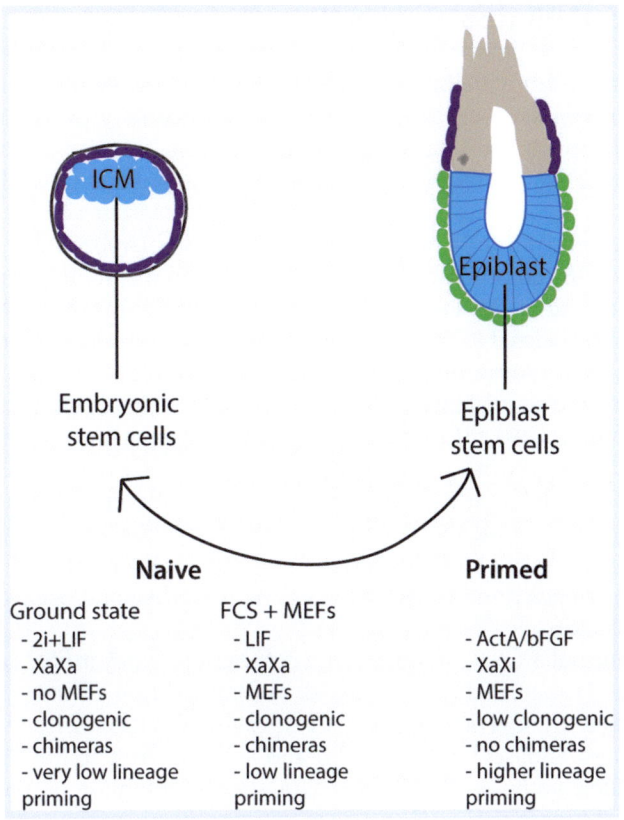

◘ Fig. 3.2 Different types of mouse embryonic stem cells. Pluripotent cell lines can be derived from the inner cell mass (ICM) of the mouse blastocyst or from the epiblast (Epi) of the early implanting embryo. These lines reflect their embryonic origin and have different signalling requirements in culture. Ground state and naïve mouse embryonic stem cells (mESCs) are LIF-dependent. Ground state mESCs are MEFs and FCS-free. Epiblast embryonic stem cells (EpiSCs) are FGF-dependent and have one inactivated X chromosome (Xi) contrary to the two active X (Xa) in ground and naïve mESCs (reflecting the X status of the ICM and epiblast). Although EpiSCs can differentiate into the three germ layers in vitro, they do not generate chimeras and have lower clonogenicity than naïve mESCs. Abbreviations: Epi epiblast, ICM inner cell mass, mESCs mouse embryonic stem cells, MEFs mouse embryonic fibroblasts, FCS foetal calf serum, EpiSCs epiblast embryonic stem cells, LIF leukaemia inhibitory factor, FGF fibroblast growth factor

tency, as they are capable of generating all the cells of the adult body in chimeras, but not TE-derived cells [14]. The characteristics of mESCs in culture also reflect their biological origin: they express ICM-associated genes such as *Pou5f1*, *Nanog* and *Sox2*, and can be differentiated in vitro into derivatives of mesoderm, endoderm and ectoderm, but they do not form TE-derived cells [14–16]. Pluripotency of mESCs is usually demonstrated through two classical in vivo experiments: (1) when injected into immunodeficient mice, mESCs are capable of generating tumours containing derivatives of the three embryonic germ layers called teratomas ("teratoma assay"); and (2) when injected into a blastocyst or after morula aggregation, mESCs can contribute to the three embryonic germ layers and the germline of the resulting chimeric embryo.

3.2.1 States of Pluripotency In Vitro

The first in vitro culture methods for mESCs involved the conditioning of medium by teratocarcinoma stem cell lines as the main source of growth factors [12, 13]. Later, it was demonstrated that the cytokine leukaemia inhibitory factor (LIF) was one of the growth factors required for mESC self-renewal and to inhibit differentiation [17, 18]. LIF signals through the JAK/STAT (Janus kinase/signal transducer and activator of transcription) pathway, particularly the JAK/STAT3 axis [19, 20]. Nevertheless, mouse embryonic fibroblasts (MEFs) and foetal calf serum (FCS) were usually still required as the source of other important factors. mESCs derived from the E3.5 ICM are referred to as naïve pluripotent stem cells. They differ from another type of pluripotent mouse stem cells derived from the E5.5 post-implantation epiblast, known as mouse epiblast stem cells (mEpiSCs) [21, 22] (◘ Fig. 3.2). Despite being pluripotent in vitro, mEpiSCs do not give rise to chimeras very efficiently; thus, they are considered to be in a primed state [23, 24]. mEpiSCs require ActivinA (ActA) and fibroblast growth factor 2 (FGF2) to maintain pluripotency rather than LIF and FCS, and mEpiSCs are highly prone to undergo apoptosis when passaged in single cells [22, 24]. Naïve mESCs cultured in the presence of MEFs and FCS show a high degree of heterogeneity and transit between the naïve-like and primed-like pluripotency states [25, 26].

In the absence of FCS, LIF alone is not sufficient to block mESC commitment to differentiation. However, this is bypassed by the addition of two small-molecule inhibitors (called 2i): CHIR99021 and PD0325901. CHIR99021 acts by inhibiting glycogen synthase kinase 3 beta (Gsk3β) and consequently activating the WNT/β-catenin signalling pathway. PD0325901 is a specific inhibitor of the extracellular signal-regulated kinase (ERK1/2)/mitogen-activated protein kinase (MAPK) signal transduction pathway [27]. mESCs cultured in LIF+2i medium can be maintained without FCS and MEFs, and are in the ground state of pluripotency. The identification of this ground state allowed not only the maintenance of mESCs under chemically defined culture conditions but also the derivation of mESCs from "non-permissive" mouse strains such as C57Bl6 [28]. Ground state mESCs are phenotypically more homogeneous than mESCs grown in FCS and MEFs, show lower levels of DNA methylation and lower expression of lineage specific-associated genes, and hence are more similar to E3.5 ICM cells than mESCs grown in FCS and MEFs [29, 30] (◘ Figs. 3.2 and 3.3).

3.2.1.1 Gene Regulatory Networks in Mouse Naïve Pluripotent mESCs

The core gene regulatory network associated with pluripotency in naïve mESCs consists of *Nanog*, *Sox2* and *Pou5f1* [31]. It is the delicate transcriptional balance between these core regulatory genes that maintains naïve pluripotency and prevents differentiation during in vitro culture [32]. In addition to these core genes, other transcription factors such as *Klf2*, *Klf4*, *Zpf42*, *Myc*, *Prdm14*, *Sall4*, *Tfcp2l1*, *Esrrb*, *Tcf3*, *Gbx2*, *Dppa3* and *Tbx3* are also involved in the maintenance of pluripotency in naïve mESCs [33–40]. The balance between self-renewal and differentiation in mESCs is also regulated by factors that include *Id1* and *Dusp9*, both downstream targets of the bone morphogenic protein (BMP) signalling pathway [41, 42].

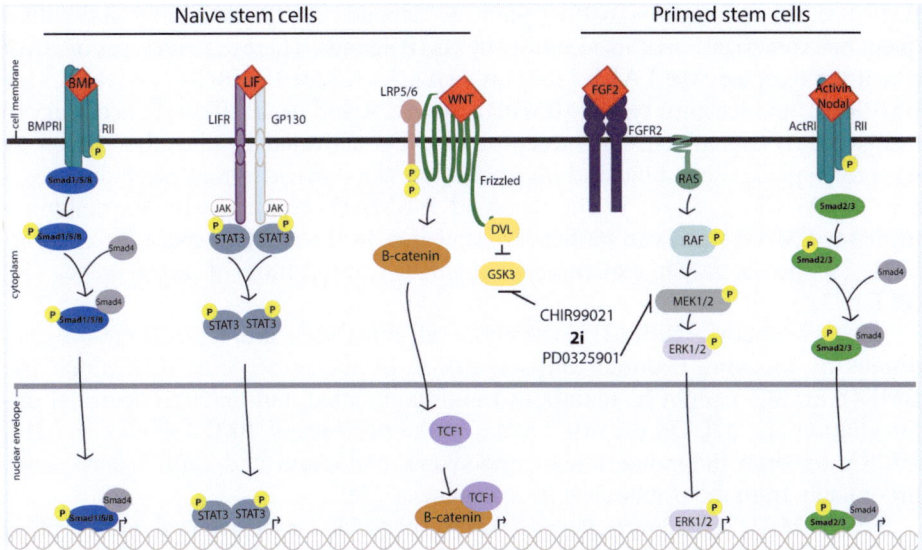

Fig. 3.3 Signalling pathways in naïve and primed pluripotency. The major signalling pathways in pluripotency include the BMP, LIF, WNT, FGF and ActivinA/Nodal signalling pathways. Typically, the ligands (red) bind and bring together their transmembrane receptors and co-receptors. Those complexes usually include a kinase that is able to phosphorylate (P) the cytoplasmic effectors, which are then able to translocate to the nucleus and regulate transcription of target genes. Abbreviations: BMP bone morphogenic protein, BMPRI BMP receptor type 1, RII receptor type 2, P phosphorylation, LIF leukaemia inhibitory factor, LIFR LIF receptor, GP130 glycoprotein 130, JAK Janus kinase, STAT signal transducer and activator of transcription, WNT wingless/integrated, LRP low-density lipoprotein receptor-related protein, DVL dishevelled, GSK glycogen synthase kinase, TCF T-cell factor, FGF fibroblast growth factor, FGFR2 FGF receptor 2, RAS rat sarcoma oncogene, RAF rapidly accelerated fibrosarcoma, MEK mitogen-activated protein kinase kinase, ERK extracellular signal-regulated kinase, ActRI activin receptor type 1

LIF Signalling in mESCs

The LIF signalling cascade is initiated with the binding of LIF to its receptor LIFR in association with its co-receptor subunit glycoprotein 130 (gp130). LIFR and gp130 heterodimers activate associated tyrosine kinases such as the family of Janus kinases (JAKs). JAKs then phosphorylate gp130 promoting the recruitment of STAT3. STAT3 is also phosphorylated by JAKs, dimerizes and translocates to the nucleus where it regulates the transcription of target genes [43, 44] (Fig. 3.3).

In mESCs, LIF triggers several different signalling pathways: the JAK/STAT3 pathway; the PI3K (phosphoinositide 3-kinase)/PKB (protein kinase B) pathway; and the SHP2 (SH2 domain-containing tyrosine phosphatase 2)/MAPK pathway [45, 46]. Nevertheless, only the STAT3 pathway is essential for LIF-mediated mESCs self-renewal [47, 48]. Downstream targets of LIF are, for example, *Myc*, *Klf4*, *Pim1/3*, *Prr13*, *Gbx2*, *Pramel7*, *Pem/Rhox5*, *Jmjd1a* and *Tfcp2l1* [36, 40, 49–54].

BMP Signalling in mESCs

BMP is part of the larger transforming growth factor (TGF)-β family. These proteins are involved in the regulation of cell proliferation, differentiation and apoptosis, thus playing key roles during embryonic development and pattern formation [55]. The

BMP-SMAD canonical signalling pathway depends on the activation and subsequent heteromerization of its receptors by BMP ligands. There are two types of BMP receptors: receptor type I ALK2 (or AcvR1A), ALK3 (or BMPRIA) and ALK6 (or BMPRIB), and receptor type II (BMPR2, AcvR2A and AcvR2B), both necessary to mediate BMP signalling. Once receptor type I is activated by phosphorylation by receptor type II, it can bind and phosphorylate the downstream intracellular receptor (R)-SMADs 1, 5 and 8. The activated R-SMADs complex with the common-mediator SMAD4 and can be then translocated to the nucleus where the complex binds to specific target sequences to regulate transcription of target genes [56] (Fig. 3.3).

BMP4 cooperates with LIF to promote self-renewal of mESCs in 2i-culture conditions by blocking neuronal differentiation. In the presence of LIF alone (no BMP4), mESCs cannot be maintained undifferentiated, but undergo neuronal differentiation [41, 57]. On the other hand, in the presence of BMP4 alone (no LIF), mESCs undergo differentiation to mesoderm, endoderm and $Cdx2+$ derivatives, presumably from the trophoblast lineage, instead [58].

BMP-SMAD signalling is, nevertheless, dispensable for self-renewal, since mESCs knockout for *Smad1* and *Smad5* self-renew at the same rate as the wild-type lines [42]. This double KO mESC line remained pluripotent, but showed high levels of DNA methylation and high propensity to differentiate, highlighting a role of BMP-SMAD in the regulation of lineage priming, rather than self-renewal.

WNT Signalling in mESCs

In the presence of WNT ligands, the transmembrane receptor frizzled (FZ) and LRP6 or LRP5 (low-density lipoprotein receptor-related protein 5 or 6) form a complex. This WNT-FZ-LRP5/6 complex recruits dishevelled (Dvl), and this event promotes the phosphorylation of LRP5/6 and activation and recruitment of the Axin complex. This inhibits Axin-mediated β-catenin phosphorylation, promoting the stabilization of β-catenin and its accumulation in the cytoplasm. β-Catenin will then translocate to the nucleus, forming complexes with TCF/LEF, regulating the expression of WNT targeted genes [59] (Fig. 3.3).

WNT/β-catenin signalling insures the maintenance of naïve pluripotency in mESCs by multiple mechanisms and tight synergy with other signalling pathways like BMP-SMAD, FGF-ERK and TGFβ-ActA [60–62]. In addition to effects on DNA methylation by regulating the expression of TET proteins [63], the activation of WNT signalling results in the upregulation of *Stat3*, *Klf2* and *Tfcp2l1* [64, 65], and the suppression of neuroectodermal differentiation by downregulation of *Tcf3* [66]. Fluctuations of β-catenin have been correlated with *Nanog* and *Pou5f1* expression in naïve mESCs [67, 68].

3.2.1.2 Gene Regulatory Networks in Mouse Primed Pluripotent mEpiSCs

The core transcriptional regulatory genes, *Pou5f1*, *Nanog* and *Sox2*, are still expressed in mEpiSCs, although there is a downregulation of *Nanog* [69]. Nevertheless, the similarities end here. Whereas genes like *Klf2*, *Klf4*, *Klf5*, *Zpf42*, *Esrrb*, *Dppa3*, *Tfcp2l1*, *Fgf4*, *Tbx3* and *Cdh1* are highly expressed in naïve mESCs, mEpiSCs express genes associated with early lineage specification like *Dnmt3b*, *Fgf5*, *Pou3f1*, *Meis1*, *Otx2*, *Sox11*, *Sox17*, *T* and *Gdf3* [70]. *Essrb* expression is reduced in mEpiSCs, due to the translocation of *Tfe3* from the nucleus to the cytoplasm during conversion to the primed state [71]. Primed pluripotency requires both the TGF-β signalling pathway, via the ligands ActA/Nodal and FGF2 (or bFGF) for self-renewal [21, 22].

FGF Signalling in mEpiSCs

The FGF family contains 22 genes divided into 6 subfamilies. Each FGF ligand binds to specific splice variants of the FGF receptor (FGFR) and uses either heparin-like glycosaminoglycans or transmembrane Klotho enzymes as co-factors. Upon ligand–receptor interaction, autophosphorylation of the intracellular region of the FGFR occurs, and this can activate four distinct pathways: JAK/STAT, PI3K, PLCγ (phosphoinositide phospholipase C) and Erk pathways [72].

FGF2 uses heparin-like glycosaminoglycans as co-factors [73]. FGF2 appears to stabilize the primed pluripotency state by dual inhibition of differentiation to neuroectoderm and blocking the reversion to a naïve state [62] (◘ Fig. 3.3). FGF2 also has an indirect effect on the maintenance of primed pluripotency by stimulating MEFs to produce ActA [74]. Interestingly, it has been shown that FGF4, which also uses heparin-like molecules as co-factors, promotes self-renewal of mEpiSCs without exogenous stimulation of ActA/Nodal [70]. mEpiSCs cultured with FGF4 were more homogeneous regarding *Pou5f1* expression [70].

ActA/Nodal Signalling in mEpiSCs

ActA and Nodal are members of the TGF-β superfamily [75]. ActA and Nodal ligands signal via the same receptors and effectors, and in the majority of the cases, the resulting signalling is the same. They bind to receptor type II AcvR2A and AcvR2B (both also BMP receptors), leading to the recruitment of the specific receptor type I ALK4 (or AcvR1B) and ALK7 (or AcvR1C) [76]. AcvR2A/2B and ALK4/7 then trigger the phosphorylation of the R-SMADs 2 and 3, which complex with the common-mediator SMAD4 and translocate to the nucleus regulating gene expression of specific targets [75] (◘ Fig. 3.3). Activation of the ActA/Nodal pathway promotes self-renewal of EpiSCs via direct activation of *Nanog*, whereas inhibition of this pathway induces neuroectodermal differentiation [62].

3.2.1.3 Epigenetics in Naïve Versus Primed Pluripotency in Mice

There are important epigenetic differences between mESCs and mEpiSCs [30, 77]. Female mESCs have two active X chromosomes, while in female mEpiSCs, X chromosome inactivation has occurred, so they present one inactive or silent X chromosome [24]. This reflects the different embryonic origins of naïve and primed pluripotency. In general, the genome of mEpiSCs is hypermethylated when compared to mESCs [29, 78]. Similarly, there is a reduced prevalence of the repressive histone mark H3K27me3 (histone 3 lysine 27 trimethylation) at promoters and fewer bivalent domains in naïve mESCs [30]. Most importantly, there are also differences in the enhancer usage of genes between naïve and primed cells. This occurs not only in differentially expressed genes but also in commonly expressed genes, such as *Pou5f1*. The distal enhancer (DE) of *Pou5f1* is used in naïve mESCs, whereas its proximal enhancer (PE) is methylated. By contrast, in primed mEpiSCs, the *Pou5f1* PE is used and the *Pou5f1* DE is methylated [79]. Also, development-associated enhancers, called seed enhancers, convert from a dormant to an active state in mEpiSCs, and this is thought to regulate lineage priming [80].

3.3 Human Embryonic Stem Cells

The first pluripotent human embryonic stem cells (hESCs) were derived in 1998 and initially maintained in culture on MEFs and in medium containing FCS [81]. Although derived from the same embryonic-stage blastocyst embryos, hESCs display primed pluripotency and thus show important differences to naïve mESCs: flattened colonies that are highly sensitive to single-cell passage; different associated pluripotent markers like SSEA-3, SSEA-4 and TRA-1-81 instead of SSEA-1; different signalling requirements in culture (dependency on FGF/TGFβ instead of LIF/STAT3) and an inactive X chromosome in female hESCs [24, 82] (◘ Fig. 3.4).

The similarities between mEpiSCs and hESCs suggest that the isolated human ICM displays a more advanced embryonic state, acquiring an EpiSC-like signature in culture. Alternatively, the primed signature reflects the fact that human ICM does not have the ability to undergo diapause (pause development), unlike the mouse, and epithelializes in culture. In agreement, an epithelialized post-ICM intermediate

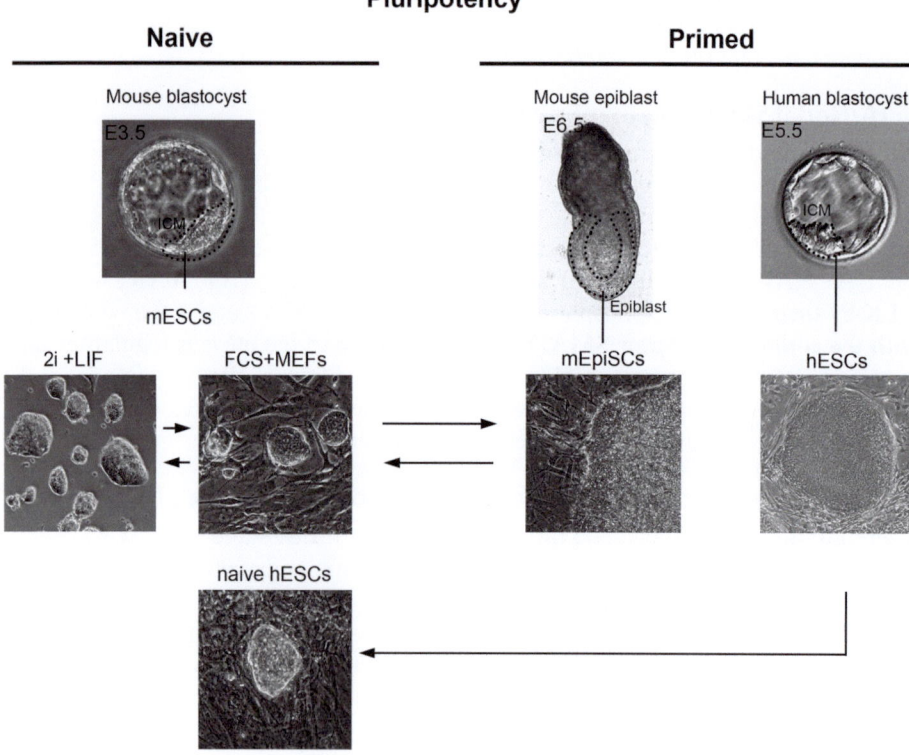

◘ Fig. 3.4 Origins of mouse and human embryonic pluripotent stem cells. Naïve mESCs (mouse embryonic stem cells) are derived from the inner cell mass (ICM) of E3.5 blastocysts. mESCs represent naïve pluripotency in 2i+LIF (ground state) and FCS and MEFs culture conditions. Primed mEpiSCs (mouse epiblast stem cells) are derived from the epiblast of E6.5 embryos. mEpiSCs can be converted to mESCs and vice versa. Primed hESCs (human embryonic stem cells) are derived from the ICM of E5–E7 human blastocyst. hESCs can be reverted to a naïve-like state. Abbreviations: ICM inner cell mass, mESCs mouse embryonic stem cells, hESCs human embryonic stem cells, MEF mouse embryonic fibroblasts, FCS foetal calf serum, mEpiSCs mouse epiblast embryonic stem cells, LIF leukaemia inhibitory factor, E embryonic day

(PICMI) has been identified during the transition from human ICM to hESCs in culture [83] (Fig. 3.4).

hESCs and EpiSCs are not strictly identical: contrarily to EpiSCs, hESCs express pre-implantation markers such as *REX1* [84] and not post-implantation markers such as *FGF5* [85]. In addition, not all female hESCs lines have gone through X chromosome inactivation, and different X chromosome states have been described in hESCs [86–89]. Also, the pattern of DNA methylation of primed hESCs resembles more the one from naïve mESCs [90, 91].

3.3.1 Primed Versus Naïve Pluripotency in hESCs

After the identification of naïve, primed and ground states in the mouse, efforts are being made to push primed hESCs into the ground state (reviewed in [92, 93]). Most of these protocols included LIF+2i conditions, but these alone were not sufficient to induce naïve pluripotency in hESCs [84, 94–97].

The first attempts to induce a naïve state in (primed) hESCs relied on the overexpression of *KLF4*, *KLF2* and *POU5F1* in the presence of LIF and 2i [95]. These naïve-like hESCs showed high levels of phosphorylated (p)STAT3 and differentiated when exposed to a JAK inhibitor that blocks phosphorylation of STAT3, similar to naïve mESCs. Also, naïve-like hESCs did not differentiate upon addition of BMP4 or inhibition of FGF2, as primed hESCs and mEpiSCs do.

The first transgene-independent naïve-like hESCs were described by Gafni and colleagues in 2013 [96]. They developed a naïve pluripotency growth medium (naïve human stem cell medium, NHSM) to use in both MEF and MEF-free conditions. This medium contained 2i+LIF together with p38 inhibitor (p38i), Jun N-terminal kinase inhibitor (JNKi), aPKCi, RHO-associated protein kinase 1 inhibitor (ROCKi) and a low dose of FGF2 and TGFβ1 (or ActA). These converted naïve-like hESCs showed downregulation of lineage priming-associated genes including *DNMT3B*, *OTX2, ZIC2* and *CD24*. Other studies followed describing the conversion of primed hESCs to naïve-like pluripotency using different cocktails of molecules [84, 94, 97], but in none of them, a complete independency of FGF2 and/or TGFβ/ActA signalling was achieved. Interestingly, naïve-like hESCs under different culture conditions exhibit dependence of the mTORC2 subunit of PI3K/PKB/mTORC pathway [98]. This suggests that naïve pluripotency is different in mESCs and hESCs, although more studies are required to understand these differences.

3.3.2 Epigenetics in Naïve Versus Primed Pluripotency in Humans

Similarly with what has been shown in naïve and primed mouse pluripotent stem cells, naïve-like hESCs show lower levels of DNA methylation that primed hESCs [99]. Nevertheless, it is hard to pinpoint epigenetic differences between naïve-like and primed hESCs. The different approaches to induce naïve-like hESCs using different molecule cocktails result in hESCs that show a wide "range" of naïve properties, and are (epi)genetically unstable [100]. Moreover, the switch from the PE to the DE usage in the *POU5F1* locus, used as a hallmark of naïve and primed mouse stem cells, may not occur in hESCs. Moreover, the long-term culture conditions cause abnormal erasure of DNA methylation in imprinting regions [101].

The X chromosome inactivation state in female naïve-like hESCs does not predict the pluripotency state, as several X chromosome activation states have been described in primed hESCs [86–89, 102]. In addition, during long-term culture, there is progressive "erosion" of silencing marks throughout the silent X chromosome in primed hESCs, which can include the absence of the *XIST* cloud, deposition of H3K27me3 and DNA methylation in certain regions of the silent X chromosome [103]. When erosion occurs in the silent X chromosome of female primed hESCs, even if differentiated the cells will not reacquire the silencing marks [104].

3.4 hESCs as a Model to Study Embryonic Development

The interest in hESCs and their pluripotent state resides in the future applications for gene therapy, drug discovery and regenerative medicine. Recently, the scientific community turned its interest to the self-organization capacity of hESCs during in vitro culture, as a model to understand early human development (◘ Fig. 3.5).

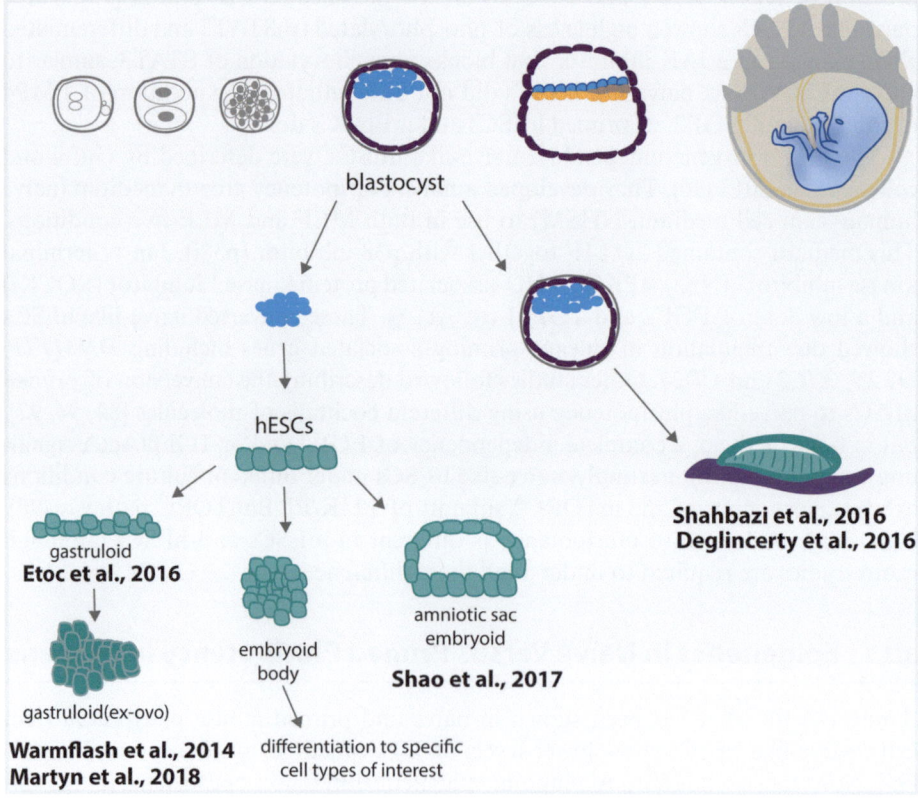

◘ **Fig. 3.5** Can human stem cells mimic some aspects of human embryology? Recent developments and protocols for the culture of hESCs (human embryonic stem cells) and human blastocysts allowed the study of human embryonic development and particularly the study of early implantation period. hESCs can self-organize in vitro into structures that mimic gastrulation (gastruloids), recapitulating some aspects of early embryonic events. hESCs can also be directed to differentiate into specific types of cells via embryoid body differentiation. Abbreviations: hESCs human embryonic stem cells

It has been shown that hESC colonies of a certain size (500 μm), in response to BMP4, pattern spontaneously into concentrically arranged zones, mimicking the arrangement of the mammalian germ layers [105, 106]. More recently, hESCs treated with WNT and ActA and grafted in a chick embryo directed the development of a secondary axis and induced a neural fate in the host, acting like the primitive streak organizer [107]. Another group has generated a synthetic human amniotic sac from hESCs, which they called post-implantation amniotic sac embryoid (PASE) [108].

Using a different approach to understand the human early development, human pre-implantation embryos have been cultured to the implantation stage (14-day limit) in the absence of maternal tissues [109, 110]. These studies open many possibilities to understand human early embryology.

3.5 Primordial Germ Cells Have an Underlying Pluripotent State

Another type of cell that seems to retain some aspects of pluripotency is the primordial germ cell (PGC). PGCs are the first embryonic cell lineage to be lineage restricted in the embryo. Due to their pluripotent-like properties, PGCs can be used to derive another type of pluripotent stem cells known as embryonic germ cells (EGCs). EGCs are derived by culturing PGCs from E8.5–E12.5 mouse embryos in the presence of LIF, FGF2 and stem cell factor (SCF) [111–113]. Like mESCs, mEGCs express the core pluripotency genes, *Pou5f1*, *Sox2* and *Nanog*, and can contribute to mouse chimeras showing germline transmission [111, 114, 115]. Since PGCs only give rise to oocytes or sperm in vivo, the derivation of EGCs is considered a reprogramming event. mEGCs can also be maintained in the ground state [116]. The derivation of human EGCs from human gonadal PGCs has also been attempted [117–119], but the long-term culture of hEGCs has not been achieved successfully.

3.5.1 Origin and Specification of PGCs in Humans and Other Animals

PGCs are highly specialized cells that give rise to gametes during adult life. They are the vehicle through which genetic and epigenetic information is passed from one generation to the next [120]. The mechanisms through which PGCs are specified, migrate and differentiate, first outside and then inside the gonads, differ between species.

There are two main mechanisms thought to govern the formation of PGCs: preformation and epigenesis (or induction). In animals that use preformation, germ cell precursors are defined by the direct inheritance of maternal factors (germ plasma) physically contained in the oocyte. In animals using epigenesis (or induction), germ cell fate is induced de novo during embryonic development [121–123]. Preformation has been documented, for example, in the fruit fly and chicken [124–126]. In the chick, cells expressing the post-migratory germ cell marker CVH (chicken vasa homologue) are already present in the centre of the blastoderm [126, 127] (◘ Fig. 3.6a). Epigenesis is so far common to all mammals studied, but it has also been described in the axolotl [128].

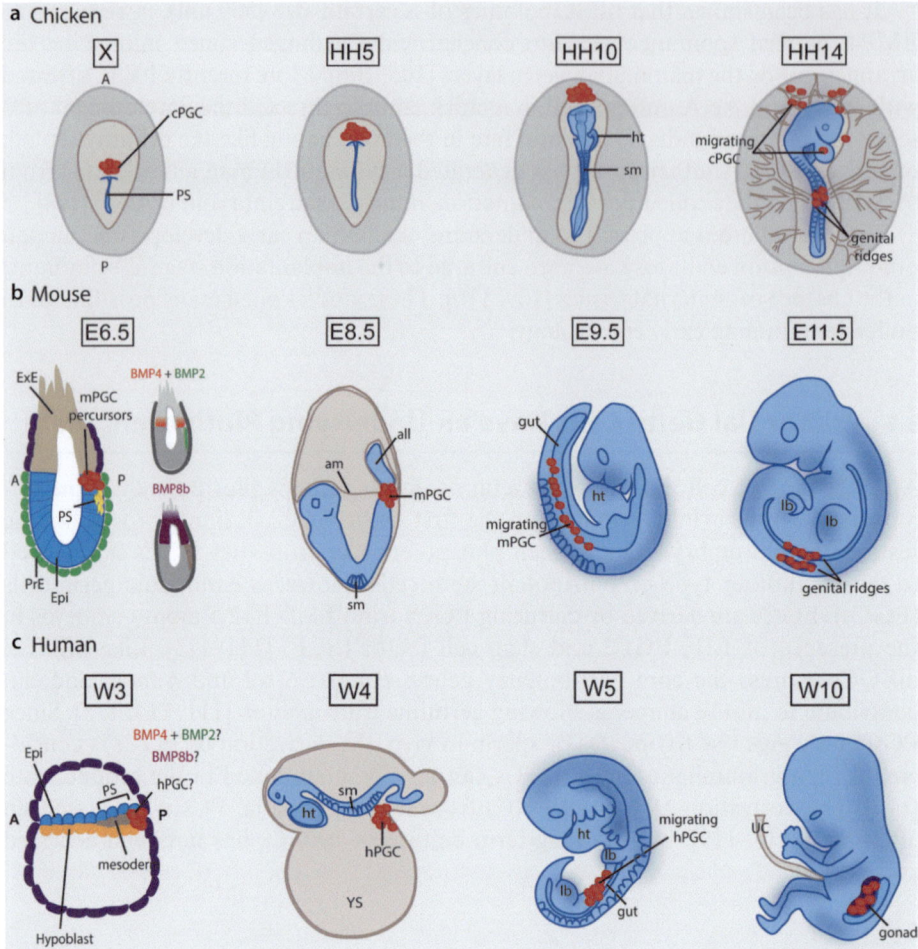

Fig. 3.6 Different origins and migratory routes of primordial germ cells. **a** In chicken, primordial germ cell (PGC) fate is passed on via germ plasma present in the egg. During gastrulation (Eyal–Giladi and Kochav stage X), the primitive streak develops, and the chicken PGCs (cPGCs) are localized ventrally and move to the anterior part of the embryo to localize to the germinal crescent, anteriorly to the head at Hamburger–Hamilton stage (HH)10. Around HH15, cPGCs migrate from the crescent to the genital ridges via the anterior vitelline veins and the aorta. **b** In mice, BMP4 and BMP8b from the extraembryonic ectoderm (ExE) and BMP2 visceral endoderm induce PGC competence at the posterior part of epiblast adjacent to the extraembryonic ectoderm at E6.5. mPGCs localized at the base of the allantois migrate via the hindgut, reaching the genital ridges at embryonic day (E)11.5. **c** In humans, the origin of the signals that induce PGC competence is unknown. hPGC competence is thought to be initiated around gastrulation (W3 of embryonic development, 12–16 days after fertilization) at the posterior part of the embryo, where the primitive streak is formed. At W5, hPGCs are located at the endoderm of the yolk sac wall near the allantois. They migrate via the gut endoderm and the dorsal mesentery to colonize the gonadal ridges around W6–7. At W10, hPGCs are encapsulated in the gonads. Abbreviations: A anterior part, P posterior part, PS primitive streak, ht heart, sm somites, ExE extraembryonic ectoderm, PrE primitive endoderm, Epi epiblast, am amnion, all allantois, lb limb bud, UC umbilical cord, PGC primordial germ cells, cPGC chicken PGCs, mPGCs mouse PGCs, hPGCs human PGCs, HH Hamburger–Hamilton stage, E embryonic day, W week, BMP bone morphogenic protein

Although the two models seem different, there is mounting evidence that in animals using preformation, an induction mechanism is also important for PGC lineage specification, and conversely, in animals with epigenesis, the oocyte may retain some maternal-inherited factors important for PGCs. Recently, a unifying model has been proposed, suggesting that all animals show a period of multipotent pre-PGCs, followed by lineage restriction by induction [123].

3.5.2 Molecular Mechanisms Regulating Specification of PGCs in Humans and Other Animals

In the mouse, competence for the generation of PGCs is set at E6.0–E6.5 in the proximal epiblast adjacent to the extraembryonic ectoderm (ExE) [129, 130] (◘ Fig. 3.6b). This initial population can be identified by expression of *Ifitm3* (or *Fragilis*). *Ifitm3* is considered the first gene to mark the onset of competence for the PGC fate [131]. Nevertheless, not all *Ifitm3+* cells are to become bona fide pre-PGCs. From this initial niche, about six *Ifitm3+* cells begin to express *Prmd1* (or *Blimp1*), and these are considered the mPGC precursors or pre-PGCs [132]. These pre-PGCs become lineage restricted, when they start expressing *Prdm14*, *Tfap2c* (or *Ap2γ*) and *Dppa3* (or *Stella*) at E7.25 and acquire the characteristic alkaline phosphatase (*Alpl* or *Tnap*) activity. This cluster of about 45 founder mPGCs is embedded in the extraembryonic mesoderm and visible at the base of the allantois [133–135]. The mPGCs are positive for *Pou5f1* (via the distal enhancer of the *Pou5f1* promoter just as the ICM, in contrast to the epiblast that uses the proximal enhancer of the *Pou5f1* promoter) and start re-expressing pluripotency-associated genes such as *Sox2, Nanog* and *Sall4* [136–139].

After implantation, the mouse embryo develops as an egg cylinder while the human embryo develops as a flat disc [11, 140]. Recent studies in pig embryos, in which peri-implantation development is closer to humans than to mice, show that competence for porcine PGC specification is set initially in the posterior end of the nascent primitive streak [141]. In the cynomolgus monkey, cyPGCs were identified prior to gastrulation, at E11, in the dorsal amnion [142]. Although both pig PGCs and cyPGCs seem to depend on BMP signalling for specification and express common germ cell markers, the PGC specification takes place at different locations suggesting that the mechanism of induction of PGC fate may not be entirely conserved among mammals [141, 142]. The timing of establishment of competence for PGC differentiation and the embryonic origin of the PGC founder population in the human embryo is presently unknown, but it is expected to occur before the initiation of gastrulation at day 14 (◘ Fig. 3.6c).

3.5.3 Migration of PGCs in Humans and Other Animals

After specification, mPGCs migrate towards the future gonads via the endoderm epithelium of the hindgut, reaching the mesentery at E9.5 and colonizing the (left and right) genital ridges at E10.5 [143–145] (◘ Fig. 3.6b). During migration and briefly after colonization of the gonads, PGCs proliferate and undergo epigenetic reprogramming characterized by genome-wide DNA demethylation, X chromosome reactivation (in the females) and erasure of genomic imprinting [143]. mPGCs start to express *c-Kit* and *SSEA1* during migration [146–149]. Contrary to the mouse, chick PGCs migrate through the vascular system to reach the genital ridges [150] (Fig. 3.6a), instead of using the gut.

During human development, human PGCs (hPGCs) have been identified the earliest in week (W)5 of gestation (or week 3 of development), recognized by their morphology and alkaline phosphatase activity. hPGCs were observed in the extragonadal region of the developing embryo, more specifically in the posterior-ventral part of the endoderm of the yolk sac wall near the allantois [151]. hPGCs have been shown to migrate via the midgut and hindgut endoderm and later via the dorsal mesentery to colonize the gonadal ridges around W8 [152] (Figs. 3.6c and 3.7).

Recently, the expression of germ cell markers associated with migratory and early post-migratory in hPGCs has been validated at W4.5 [153]. In this study, the migration of hPGCs in a human embryo was analysed, and specific expression of *NANOG*, *POU5F1*, *TFAP2C* and *PRDM1* in hPGCs in the AGM (aorta–mesonephros–gonadal region) was shown. On the other hand, the well-known mouse germ cell markers, like *ALPL* and *KIT*, and others, like *SOX17*, *TUBB3* and *ITGA6*, were expressed in other cell types in the AGM, highlighting that to identify hPGCs at this developmental stage, using a suitable combination of markers is essential [153].

3.5.4 Arrival and Colonization of the Gonad

After initial colonization of the gonad by E10.5, mPGCs undergo mitotic division until E12.5 [154]. By this time, *Dazl* and *Ddx4* (or *Vasa/Mvh*) are upregulated [155, 156], and mPGCs lose expression of early germ cell and pluripotency-associated markers like *Prdm1*, *Dppa3*, *Pou5f1*, *Sox2*, *Nanog* and *Alpl* [157–160].

In humans, long after colonization of the human gonadal ridges, around W9 of gestation, *DAZL* and *DDX4* start to become upregulated, while *POU5F1* and *NANOG* expression decreases, becoming mutually exclusive [161–163] (Fig. 3.7). While in the mouse, meiosis entry is relatively synchronized and occurs in a short-lived wave of about 12 hours [157, 159], in human this process is asynchronous, taking place from W17 to birth [162, 164]. More recently, with advances in technology and access to human foetal gonads with ages varying from W7 to W20, the transcriptome and epigenome of hPGCs have been investigated by FACS-sorting hPGCs from embryonic somatic tissue using surface markers and performing single-cell transcriptomics [163–166].

3.5.5 Protecting PGC Genome Integrity

During development, mPGCs undergo a series of epigenetic reprogramming waves where epigenetic marks such as global DNA methylation and genomic imprints are erased and re-established later in a sex-specific manner [167, 168]. Throughout this reprogramming period, when global DNA methylation is low, the genome is particularly vulnerable to random integration by repetitive transposable elements (TrE) that are usually repressed by DNA methylation [169]. In addition to DNA methylation,

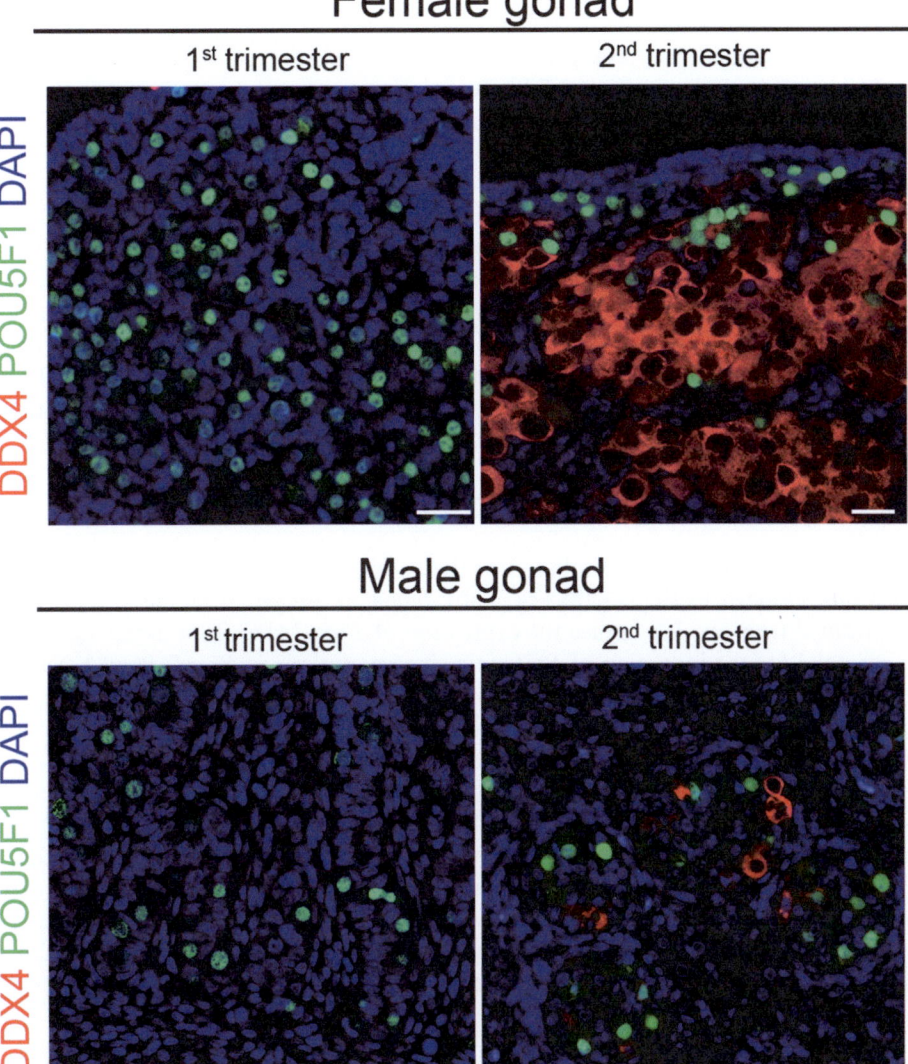

Fig. 3.7 Germ cells in first- and second-trimester human gonads. Histological sections of female and male gonads immunostained for the pre-meiotic germ cell marker DDX4 (red) and early germ cell marker POU5F1 (green) in the first- and second-trimester gonad. Nuclei are counterstained with DAPI (blue). Scale bars are 20 μm

another strategy to block the integration of TrE in the genome is targeting of TrE transcripts for degradation [169]. Targeting of TrE for degradation is achieved in complexes of Piwil (P-element induced wimpy testis-like) proteins with small non-coding RNAs that are germline-specific, the piRNAs [170–172]. These piRNA-induced silencing complexes (piRISC) recognize and cleave target TrE transcripts with complementary sequences to the loaded piRNAs. Mice have three Piwi-like paralogues (*Piwil1/Miwi*, *Piwil2/Mili* and *Piwil4/Miwil2*), while humans have one extra Piwil gene, *PIWIL3* [173, 174]. Mutations in *Piwil*, in mice, have been associated with defects in meiosis, specifically in the male germline, while females remain fertile, and mutated oocytes are able to resume meiosis normally (reviewed in [175]). Retrotransposon silencing via piRISC usually occurs in a peri-nuclear cytoplasmic structure rich in mitochondria and endoplasmic reticulum surrounding a Golgi aggregate, known as intermitochondrial cement [176–178].

Specific haplotypes in *PIWIL4* and *PIWIL3* together with hypermethylation of the *PIWIL2* promoter have been associated with infertility in humans [179–181]. Due to the prominent role in human gametogenesis, the expression of the different PIWIL during male and female germline development has recently been systematically investigated [173, 174, 182]. *PIWIL1*, *PIWIL2* and *PIWIL4* have a mutually exclusive pattern of subcellular localization, particularly in female oocytes. In contrast to mice, in humans *PIWIL4*, but not *PIWIL2*, is localized to the intermitochondrial cement [182] highlighting important differences between mouse and human.

3.5.6 Protocols for In Vitro Germ Cell Development

In mouse, PGC induction is initiated by BMPs originating in the ExE and proximal visceral endoderm (VE) [183] (◘ Fig. 3.6b). *Bmp4* and *Bmp8b* from the ExE together with *Bmp2* from the VE induce the expression of *Prdm1* [183–185]. BMP signalling is essential for the induction of PGC fate since mutations in *Bmp4*, *Bmp8b*, *Smad1* and *Smad5* result in impaired PGC development in vivo [186–188]. Similarly, protocols aiming to generate mPGCs from pluripotent stem cells in vitro require, among other factors, the addition of Bmp4 and Bmp8b [189]. In mouse, PGC-like cells (mPGCLCs) with full competence to generate functional gametes after transplantation to mice in vivo have successfully been induced from mESCs [189, 190]. The mPGCLCs were able to undergo gametogenesis leading to the formation of functional sperm or oocytes, but meiosis was only accomplished due to co-culture with gonadal tissue [189, 190]. More recently, meiosis has been completed in vitro, without the need to transplant the aggregates of gonadal tissue containing the mPGCLCs into mice, resulting in differentiation to both female functional gametes [191] and male functional gametes [192].

The studies in mice have paved the way to protocols to generate human PGC-like cells (hPGCLCs) [141, 193, 194]. These protocols rely on the comparison of gene expression between the newly generated hPGCLC and gonadal hPGCs. Primed hESCs cultured in the presence of FGF2 have low germline competence, but when grown in 4i-medium (four inhibitors: CHIR99021, PD0325901, SB203580 and SP600125) containing BMP4 or BMP2, hESCs efficiently differentiated to hPGCLCs [141, 193, 194]. W7 hPGCs and hPGCLCs have relatively similar global transcriptional profiles with the expression of *PRDM1, ALPL, DDPA3, TFAP2C, NANOS3, KIT, NANOG, POU5F1, KLF4* and, surprisingly, some other lineage-marker genes like endoderm marker *SOX17*. Surprisingly, *SOX17* and *KLF4* are specifically expressed in hPGCLCs, but not in mPGCLCs. By contrast, *SOX2* and *PRDM14* are specifically expressed in mPGCLCs and not in hPGCLCs [141, 165, 166, 193, 195, 196]. In 4i condition, hESCs develop early mesodermal characteristics and differentiate into hPGCLCs with high efficiency [193, 194]. This is similar to mPGCs that transiently upregulated mesoderm markers, such as *T*, before PGC specification [197]. *SOX17, PRDM1* and *TFAP2C* were also upregulated during the in vitro acquisition of competence for cyPGC fate from cynomolgus monkey pluripotent stem cells [141], suggesting that there are conserved aspects between primates.

Developing protocols to mimic gametogenesis starting from human pluripotent stem cells is currently challenging, but it will surely open novel avenues to understand causes and develop treatments for human infertility and perhaps even revolutionize the way we reproduce in the future.

Take-Home Message

This chapter covers the following topics:
- We have introduced early embryonic development.
- We have explained the differences between different pluripotency stages.
- We have raised awareness of differences between animal models.
- We presented the different signalling pathways involved in pluripotency.
- We have introduced the development of germ cells in vivo and in vitro.

Review Questions for This Chapter
1. Discuss the differences between totipotency and pluripotency.
2. Discuss the differences between naïve and primed pluripotent stem cells.
3. Describe the differences between hESCs and mESCs.
4. Compare the signalling pathways important for naïve and primed pluripotency.
5. Explain the BMP signalling pathway.
6. Enumerate genes associated with naïve and primed pluripotency.
7. Describe the events that take place during pre-implantation.
8. Describe the formation of PGCs, including specification, migration, epigenetic reprogramming and genomic integrity.

References

1. Cockburn K, Rossant J. Making the blastocyst: lessons from the mouse. J Clin Invest. 2010;120(4):995–1003.
2. Gasperowicz M, Natale DRC. Establishing three blastocyst lineages—then what? Biol Reprod. 2011;84(4):621–30.
3. Niwa H, Toyooka Y, Shimosato D, Strumpf D, Takahashi K, Yagi R, et al. Interaction between Oct3/4 and Cdx2 determines trophectoderm differentiation. Cell. 2005;123(5):917–29.
4. Strumpf D, Mao CA, Yamanaka Y, Ralston A, Chawengsaksophak K, Beck F, et al. Cdx2 is required for correct cell fate specification and differentiation of trophectoderm in the mouse blastocyst. Development. 2005;132(9):2093–102.
5. Suwińska A, Czołowska R, Ożdżeński W, Tarkowski AK. Blastomeres of the mouse embryo lose totipotency after the fifth cleavage division: expression of Cdx2 and Oct4 and developmental potential of inner and outer blastomeres of 16- and 32-cell embryos. Dev Biol. 2008;322(1):133–44.
6. Chazaud C, Yamanaka Y, Pawson T, Rossant J. Early lineage segregation between epiblast and primitive endoderm in mouse blastocysts through the Grb2-MAPK pathway. Dev Cell. 2006;10(5):615–24.
7. Schrode N, Saiz N, Di Talia S, Hadjantonakis AK. GATA6 levels modulate primitive endoderm cell fate choice and timing in the mouse blastocyst. Dev Cell. 2014;29(4):454–67.

8. Kojima Y, Kaufman-Francis K, Studdert JB, Steiner KA, Power MD, Loebel DAF, et al. The transcriptional and functional properties of mouse epiblast stem cells resemble the anterior primitive streak. Cell Stem Cell. 2014;14(1):107–20.
9. Najm FJ, Chenoweth JG, Anderson PD, Nadeau JH, Redline RW, McKay RDG, et al. Isolation of epiblast stem cells from preimplantation mouse embryos. Cell Stem Cell. 2011;8(3):318–25.
10. Osorno R, Tsakiridis A, Wong F, Cambray N, Economou C, Wilkie R, et al. The developmental dismantling of pluripotency is reversed by ectopic Oct4 expression. Development. 2012;139(13):2288–98.
11. Tam PPL, Behringer RR. Mouse gastrulation: the formation of a mammalian body plan. Mech Dev. 1997;68(1–2):3–25.
12. Evans MJ, Kaufman MH. Establishment in culture of pluripotential cells from mouse embryos. Nature. 1981;292(5819):154–6.
13. Martin GR. Isolation of a pluripotent cell line from early mouse embryos cultured in medium conditioned by teratocarcinoma stem cells. Proc Natl Acad Sci U S A. 1981;78(12):7634–8.
14. Smith A. A glossary for stem-cell biology. Nature. 2006;441(7097):1060.
15. Loebel DAF, Watson CM, De Young RA, Tam PPL. Lineage choice and differentiation in mouse embryos and embryonic stem cells. Dev Biol. 2003;264(1):1–14.
16. Tang F, Barbacioru C, Bao S, Lee C, Nordman E, Wang X, et al. Tracing the derivation of embryonic stem cells from the inner cell mass by single-cell RNA-Seq analysis. Cell Stem Cell. 2010;6(5–2):468–78.
17. Smith AG, Heath JK, Donaldson DD, Wong GG, Moreau J, Stahl M, et al. Inhibition of pluripotential embryonic stem cell differentiation by purified polypeptides. Nature. 1988;336(6200):688–90.
18. Williams RL, Hilton DJ, Pease S, Willson TA, Stewart CL, Gearing DP, et al. Myeloid leukaemia inhibitory factor maintains the developmental potential of embryonic stem cells. Nature. 1988;336(6200):684–7.
19. Boeuf H, Hauss C, Graeve FD, Baran N, Kedinger C. Leukemia inhibitory factor–dependent transcriptional activation in embryonic stem cells. J Cell Biol. 1997;138(6):1207–17.
20. Niwa H, Burdon T, Chambers I, Smith A. Self-renewal of pluripotent embryonic stem cells is mediated via activation of STAT3. Genes Dev. 1998;12(13):2048–60.
21. Tesar PJ, Chenoweth JG, Brook FA, Davies TJ, Evans EP, Mack DL, et al. New cell lines from mouse epiblast share defining features with human embryonic stem cells. Nature. 2007;448:196.
22. Brons IGM, Smithers LE, Trotter MWB, Rugg-Gunn P, Sun B, Chuva de Sousa Lopes SM, et al. Derivation of pluripotent epiblast stem cells from mammalian embryos. Nature. 2007;448(7150):191–5.
23. Han DW, Tapia N, Joo JY, Greber B, Araúzo-Bravo MJ, Bernemann C, et al. Epiblast stem cell subpopulations represent mouse embryos of distinct pregastrulation stages. Cell. 2010;143(4):617–27.
24. Nichols J, Smith A. Naive and primed pluripotent states. Cell Stem Cell. 2009;4(6):487–92.
25. Sasai M, Kawabata Y, Makishi K, Itoh K, Terada TP. Time scales in epigenetic dynamics and phenotypic heterogeneity of embryonic stem cells. PLoS Comput Biol. 2013;9(12):e1003380.
26. Trott J, Martinez Arias A. Single cell lineage analysis of mouse embryonic stem cells at the exit from pluripotency. Biol Open. 2013;2(10):1049–56.
27. Ying QL, Wray J, Nichols J, Batlle-Morera L, Doble B, Woodgett J, et al. The ground state of embryonic stem cell self-renewal. Nature. 2008;453(7194):519–23.
28. Blair K, Wray J, Smith A. The liberation of embryonic stem cells. PLoS Genet. 2011;7(4):e1002019.
29. Habibi E, Brinkman AB, Arand J, Kroeze LI, Kerstens HHD, Matarese F, et al. Whole-genome bisulfite sequencing of two distinct interconvertible DNA methylomes of mouse embryonic stem cells. Cell Stem Cell. 2013;13(3):360–9.
30. Marks H, Kalkan T, Menafra R, Denissov S, Jones K, Hofemeister H, et al. The transcriptional and epigenomic foundations of ground state pluripotency. Cell. 2012;149(3):590–604.
31. Chambers I, Tomlinson SR. The transcriptional foundation of pluripotency. Development. 2009;136(14):2311–22.
32. Li M, Belmonte JCI. Ground rules of the pluripotency gene regulatory network. Nat Rev Genet. 2017;18(3):180–91.
33. Jiang J, Chan YS, Loh YH, Cai J, Tong GQ, Lim CA, et al. A core Klf circuitry regulates self-renewal of embryonic stem cells. Nat Cell Biol. 2008;10(3):353–60.
34. Martello G, Sugimoto T, Diamanti E, Joshi A, Hannah R, Ohtsuka S, et al. Esrrb is a pivotal target of the Gsk3/Tcf3 axis regulating embryonic stem cell self-renewal. Cell Stem Cell. 2012;11(4):491–504.
35. Scotland KB, Chen S, Sylvester R, Gudas LJ. Analysis of Rex1 (zfp42) function in embryonic stem cell differentiation. Dev Dyn. 2009;238(8):1863–77.

36. Tai CI, Ying QL. Gbx2, a LIF/Stat3 target, promotes reprogramming to and retention of the pluripotent ground state. J Cell Sci. 2013;126(5):1093–8.
37. Varlakhanova NV, Cotterman RF, deVries WN, Morgan J, Donahue LR, Murray S, et al. myc maintains embryonic stem cell pluripotency and self-renewal. Differentiation. 2010;80(1):9–19.
38. Waghray A, Saiz N, Jayaprakash AD, Freire AG, Papatsenko D, Pereira CF, et al. Tbx3 controls Dppa3 levels and exit from pluripotency toward mesoderm. Stem Cell Rep. 2015;5(1):97–110.
39. Yamaji M, Ueda J, Hayashi K, Ohta H, Yabuta Y, Kurimoto K, et al. PRDM14 ensures naive pluripotency through dual regulation of signaling and epigenetic pathways in mouse embryonic stem cells. Cell Stem Cell. 2013;12(3):368–82.
40. Ye S, Li P, Tong C, Ying QL. Embryonic stem cell self-renewal pathways converge on the transcription factor Tfcp2l1. EMBO J. 2013;32(19):2548–60.
41. Li Z, Fei T, Zhang J, Zhu G, Wang L, Lu D, et al. BMP4 signaling acts via dual-specificity phosphatase 9 to control ERK activity in mouse embryonic stem cells. Cell Stem Cell. 2012;10(2):171–82.
42. Gomes Fernandes M, Dries R, Roost MS, Semrau S, de Melo Bernardo A, Davis RP, et al. BMP-SMAD signaling regulates lineage priming, but is dispensable for self-renewal in mouse embryonic stem cells. Stem Cell Rep. 2016;6(1):85–94.
43. Sasse J, Hemmann U, Schwartz C, Schniertshauer U, Heesel B, Landgraf C, et al. Mutational analysis of acute-phase response factor/Stat3 activation and dimerization. Mol Cell Biol. 1997;17(8):4677–86.
44. Inoue M, Minami M, Matsumoto M, Kishimoto T, Akira S. The amino acid residues immediately carboxyl-terminal to the tyrosine phosphorylation site contribute to interleukin 6-specific activation of signal transducer and activator of transcription 3. J Biol Chem. 1997;272(14):9550–5.
45. Burdon T, Chambers I, Stracey C, Niwa H, Smith A. Signaling mechanisms regulating self-renewal and differentiation of pluripotent embryonic stem cells. Cells Tissues Organs. 1999;165(3–4):131–43.
46. Hirai H, Karian P, Kikyo N. Regulation of embryonic stem cell self-renewal and pluripotency by leukaemia inhibitory factor. Biochem J. 2011;438(1):11–23.
47. Raz R, Lee CK, Cannizzaro LA, d'Eustachio P, Levy DE. Essential role of STAT3 for embryonic stem cell pluripotency. Proc Natl Acad Sci. 1999;96(6):2846–51.
48. Niwa H, Ogawa K, Shimosato D, Adachi K. A parallel circuit of LIF signalling pathways maintains pluripotency of mouse ES cells. Nature. 2009;460:118.
49. Aksoy I, Sakabedoyan C, Bourillot PY, Malashicheva AB, Mancip J, Knoblauch K, Afanassieff M, Savatier P. Self-renewal of murine embryonic stem cells is supported by the serine/threonine kinases Pim-1 and Pim-3. Stem Cells. 2007;25(12):2996–3004.
50. Hall J, Guo G, Wray J, Eyres I, Nichols J, Grotewold L, et al. Oct4 and LIF/Stat3 additively induce Kruppel factors to sustain embryonic stem cell self-renewal. Cell Stem Cell. 2009;5(6):597–609.
51. Cartwright P, McLean C, Sheppard A, Rivett D, Jones K, Dalton S. LIF/STAT3 controls ES cell self-renewal and pluripotency by a Myc-dependent mechanism. Development. 2005;132(5):885–96.
52. Martello G, Bertone P, Smith A. Identification of the missing pluripotency mediator downstream of leukaemia inhibitory factor. EMBO J. 2013;32(19):2561–74.
53. Casanova EA, Shakhova O, Patel SS, Asner IN, Pelczar P, Weber FA, et al. Pramel7 mediates LIF/STAT3-dependent self-renewal in embryonic stem cells. Stem Cells. 2011;29(3):474–85.
54. Li Y, McClintick J, Zhong L, Edenberg HJ, Yoder MC, Chan RJ. Murine embryonic stem cell differentiation is promoted by SOCS-3 and inhibited by the zinc finger transcription factor Klf4. Blood. 2005;105(2):635–7.
55. Massagué J, Attisano L, Wrana JL. The TGF-β family and its composite receptors. Trends Cell Biol. 1994;4(5):172–8.
56. Nohe A, Keating E, Knaus P, Petersen NO. Signal transduction of bone morphogenetic protein receptors. Cell Signal. 2004;16(3):291–9.
57. Ying QL, Nichols J, Chambers I, Smith A. BMP induction of Id proteins suppresses differentiation and sustains embryonic stem cell self-renewal in collaboration with STAT3. Cell. 2003;115(3):281–92.
58. Zhang J, Li L. BMP signaling and stem cell regulation. Dev Biol. 2005;284(1):1–11.
59. Nusse R, Clevers H. Wnt/β-catenin signaling, disease, and emerging therapeutic modalities. Cell. 2017;169(6):985–99.
60. Illich DJ, Zhang M, Ursu A, Osorno R, Kim K-P, Yoon J, et al. Distinct signaling requirements for the establishment of ESC pluripotency in late-stage EpiSCs. Cell Rep. 2016;15(4):787–800.

61. Kurek D, Neagu A, Tastemel M, Tüysüz N, Lehmann J, van de Werken HJ, et al. Endogenous WNT signals mediate BMP-induced and spontaneous differentiation of epiblast stem cells and human embryonic stem cells. Stem Cell Rep. 2015;4(1):114–28.
62. Greber B, Wu G, Bernemann C, Joo JY, Han DW, Ko K, et al. Conserved and divergent roles of FGF signaling in mouse epiblast stem cells and human embryonic stem cells. Cell Stem Cell. 2010;6(3):215–26.
63. Ko M, An J, Bandukwala HS, Chavez L, Äijö T, Pastor WA, et al. Modulation of TET2 expression and 5-methylcytosine oxidation by the CXXC domain protein IDAX. Nature. 2013;497(7447):122–6.
64. Hao J, Li TG, Qi X, Zhao DF, Zhao GQ. WNT/beta-catenin pathway up-regulates Stat3 and converges on LIF to prevent differentiation of mouse embryonic stem cells. Dev Biol. 2006;290(1):81–91.
65. Qiu D, Ye S, Ruiz B, Zhou X, Liu D, Zhang Q, et al. Klf2 and Tfcp2l1, two Wnt/β-catenin targets, act synergistically to induce and maintain naive pluripotency. Stem Cell Rep. 2015;5(3):314–22.
66. Atlasi Y, Noori R, Gaspar C, Franken P, Sacchetti A, Rafati H, et al. Wnt signaling regulates the lineage differentiation potential of mouse embryonic stem cells through Tcf3 Down-regulation. PLoS Genet. 2013;9(5):e1003424.
67. Marucci L, Pedone E, Di Vicino U, Sanuy-Escribano B, Isalan M, Cosma MP. β-catenin fluctuates in mouse ESCs and is essential for Nanog-mediated reprogramming of somatic cells to pluripotency. Cell Rep. 2014;8(6):1686–96.
68. Takao Y, Yokota T, Koide H. β-catenin up-regulates Nanog expression through interaction with Oct-3/4 in embryonic stem cells. Biochem Biophys Res Commun. 2007;353:699–705.
69. Hackett JA, Surani MA. Regulatory principles of pluripotency: from the ground state up. Cell Stem Cell. 2014;15(4):416–30.
70. Joo JY, Choi HW, Kim MJ, Zaehres H, Tapia N, Stehling M, et al. Establishment of a primed pluripotent epiblast stem cell in FGF4-based conditions. Sci Rep. 2014;4:7477.
71. Betschinger J, Nichols J, Dietmann S, Corrin PD, Paddison PJ, Smith A. Exit from pluripotency is gated by intracellular redistribution of the bHLH transcription factor Tfe3. Cell. 2013;153(2):335–47.
72. Dailey L, Ambrosetti D, Mansukhani A, Basilico C. Mechanisms underlying differential responses to FGF signaling. Cytokine Growth Factor Rev. 2005;16(2):233–47.
73. Ornitz DM, Itoh N. The fibroblast growth factor signaling pathway. Wiley Interdiscip Rev Dev Biol. 2015;4(3):215–66.
74. Boris G, Hans L, James A. Fibroblast growth factor 2 modulates transforming growth factor β signaling in mouse embryonic fibroblasts and human ESCs (hESCs) to support hESC self-renewal. Stem Cells. 2007;25(2):455–64.
75. Wrana JL, Attisano L, Wieser R, Ventura F, Massagué J. Mechanism of activation of the TGF-β receptor. Nature. 1994;370:341.
76. Tsuchida K, Nakatani M, Yamakawa N, Hashimoto O, Hasegawa Y, Sugino H. Activin isoforms signal through type I receptor serine/threonine kinase ALK7. Mol Cell Endocrinol. 2004;220(1–2):59–65.
77. Ghimire S, Van der Jeught M, Neupane J, Roost MS, Anckaert J, Popovic M, et al. Comparative analysis of naive, primed and ground state pluripotency in mouse embryonic stem cells originating from the same genetic background. Sci Rep. 2018;8(1):5884.
78. Ficz G, Hore TA, Santos F, Lee HJ, Dean W, Arand J, et al. FGF signaling inhibition in ESCs drives rapid genome-wide demethylation to the epigenetic ground state of pluripotency. Cell Stem Cell. 2013;13(3):351–9.
79. Choi HW, Joo JY, Hong YJ, Kim JS, Song H, Lee JW, et al. Distinct enhancer activity of Oct4 in naive and primed mouse pluripotency. Stem Cell Rep. 2016;7(5):911–26.
80. Factor DC, Corradin O, Zentner GE, Saiakhova A, Song L, Chenoweth JG, et al. Epigenomic comparison reveals activation of "seed" enhancers during transition from naive to primed pluripotency. Cell Stem Cell. 2014;14(6):854–63.
81. Thomson JA, Itskovitz-Eldor J, Shapiro SS, Waknitz MA, Swiergiel JJ, Marshall VS, et al. Embryonic stem cell lines derived from human blastocysts. Science. 1998;282(5391):1145–7.
82. Vallier L, Alexander M, Pedersen RA. Activin/Nodal and FGF pathways cooperate to maintain pluripotency of human embryonic stem cells. J Cell Sci. 2005;118(19):4495–509.
83. O'Leary T, Heindryckx B, Lierman S, van Bruggen D, Goeman JJ, Vandewoestyne M, et al. Tracking the progression of the human inner cell mass during embryonic stem cell derivation. Nat Biotechnol. 2012;30:278.

84. Chan YS, Göke J, Ng JH, Lu X, Gonzales KAU, Tan CP, et al. Induction of a human pluripotent state with distinct regulatory circuitry that resembles preimplantation epiblast. Cell Stem Cell. 2013;13(6):663–75.
85. Vallier L, Reynolds D, Pedersen RA. Nodal inhibits differentiation of human embryonic stem cells along the neuroectodermal default pathway. Dev Biol. 2004;275(2):403–21.
86. Hoffman LM, Hall L, Batten JL, Young H, Pardasani D, Baetge EE, et al. X-inactivation status varies in human embryonic stem cell lines. Stem Cells. 2005;23(10):1468–78.
87. Lengner CJ, Gimelbrant AA, Erwin JA, Cheng AW, Guenther MG, Welstead GG, et al. Derivation of pre-X inactivation human embryonic stem cells under physiological oxygen concentrations. Cell. 2010;141(5):872–83.
88. Shen Y, Matsuno Y, Fouse SD, Rao N, Root S, Xu R, et al. X-inactivation in female human embryonic stem cells is in a nonrandom pattern and prone to epigenetic alterations. Proc Natl Acad Sci. 2008;105(12):4709–14.
89. Silva SS, Rowntree RK, Mekhoubad S, Lee JT. X-chromosome inactivation and epigenetic fluidity in human embryonic stem cells. Proc Natl Acad Sci. 2008;105(12):4820–5.
90. Hackett JA, Dietmann S, Murakami K, Down TA, Leitch HG, Surani MA. Synergistic mechanisms of DNA demethylation during transition to ground-state pluripotency. Stem Cell Rep. 2013;1(6):518–31.
91. Shipony Z, Mukamel Z, Cohen NM, Landan G, Chomsky E, Zeliger SR, et al. Dynamic and static maintenance of epigenetic memory in pluripotent and somatic cells. Nature. 2014;513(7516): 115–9.
92. Wu J, Izpisua Belmonte JC. Dynamic pluripotent stem cell states and their applications. Cell Stem Cell. 2015;17(5):509–25.
93. Van der Jeught M, O'Leary T, Duggal G, De Sutter P, Chuva de Sousa Lopes SM, Heindryckx B. The post-inner cell mass intermediate: implications for stem cell biology and assisted reproductive technology. Hum Reprod Update. 2015;21(5):616–26.
94. Duggal G, Warrier S, Ghimire S, Broekaert D, Van der Jeught M, Lierman S, et al. Alternative routes to induce naïve pluripotency in human embryonic stem cells. Stem Cells. 2015;33(9):2686–98.
95. Hanna J, Cheng AW, Saha K, Kim J, Lengner CJ, Soldner F, et al. Human embryonic stem cells with biological and epigenetic characteristics similar to those of mouse ESCs. Proc Natl Acad Sci. 2010;107(20):9222–7.
96. Gafni O, Weinberger L, Mansour AA, Manor YS, Chomsky E, Ben-Yosef D, et al. Derivation of novel human ground state naive pluripotent stem cells. Nature. 2013;504:282.
97. Theunissen TW, Powell BE, Wang H, Mitalipova M, Faddah DA, Reddy J, et al. Systematic identification of culture conditions for induction and maintenance of naive human pluripotency. Cell Stem Cell. 2014;15(4):471–87.
98. Warrier S, Van der Jeught M, Duggal G, Tilleman L, Sutherland E, Taelman J, et al. Direct comparison of distinct naive pluripotent states in human embryonic stem cells. Nat Commun. 2017;8:15055.
99. Pastor WA, Chen D, Liu W, Kim R, Sahakyan A, Lukianchikov A, et al. Naïve human pluripotent cells feature a methylation landscape devoid of blastocyst or germline memory. Cell Stem Cell. 2016;18(3):323–9.
100. Takahashi S, Kobayashi S, Hiratani I. Epigenetic differences between naïve and primed pluripotent stem cells. Cell Mol Life Sci. 2018;75(7):1191–203.
101. Theunissen TW, Friedli M, He Y, Planet E, O'Neil RC, Markoulaki S, et al. Molecular criteria for defining the naive human pluripotent state. Cell Stem Cell. 2016;19(4):502–15.
102. Geens M, Chuva de Sousa Lopes SM. X chromosome inactivation in human pluripotent stem cells as a model for human development: back to the drawing board? Hum Reprod Update. 2017;23(5):520–32.
103. Mekhoubad S, Bock C, de Boer AS, Kiskinis E, Meissner A, Eggan K. Erosion of dosage compensation impacts human iPSC disease modeling. Cell Stem Cell. 2012;10(5):595–609.
104. Vallot C, Ouimette JF, Makhlouf M, Féraud O, Pontis J, Côme J, et al. Erosion of X chromosome inactivation in human pluripotent cells initiates with XACT coating and depends on a specific heterochromatin landscape. Cell Stem Cell. 2015;16(5):533–46.
105. Etoc F, Metzger J, Ruzo A, Kirst C, Yoney A, Ozair MZ, et al. A balance between secreted inhibitors and edge sensing controls gastruloid self-organization. Dev Cell. 2016;39(3):302–15.
106. Warmflash A, Sorre B, Etoc F, Siggia ED, Brivanlou AH. A method to recapitulate early embryonic spatial patterning in human embryonic stem cells. Nat Methods. 2014;11:847.

107. Martyn I, Kanno TY, Ruzo A, Siggia ED, Brivanlou AH. Self-organization of a human organizer by combined Wnt and Nodal signalling. Nature. 2018;558(7708):132–5.
108. Shao Y, Taniguchi K, Townshend RF, Miki T, Gumucio DL, Fu J. A pluripotent stem cell-based model for post-implantation human amniotic sac development. Nat Commun. 2017;8(1):208.
109. Shahbazi MN, Jedrusik A, Vuoristo S, Recher G, Hupalowska A, Bolton V, et al. Self-organization of the human embryo in the absence of maternal tissues. Nat Cell Biol. 2016;18:700.
110. Deglincerti A, Croft GF, Pietila LN, Zernicka-Goetz M, Siggia ED, Brivanlou AH. Self-organization of the in vitro attached human embryo. Nature. 2016;533:251.
111. Durcova-Hills G, Ainscough JFX, McLaren A. Pluripotential stem cells derived from migrating primordial germ cells. Differentiation. 2001;68(4):220–6.
112. Matsui Y, Toksoz D, Nishikawa S, Nishikawa SI, Williams D, Zsebo K, et al. Effect of Steel factor and leukaemia inhibitory factor on murine primordial germ cells in culture. Nature. 1991;353:750.
113. Resnick JL, Bixler LS, Cheng L, Donovan PJ. Long-term proliferation of mouse primordial germ cells in culture. Nature. 1992;359:550.
114. Labosky PA, Barlow DP, Hogan BL. Mouse embryonic germ (EG) cell lines: transmission through the germline and differences in the methylation imprint of insulin-like growth factor 2 receptor (Igf2r) gene compared with embryonic stem (ES) cell lines. Development. 1994;120(11):3197–204.
115. Stewart CL, Gadi I, Bhatt H. Stem cells from primordial germ cells can reenter the germ line. Dev Biol. 1994;161(2):626–8.
116. Leitch HG, Blair K, Mansfield W, Ayetey H, Humphreys P, Nichols J, et al. Embryonic germ cells from mice and rats exhibit properties consistent with a generic pluripotent ground state. Development. 2010;137(14):2279–87.
117. Shamblott MJ, Axelman J, Wang S, Bugg EM, Littlefield JW, Donovan PJ, et al. Derivation of pluripotent stem cells from cultured human primordial germ cells. Proc Natl Acad Sci. 1998;95(23):13726–31.
118. Turnpenny L, Brickwood S, Spalluto CM, Piper K, Cameron IT, Wilson DI, Hanley NA. Derivation of human embryonic germ cells: an alternative source of pluripotent stem cells. Stem Cells. 2003;21(5):598–609.
119. Hua J, Yu H, Liu S, Dou Z, Sun Y, Jing X, et al. Derivation and characterization of human embryonic germ cells: serum-free culture and differentiation potential. Reprod Biomed Online. 2009;19:238–49.
120. Nikolic A, Volarevic V, Armstrong L, Lako M, Stojkovic M. Primordial germ cells: current knowledge and perspectives. Stem Cells Int. 2016;2016:1741072.
121. Lesch BJ, Page DC. Genetics of germ cell development. Nat Rev Genet. 2012;13(11):781–94.
122. Strome S, Updike D. Specifying and protecting germ cell fate. Nat Rev Mol Cell Biol. 2015;16(7):406–16.
123. Bertocchini F, Chuva de Sousa Lopes SM. Germline development in amniotes: a paradigm shift in primordial germ cell specification. BioEssays. 2016;38(8):791–800.
124. Gavis ER, Lehmann R. Localization of nanos RNA controls embryonic polarity. Cell. 1992;71(2):301–13.
125. Sinsimer KS, Jain RA, Chatterjee S, Gavis ER. A late phase of germ plasm accumulation during Drosophila oogenesis requires Lost and Rumpelstiltskin. Development. 2011;138(16):3431–40.
126. Tsunekawa N, Naito M, Sakai Y, Nishida T, Noce T. Isolation of chicken vasa homolog gene and tracing the origin of primordial germ cells. Development. 2000;127(12):2741–50.
127. Eyal-Giladi H, Kochav S. From cleavage to primitive streak formation: a complementary normal table and a new look at the first stages of the development of the chick. Dev Biol. 1976;49(2):321–37.
128. Johnson AD, Richardson E, Bachvarova RF, Crother BI. Evolution of the germ line–soma relationship in vertebrate embryos. Reproduction. 2011;141(3):291–300.
129. Lawson KA, Meneses JJ, Pedersen RA. Clonal analysis of epiblast fate during germ layer formation in the mouse embryo. Development. 1991;113(3):891–911.
130. Chuva de Sousa Lopes SM, Hayashi K, Surani MA. Proximal visceral endoderm and extraembryonic ectoderm regulate the formation of primordial germ cell precursors. BMC Dev Biol. 2007;7(1):140.
131. Saitou M, Barton SC, Surani MA. A molecular programme for the specification of germ cell fate in mice. Nature. 2002;418(6895):293–300.
132. Saitou M, Payer B, O'Carroll D, Ohinata Y, Surani MA. Blimp1 and the emergence of the germ line during development in the mouse. Cell Cycle. 2005;4(12):1736–40.

133. Hayashi K, Chuva de Sousa Lopes SM, Surani MA. Germ cell specification in mice. Science. 2007;316(5823):394–6.
134. Magnúsdóttir E, Dietmann S, Murakami K, Günesdogan U, Tang F, Bao S, et al. A tripartite transcription factor network regulates primordial germ cell specification in mice. Nat Cell Biol. 2013;15(8):905–15.
135. Yamaji M, Seki Y, Kurimoto K, Yabuta Y, Yuasa M, Shigeta M, et al. Critical function of Prdm14 for the establishment of the germ cell lineage in mice. Nat Genet. 2008;40(8):1016–22.
136. Ohinata Y, Payer B, O'Carroll D, Ancelin K, Ono Y, Sano M, et al. Blimp1 is a critical determinant of the germ cell lineage in mice. Nature. 2005;436:207–13.
137. Yabuta Y, Kurimoto K, Ohinata Y, Seki Y, Saitou M. Gene expression dynamics during germline specification in mice identified by quantitative single-cell gene expression profiling. Biol Reprod. 2006;75(5):705–16.
138. Yamaguchi S, Kimura H, Tada M, Nakatsuji N, Tada T. Nanog expression in mouse germ cell development. Gene Expr Patterns. 2005;5(5):639–46.
139. Yamaguchi YL, Tanaka SS, Kumagai M, Fujimoto Y, Terabayashi T, Matsui Y, et al. Sall4 is essential for mouse primordial germ cell specification by suppressing somatic cell program genes. Stem Cells. 2015;33(1):289–300.
140. Hertig AT, Rock J, Adams EC. A description of 34 human ova within the first 17 days of development. Am J Anat. 1956;98(3):435–93.
141. Kobayashi T, Zhang H, Tang WWC, Irie N, Withey S, Klisch D, et al. Principles of early human development and germ cell program from conserved model systems. Nature. 2017;546(7658): 416–20.
142. Sasaki K, Nakamura T, Okamoto I, Yabuta Y, Iwatani C, Tsuchiya H, et al. The germ cell fate of cynomolgus monkeys is specified in the nascent amnion. Dev Cell. 2016;39(2):169–85.
143. Denis H. A parallel between development and evolution: germ cell recruitment by the gonads. BioEssays. 1994;16(12):933–8.
144. Molyneaux KA, Stallock J, Schaible K, Wylie C. Time-lapse analysis of living mouse germ cell migration. Dev Biol. 2001;240(2):488–98.
145. Richardson BE, Lehmann R. Mechanisms guiding primordial germ cell migration: strategies from different organisms. Nat Rev Mol Cell Biol. 2010;11(1):37–49.
146. Anderson R, Copeland TK, Schöler H, Heasman J, Wylie C. The onset of germ cell migration in the mouse embryo. Mech Dev. 2000;91(1–2):61–8.
147. Mintz B, Russell ES. Gene-induced embryological modifications of primordial germ cells in the mouse. J Exp Zool. 1957;134(2):207–37.
148. Pelosi E, Forabosco A, Schlessinger D. Germ cell formation from embryonic stem cells and the use of somatic cell nuclei in oocytes. Ann N Y Acad Sci. 2011;1221(1):18–26.
149. Pesce M, Di Carlo A, De Felici M. The c-kit receptor is involved in the adhesion of mouse primordial germ cells to somatic cells in culture. Mech Dev. 1997;68(1–2):37–44.
150. De Melo Bernardo A, Sprenkels K, Rodrigues G, Noce T, Chuva de Sousa Lopes SM. Chicken primordial germ cells use the anterior vitelline veins to enter the embryonic circulation. Biol Open. 2012;1:1146–52.
151. Witschi E. Migration of the germ cells of human embryos from the yolk sac to the primitive gonadal folds. Contrib Embryol. 1948;209:67–80.
152. De Felici M. Origin, migration, and proliferation of human primordial germ cells. In: Coticchio G, Albertini DF, De Santis L, editors. Oogenesis. London: Springer London; 2013. p. 19–37.
153. Gomes Fernandes M, Bialecka M, Salvatori DCF, Chuva de Sousa Lopes SM. Characterization of migratory primordial germ cells in the aorta-gonad-mesonephros of a 4.5-week-old human embryo: a toolbox to evaluate in vitro early gametogenesis. Mol Hum Reprod. 2018;24(5):233–43.
154. Tevosian SG, Albrecht KH, Crispino JD, Fujiwara Y, Eicher EM, Orkin SH. Gonadal differentiation, sex determination and normal Sry expression in mice require direct interaction between transcription partners GATA4 and FOG2. Development. 2002;129(19):4627–34.
155. Haston KM, Tung JY, Reijo Pera RA. Dazl functions in maintenance of pluripotency and genetic and epigenetic programs of differentiation in mouse primordial germ cells in vivo and in vitro. PLoS One. 2009;4(5):e5654.
156. Tanaka SS, Toyooka Y, Akasu R, Katoh-Fukui Y, Nakahara Y, Suzuki R, et al. The mouse homolog of Drosophila Vasa is required for the development of male germ cells. Genes Dev. 2000;14(7):841–53.

157. Bullejos M, Koopman P. Germ cells enter meiosis in a rostro-caudal wave during development of the mouse ovary. Mol Reprod Dev. 2004;68(4):422–8.
158. Hu Y-C, Nicholls PK, Soh YQS, Daniele JR, Junker JP, van Oudenaarden A, et al. Licensing of primordial germ cells for gametogenesis depends on genital ridge signaling. PLoS Genet. 2015;11(3):e1005019.
159. Kimura T, Yomogida K, Iwai N, Kato Y, Nakano T. Molecular cloning and genomic organization of mouse homologue of Drosophila germ cell-less and its expression in germ lineage cells. Biochem Biophys Res Commun. 1999;262(1):223–30.
160. Pesce M, Wang X, Wolgemuth DJ, Schöler HR. Differential expression of the Oct-4 transcription factor during mouse germ cell differentiation. Mech Dev. 1998;71(1–2):89–98.
161. Anderson RA, Fulton N, Cowan G, Coutts S, Saunders PT. Conserved and divergent patterns of expression of DAZL, VASA and OCT4 in the germ cells of the human fetal ovary and testis. BMC Dev Biol. 2007;7(1):136.
162. Heeren AM, He N, de Souza AF, Goercharn-Ramlal A, van Iperen L, Roost MS, et al. On the development of extragonadal and gonadal human germ cells. Biol Open. 2016;5(2):185–94.
163. Vértesy Á, Arindrarto W, Roost MS, Reinius B, Torrens-Juaneda V, Bialecka M, et al. Parental haplotype-specific single-cell transcriptomics reveal incomplete epigenetic reprogramming in human female germ cells. Nat Commun. 2018;9(1):1873.
164. Gkountela S, Li Z, Vincent JJ, Zhang KX, Chen A, Pellegrini M, et al. The ontogeny of cKIT+ human primordial germ cells proves to be a resource for human germ line reprogramming, imprint erasure and in vitro differentiation. Nat Cell Biol. 2013;15(1):113–22.
165. Gkountela S, Zhang KX, Shafiq TA, Liao W-W, Hargan-Calvopiña J, Chen PY, et al. DNA demethylation dynamics in the human prenatal germline. Cell. 2015;161(6):1425–36.
166. Guo F, Yan L, Guo H, Li L, Hu B, Zhao Y, et al. The transcriptome and DNA methylome landscapes of human primordial germ cells. Cell. 2015;161(6):1437–52.
167. Cantone I, Fisher AG. Epigenetic programming and reprogramming during development. Nat Struct Mol Biol. 2013;20(3):282–9.
168. Kimmins S, Sassone-Corsi P. Chromatin remodelling and epigenetic features of germ cells. Nature. 2005;434(7033):583–9.
169. Yang F, Wang PJ. Multiple LINEs of retrotransposon silencing mechanisms in the mammalian germline. Semin Cell Dev Biol. 2016;59:118–25.
170. Aravin AA, Sachidanandam R, Bourc'his D, Schaefer C, Pezic D, Fejes Toth K, et al. A piRNA pathway primed by individual transposons is linked to de novo DNA methylation in mice. Mol Cell. 2008;31(6):785–99.
171. Kuramochi-Miyagawa S, Watanabe T, Gotoh K, Totoki Y, Toyoda A, Ikawa M, et al. DNA methylation of retrotransposon genes is regulated by Piwi family members MILI and MIWI2 in murine fetal testes. Genes Dev. 2008;22(7):908–17.
172. Siomi MC, Sato K, Pezic D, Aravin AA. PIWI-interacting small RNAs: the vanguard of genome defence. Nat Rev Mol Cell Biol. 2011;12:246.
173. Roovers EF, Rosenkranz D, Mahdipour M, Han CT, He N, Chuva de Sousa Lopes SM, et al. Piwi proteins and piRNAs in mammalian oocytes and early embryos. Cell Rep. 2015;10(12):2069–82.
174. Williams Z, Morozov P, Mihailovic A, Lin C, Puvvula Pavan K, Juranek S, et al. Discovery and characterization of piRNAs in the human fetal ovary. Cell Rep. 2015;13(4):854–63.
175. Juliano C, Wang J, Lin H. Uniting germline and stem cells: the function of Piwi proteins and the piRNA pathway in diverse organisms. Annu Rev Genet. 2011;45:447–69. https://doi.org/10.1146/annurev-genet-110410-32541.
176. Aravin AA, van der Heijden GW, Castañeda J, Vagin VV, Hannon GJ, Bortvin A. Cytoplasmic compartmentalization of the fetal piRNA pathway in mice. PLoS Genet. 2009;5(12):e1000764.
177. Lim AK, Tao L, Kai T. piRNAs mediate posttranscriptional retroelement silencing and localization to pi-bodies in the Drosophila germline. J Cell Biol. 2009;186(3):333–42.
178. Pepling ME, Wilhelm JE, O'Hara AL, Gephardt GW, Spradling AC. Mouse oocytes within germ cell cysts and primordial follicles contain a Balbiani body. Proc Natl Acad Sci. 2007;104(1):187–92.
179. Gu A, Ji G, Shi X, Long Y, Xia Y, Song L, et al. Genetic variants in Piwi-interacting RNA pathway genes confer susceptibility to spermatogenic failure in a Chinese population. Hum Reprod. 2010;25(12):2955–61.

180. Heyn H, Ferreira HJ, Bassas L, Bonache S, Sayols S, Sandoval J, et al. Epigenetic disruption of the PIWI pathway in human spermatogenic disorders. PLoS One. 2012;7(10):e47892.
181. Muñoz X, Navarro M, Mata A, Bassas L, Larriba S. Association of PIWIL4 genetic variants with germ cell maturation arrest in infertile Spanish men. Asian J Androl. 2014;16(6):931–3.
182. Gomes Fernandes M, He N, Wang F, Van Iperen L, Eguizabal C, Matorras R, et al. Human-specific subcellular compartmentalization of P-element induced wimpy testis-like (PIWIL) granules during germ cell development and spermatogenesis. Hum Reprod. 2018;33(2):258–69.
183. Chuva de Sousa Lopes SM, Roelen BA, Monteiro RM, Emmens R, Lin HY, Li E, et al. BMP signaling mediated by ALK2 in the visceral endoderm is necessary for the generation of primordial germ cells in the mouse embryo. Genes Dev. 2004;18:1838–49.
184. Ohinata Y, Ohta H, Shigeta M, Yamanaka K, Wakayama T, Saitou M. A signaling principle for the specification of the germ cell lineage in mice. Cell. 2008;137(3):571–84.
185. Ying Y, Zhao G-Q. Cooperation of endoderm-derived BMP2 and extraembryonic ectoderm-derived BMP4 in primordial germ cell generation in the mouse. Dev Biol. 2001;232(2):484–92.
186. Chang H, Matzuk MM. Smad5 is required for mouse primordial germ cell development. Mech Dev. 2001;104(1–2):61–7.
187. Hayashi K, Kobayashi T, Umino T, Goitsuka R, Matsui Y, Kitamura D. SMAD1 signaling is critical for initial commitment of germ cell lineage from mouse epiblast. Mech Dev. 2002;118(1–2):99–109.
188. Lawson KA, Dunn NR, Roelen BAJ, Zeinstra LM, Davis AM, Wright CVE, et al. Bmp4 is required for the generation of primordial germ cells in the mouse embryo. Genes Dev. 1999;13(4):424–36.
189. Hayashi K, Ohta H, Kurimoto K, Aramaki S, Saitou M. Reconstitution of the mouse germ cell specification pathway in culture by pluripotent stem cells. Cell. 2011;146(4):519–32.
190. Hayashi K, Ogushi S, Kurimoto K, Shimamoto S, Ohta H, Saitou M. Offspring from oocytes derived from in vitro primordial germ cell–like cells in mice. Science. 2012;338(6109):971–5.
191. Hikabe O, Hamazaki N, Nagamatsu G, Obata Y, Hirao Y, Hamada N, et al. Reconstitution in vitro of the entire cycle of the mouse female germ line. Nature. 2016;539:299.
192. Zhou Q, Wang M, Yuan Y, Wang X, Fu R, Wan H, et al. Complete meiosis from embryonic stem cell-derived germ cells in vitro. Cell Stem Cell. 2016;18(3):330–40.
193. Irie N, Weinberger L, Tang WWC, Kobayashi T, Viukov S, Manor YS, et al. SOX17 is a critical specifier of human primordial germ cell fate. Cell. 2015;160(1):253–68.
194. Sasaki K, Yokobayashi S, Nakamura T, Okamoto I, Yabuta Y, Kurimoto K, et al. Robust in vitro induction of human germ cell fate from pluripotent stem cells. Cell Stem Cell. 2015;17(2):178–94.
195. Guo H, Hu B, Yan L, Yong J, Wu Y, Gao Y, et al. DNA methylation and chromatin accessibility profiling of mouse and human fetal germ cells. Cell Res. 2017;27(2):165–83.
196. Perrett RM, Turnpenny L, Eckert JJ, O'Shea M, Sonne SB, Cameron IT, et al. The early human germ cell lineage does not express SOX2 during in vivo development or upon in vitro culture. Biol Reprod. 2008;78(5):852–8.
197. Aramaki S, Hayashi K, Kurimoto K, Ohta H, Yabuta Y, Iwanari H, et al. A mesodermal factor, T, specifies mouse germ cell fate by directly activating germline determinants. Dev Cell. 2013;27(5):516–29.

Human Induced Pluripotent Stem (hiPS) Cells: Generation and Applications

Christian Freund

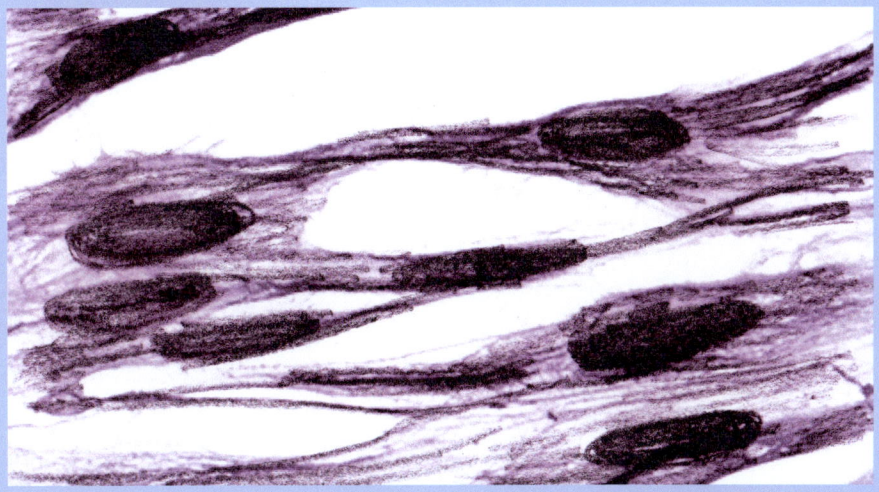

Contents

4.1 Introduction – 59

4.2 Historical Background and Generation of the First iPS Cells – 61

4.3 Reprogramming Vectors – 61

4.4 Somatic Tissue Sources for Reprogramming – 63

Website LUMC hiPSC core facility:
▶ https://www.lumc.nl/research/facilities/hipsc-core-facility/hipsc-for-lumc-researchers-and-external-parties/.

© Springer Nature Switzerland AG 2020
G. Rodrigues, B. A. J. Roelen (eds.), *Concepts and Applications of Stem Cell Biology*,
Learning Materials in Biosciences, https://doi.org/10.1007/978-3-030-43939-2_4

4.5		Generation and Validation of hiPS Cells and Mechanisms of Reprogramming – 64
4.6		Differentiation of hiPS Cells – 66
4.7		Application of hiPS Cells in Medical Research and Therapy – 67
4.7.1		Toxicology Testing – 67
4.7.2		Disease Modeling – 67
4.7.3		Drug Testing – 68
4.7.4		Regenerative Medicine – 68

References – 69

4 Human Induced Pluripotent Stem (hiPS) Cells: Generation and Applications

What Will You Learn in This Chapter?

In 2006/2007, it was discovered that somatic cells can be reverted to an embryonic stem cell-like state. This chapter describes how these so-called induced pluripotent stem (iPS) cells were first generated and briefly mentions a selection of historical findings which led to this groundbreaking discovery. Then, you learn how human iPS (hiPS) cells are generated from a practical point of view: what are the advantages and disadvantages of using certain somatic cell types and reprogramming vectors and what does a typical reprogramming experiment look like? Next the underlying mechanisms of reprogramming are explained briefly as well as what it takes to validate a hiPS cell. This is followed by a section on the differentiation of hiPS cells into heart muscle cells (cardiomyocytes) based on protocols developed in our own laboratory. The last part of the chapter gives examples of how differentiated derivatives of hiPS cells are currently being successfully used for toxicology screening, disease modeling, and drug screening. Finally, you learn about the potential use of hiPS cells in regenerative medicine.

4.1 Introduction

In 2007, Japanese researchers discovered that differentiated human skin fibroblasts can be reconverted to an embryonic stem cell-like state. The resulting dedifferentiated cells were referred to as "induced pluripotent stem" (iPS) cells. Human iPS (hiPS) cells have two main features which make them unique tools for research and medical applications: Firstly, hiPS cells are able to self-renew, meaning that under certain cell culture conditions these cells maintain an undifferentiated state indefinitely. Secondly, hiPS cells are pluripotent: Thus, under appropriate culture conditions, they can be differentiated into virtually all 220 somatic cell types of the human body, e.g. neurons (ectoderm), endothelial cells (mesoderm) or liver-like cells (endoderm) (◘ Fig. 4.1). hiPS cells share self-renewal and pluripotency with human embryonic stem (hES) cells, which are derived from the inner cell mass of early human embryos. However, to obtain hES cells, the embryo has to be destroyed, which is why the usage of hES cells for research is considered ethically controversial. Depending on the country, their generation and use is restricted or sometimes even banned. By contrast, hiPS cells can be generated from a small tissue biopsy (e.g. skin or blood) without any ethical constraints.

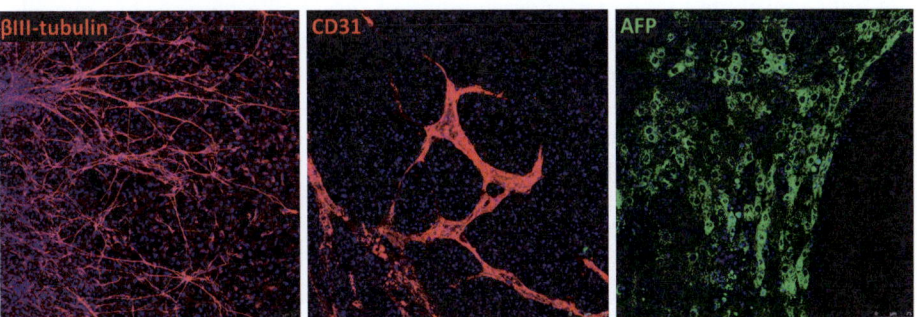

◘ **Fig. 4.1** Differentiated derivatives of hiPS cells. Confocal images showing hiPS cell-derived neurons (left), endothelial cells (middle) and hepatocyte-like cells (right) after staining with fluorescent dye-coupled antibodies against β-III tubulin, CD31 (Pecam), and α-fetoprotein (AFP), respectively. Nuclei were stained with DAPI (blue)

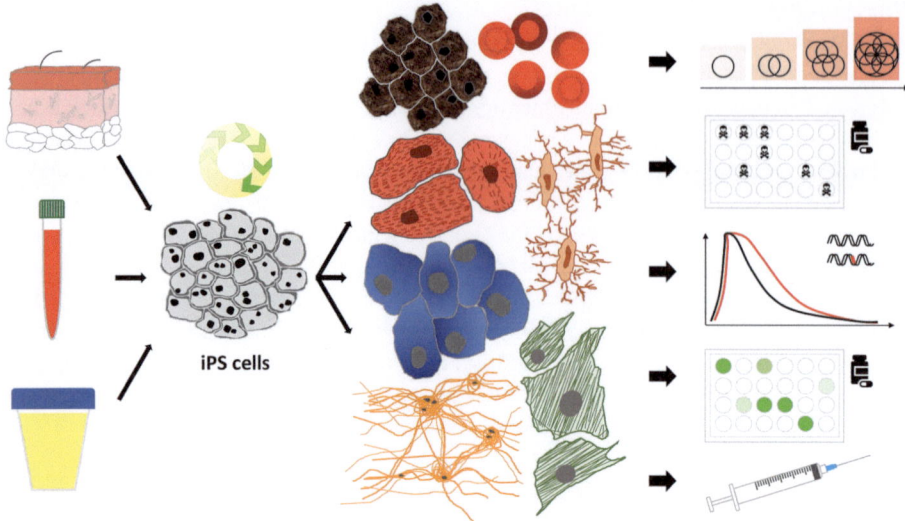

Fig. 4.2 hiPS cells: From generation to application. hiPS cells can be generated from various tissue sources such as skin, blood, and urine (left). The main features of hiPS cells are self-renewal and their capacity to differentiate into derivatives of all three germ layers (pluripotency), e.g., retinal pigment epithelial cells (dark brown), blood cells (red), cardiomyocytes (light red), liver cells (blue), bone cells (light brown), smooth muscle cells (green), and neurons (orange). Differentiated cell types from healthy individuals or patients with a genetic disease can be used for studying human development, toxicology, and drug testing as well as for clinical applications (right)

Following biopsy, the somatic cells are converted ('reprogrammed') into hiPS cells by simple overexpression of four transcription factors known to control pluripotency.

Differentiation of healthy hiPS cells toward somatic lineages is a unique tool to study human developmental processes (◘ Fig. 4.2). Furthermore, hiPS cell derivatives can be used for toxicology tests. For example, hiPS cell-derived cardiomyocytes are utilized to analyze the side effects of chemotherapeutic drugs. Importantly, as the genotype of a hiPS cell is identical to the corresponding donor, these cells are great tools for investigating human genetic diseases in vitro, such as cardiac arrhythmias caused by mutations in a cardiac ion channel gene. hiPS cell-based disease models do not only contribute to a better understanding of disease mechanisms but may also significantly improve the screening and validation of drugs. In the past, disease modeling and initial phases of drug testing heavily relied on animal experiments, especially with genetically modified rodents. However, some important differences between humans and mice exist for example in terms of heart physiology (500 heartbeats per minute versus 80) or in the cellular composition of their organs. Furthermore, certain genetic disease variants and the associated phenotypes observed in humans are absent in rodents. Therefore, hiPS cell-based disease modeling and drug testing represent a significant advancement. Finally, in the future, somatic cells differentiated from hiPS cells are expected to play an important role in regenerative medicine, either by replenishing cells lost due to aging or injury (e.g., after macular degeneration or cardiac infarction) or in the case of certain genetic diseases by replacing genetically defective cells with healthy counterparts after genetic repair of the mutation in hiPS cells.

4.2 Historical Background and Generation of the First iPS Cells

The discovery by Shinya Yamanaka and Kazutoshi Takahashi in 2006/2007 that differentiated unipotent cells with a restricted lifespan can be reprogrammed into pluripotent, immortal stem cells [1, 2] by overexpression of four transcription factors is based on seminal findings made by various researchers in earlier decades. In the 1960s, John Gurdon showed that upon transplantation into an enucleated oocyte, a nucleus from a differentiated tadpole cell gave rise to an entire frog (somatic cell nuclear transfer, SCNT, or cloning; [3]). The first mammal cloned from an adult somatic cell was the sheep Dolly [4]. These results indicated that the nucleus from a differentiated cell contained all necessary information to give rise to a whole organism and that the identity of a differentiated cell was determined by reversible epigenetic modifications rather than by irreversible changes in the DNA sequence. The fact, that the fate of a differentiated cell was not necessarily permanent, was shown by Davis et al. in 1987: Overexpression of skeletal muscle transcription factor MyoD in fibroblasts led to the formation of myoblasts [5]. Similarly, primary B cells could be converted into macrophages by overexpression of the myeloid transcription factor C/EBPα [6]. Evans & Kaufman and Martin achieved another milestone in 1981, when they succeeded in the isolation and culture of the first mouse embryonic stem (mES) cells [7, 8]. The first human embryonic stem (hES) cell lines were established in 1998 [9]. The generation of mouse and human ES cells triggered research on factors controlling pluripotency and differentiation and led to the development of protocols for both the maintenance of undifferentiated cells and for their differentiation into various cell types.

The knowledge about cellular plasticity, transcriptional networks regulating pluripotency, and culture conditions for maintenance of ES cells paved the way for Yamanaka's and Takahashi's discovery of cellular reprogramming by exogenous factors. In order to identify candidates which are able to transform a somatic cell into a pluripotent cell, they used genetically engineered mouse embryonic fibroblasts (MEFs) in which the ES cell-specific *Fbxo15* gene was coupled to a neomycin resistance cassette [2]. Only MEFs reconverted into pluripotent cells expressing *Fbxo15* would be resistant against the drug. Initially, 24 factors known to play a role in pluripotency were tested. After having obtained the first iPS cells with all 24 candidates, the list could be narrowed down to a core set of four transcription factors: *OCT3/4*, *SOX2*, *c-MYC*, and *KLF4*. These reprogramming factors or so-called Yamanaka-factors are sufficient to generate mouse iPS cells. A year later, Takahashi and Yamanaka generated iPS cells from adult human skin fibroblasts using the same four pivotal factors [1]. In parallel, the group of James Thomson was able to obtain hiPS cells with *NANOG* and *Lin28* replacing *c-MYC* and *KLF4* [10]. hiPS cells generated in both ways displayed the hallmarks of pluripotent stem cells: They were able to self-renew and to be differentiated into derivatives of all three germ layers. In 2012, Gurdon and Yamanaka received the Nobel Prize for their groundbreaking work on cellular reprogramming.

4.3 Reprogramming Vectors

For the generation of iPS cells, the coding sequences of all four Yamanaka factors have to be introduced into the somatic target cell. Reprogramming vectors can be divided into two major groups: integrating vectors such as retroviruses and lentiviruses, which persist in the iPS cells after reprogramming is completed. Importantly, the expression of the reprogramming factors must be silenced once hiPS cells have been established. By contrast, vectors like Sendai virus (SeV), episomal plasmids, and synthetic RNA are non-integrating and only transiently remain in the target cell. For this reason, non-integrating reprogramming vectors are used preferentially, but other aspects such as production costs, storability, laboratory biosafety requirements, range of target cells, and reprogramming efficiency have implications on the choice of the vector for daily use.

For their initial experiments to generate mouse iPS cells, Takahashi and Yamanaka used four retroviruses, each carrying one of the four Yamanaka factor sequences. For human iPS cell generation, an additional lentivirus encoding the murine retroviral receptor was necessary. The resulting hiPS cell lines had a minimum of five viral integrations; however, in practice, the number of integrations was sometimes up to 20. A major drawback of retroviruses is their integration into the host genome, preferentially at sites of actively transcribed genes, which may result in an altered phenotype when using hiPS cells for disease modeling. A further problem may arise from incomplete silencing of the reprogramming factors which might compromise the differentiation capacity of the cells [11].

Currently, lentiviruses are commonly used for the generation of hiPS cells destined for in vitro applications. Their production is easy and cost-effective, virus stocks can be stored frozen, and they require a lower biosafety level than retroviruses. A lentivirus generated by Warlich et al. carries the sequences of all four Yamanaka factors attached to a red fluorescent dye [12]. The latter enables monitoring of infection efficiency of somatic cells as well as silencing of the reprogramming factors in established hiPS cells. Furthermore, a single viral integration is sufficient for successful reprogramming, and if required, the transgenes can be removed by using the Flp enzyme, as Flp target sequences have been introduced at the 5′ and 3′ end.

Non-integrating vectors currently used for generation of hiPS cells include episomal vectors, SeV, and synthetic RNA. Episomal vectors are easily produced in bacteria and can efficiently generate hiPS cells [13]. However, they may occasionally integrate into the host genome, and therefore extensive screening for integration-free hiPS cell lines is obligatory prior to application in humans [14].

SeV can infect a wide range of human target cells [15]. These viruses remain exclusively in the cytoplasm of the infected cell and are therefore completely non-integrating ('zero footprint'). However, SeV vectors are relatively difficult to produce, and commercially available SeVs are expensive. In addition, in the Netherlands, hiPS cells generated with SeVs require a higher biosafety level, unless the absence of the vector has been proven at RNA and protein level.

Synthetic RNA is another non-integrating reprogramming vector, yet it normally is rapidly degraded by ribonucleases in the transfected cells. Initial protocols therefore required daily transfections in a two-week period, making synthetic RNA reprogramming laborious, expensive, and prone to error. A major improvement was achieved by Yoshioka et al. whose reprogramming method with modified synthetic RNA only requires a single transfection together with blocking RNA degradation [16]. Nevertheless, RNA reprogramming remains expensive and might not reprogram all somatic cell types with sufficient efficiency.

At our hiPS cell core facility lentiviruses, episomal vectors, SeV, and synthetic RNA are routinely used for generation of research-grade hiPS cells derived from various tissue sources [17, 18].

Various other reprogramming methods have been proposed as alternatives. One approach aims at the complete replacement of exogenous reprogramming factors using chemical compounds. Generation of iPS cells with small molecules in the absence of transgenes has been achieved using mouse somatic cells [19]. For reprogramming human somatic cells, Sox2 can be replaced by a small molecule targeting the TGF-β pathway [20], but efficient reprogramming solely with chemical compounds has not been reported yet. Finally, the directed DNA binding capacity of CRISPR/Cas9 has recently been employed to successfully reprogram human cells by activating transcription of endogenous pluripotency genes [21]. Although potentially interesting, for widespread application, the use of CRISPR/Cas9 for reprogramming cells requires further verification. In conclusion, various satisfactory reprogramming systems have been developed, but an efficient method that combines cost-efficiency, low biosafety requirements, a wide range of target cells, absence of alteration of the host genome, and high reprogramming efficiency is still pending.

4.4 Somatic Tissue Sources for Reprogramming

In theory, any somatic cell type can be used for generation of hiPS cells, even neural stem cells from brain [22]. However, in practice, the choice of human donor tissue is strongly influenced by the accessibility/invasiveness of the biopsy taken. Additional criteria include the possibility to store tissue material prior to cell isolation, the feasibility, and cost-effectiveness of cell culture, the proliferation capacity of the isolated somatic cells and their ability to be reprogrammed with common vector systems. The vast majority of hiPS cells are therefore generated from skin, blood, or urine.

Skin fibroblasts can easily be isolated and expanded from 4 mm punch biopsies. We previously found out that skin biopsies can be stored in saline buffer at 4 °C for up to 2 weeks prior to fibroblast isolation which would allow long-distance shipment of rare donor material [17]. Importantly, skin fibroblasts can be reprogrammed with any of the standard reprogramming vector systems (e.g. lentivirus, Sendai virus, episomal vectors, synthetic RNA). However, punch biopsies are painful and at least in the Netherlands the procedure is not applicable to minors. By contrast, milk teeth from children constitute a completely noninvasive tissue source, and dental pulp cells are readily reprogrammed into hiPS cells [17]. The disadvantage of using teeth lies in a higher risk of cell culture contamination with bacteria and fungi which is why they have to be processed immediately.

Blood samples can generally be stored for a maximum of 24 hours at room temperature before processing. Peripheral blood contains various cell types suitable for reprogramming. In the beginning, we focused on blood outgrowth endothelial cells (BOECs). However, these are rare and thus require high blood volumes (80 ml), which in the Netherlands is contraindicated for children. To obtain sufficient numbers of BOECs for reprogramming may take weeks and the reprogramming efficiency of this cell type with lentivirus is low [17]. By contrast, as little as 10 ml peripheral blood is needed to isolate sufficient peripheral blood mononuclear cells (PBMCs) by a simple density gradient centrifugation. PBMCs are a mixture of various cell types such as T and B cells, macrophages, and erythroid progenitor cells. The latter are rare but can be easily expanded and reprogrammed with episomal vectors or lentivirus. In our hiPS cell core facility, we now routinely use erythroblasts for reprogramming.

Interestingly also urine is a source of somatic cells for reprogramming: About 7000 renal epithelial (RE) cells are excreted daily via the urinary tract [23]. RE cells can be expanded in a specific cell culture media and are readily transformed into hiPS cells with various vector systems. Collecting urine samples is truly noninvasive, but they have to be processed immediately after collection.

When patients undergo surgical procedures, tissue material that normally is not accessible with regular biopsies may be obtained for reprogramming. For example, we were able to isolate fibroblasts and generated hiPS cells from nasal epithelium removed from patients suffering from recurrent nosebleeds due to a defect in TGF-β signaling (hereditary hemorrhagic telangiectasia). Furthermore, we isolated and reprogrammed chondrocytes from cartilage tissue leftover from hip replacement surgery.

It has been reported that hiPS cells retain epigenetic marks that are specific of the somatic cell type they were derived from ('epigenetic memory' [24]). As a consequence, hiPS cells may be differentiated more easily into their cell type of origin which is advantageous for cell types without efficient differentiation protocols. Furthermore, certain internal organs might be better cell sources for reprogramming than skin: Exposure to UV radiation from sunlight is known to cause DNA damage and skin cells may acquire more spontaneous mutations than blood cells. Finally, for disease modeling, the choice of donor material also depends on whether a disease-causing mutation is present in all tissues and cells or not. In so-called mosaic patients, for instance, only certain somatic cell types carry the mutation, whereas the healthy gene is present in others. Here, mutated and normal hiPS cells can be obtained from the same individual and, thus, have the same genetic background (isogenic), which certainly represents an ideal situation for studying the effect of the disease-specific mutation.

In conclusion, various somatic cell types from skin, blood, and urine can be readily used for reprogramming, but the choice is often limited due to the invasiveness of biopsy-taking, the feasibility of cell culture, and the reprogramming efficiency. Of note, all tissue biopsies, with the exception of leftover surgical material, must be taken with a proper informed consent in which the donor agrees with the use of the tissue material for reprogramming and downstream applications.

4.5 Generation and Validation of hiPS Cells and Mechanisms of Reprogramming

For a typical reprogramming experiment, $1 \times 10^5 - 5 \times 10^5$ somatic cells, e.g., skin fibroblasts are infected with virus or transfected with plasmids or RNA and are allowed to expand in somatic cell media for 1 week (◘ Fig. 4.3). They are then plated on irradiated (cell cycle arrested) mouse embryonic feeder cells in culture medium containing fetal calf serum or serum replacement. Alternatively, defined conditions such as recombinant extracellular matrix proteins and animal component-free media can be used [25]. First colonies of hiPS cells emerge after 3 weeks and are selected ('picking') and expanded separately to establish various clonal hiPS cell lines from the same donor.

Human Induced Pluripotent Stem (hiPS) Cells: Generation and Applications

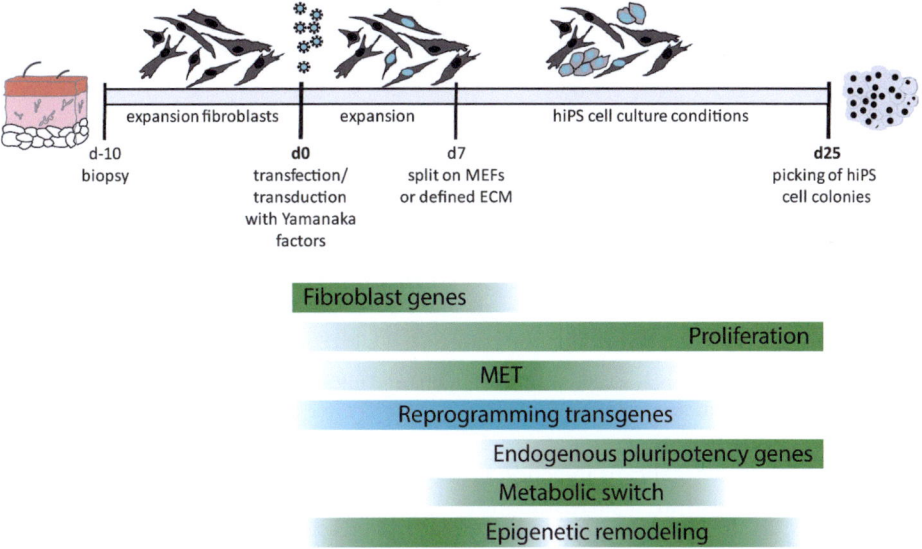

Fig. 4.3 Schematic of a reprogramming experiment. Timecourse in days (d), starting with the isolation of fibroblasts from a skin biopsy, infection of fibroblasts with viral vectors encoding the Yamanaka factors up to picking of hiPS cell colonies. Events during reprogramming are shown below. MET: mesenchymal-to-epithelial transition

Recently, reprogramming was demonstrated in a miniaturized cell culture system: Gagliano and colleagues used a microfluidic chamber allowing for a significant downscaling of somatic cell numbers and reprogramming compounds as well as the generation of multiple hiPS cell lines simultaneously [26].

Although the mechanisms of reprogramming are still not fully understood, gene expression analysis of intermediate stages has shed some light on the underlying processes [27, 28] (Fig. 4.3). Whereas transgenic *c-MYC* plays a role in the early reprogramming phase, transgenic *OCT3/4* and *SOX2* are thought to act mainly at later stages of reprogramming. When skin fibroblasts are used as somatic cell source, the expression of the Yamanaka factors initially leads to the suppression of genes controlling fibroblast identity. Transforming fibroblasts increase their proliferation rate and transition from a mesenchymal (migratory, loose cell–cell contacts) to an epithelial state (stationary, tight intercellular contacts). This process is known as mesenchymal-to-epithelial transition (MET). In addition, a metabolic switch from oxidative phosphorylation to glycolysis for cellular energy production occurs. Importantly the whole reprogramming process is accompanied by a massive epigenetic remodeling. Furthermore, the cellular morphology changes drastically: The large spindle-like fibroblast is converted into a small rounded hiPS cell of which the cytoplasm is almost completely occupied by a nucleus that contains large nucleoli (Fig. 4.4). The expression of endogenous pluripotency genes such as *OCT3/4* and *SOX2* maintains the emerging hiPS cells in an undifferentiated state, whereas the transgenic Yamanaka factors become transcriptionally silenced or are eliminated.

Although a hiPS cell colony can be clearly identified solely based on morphology, for each hiPS cell line, complete reprogramming and hES cell properties, such as self-renewal and pluripotency, have to be verified. For mouse iPS cells, the most stringent

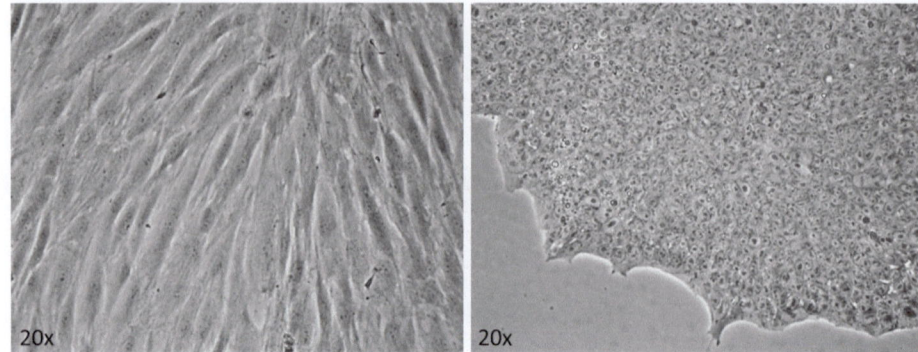

Fig. 4.4 Cellular morphologies: Skin fibroblasts (left) and hiPS cells generated from skin fibroblasts (right) at 20× magnification

assay to prove pluripotency is injection into blastocysts. Upon transplantation of the chimeric blastocysts into a surrogate mother, the organs of the embryo forming will partially contain cells that originate from the injected pluripotent stem cells (chimera). In case of injecting miPS cells into a tetraploid blastocyst which can only give rise to extraembryonic tissues, the entire embryo evolves from the iPS cells [29]. In most countries, analogous experiments injecting human iPS cells into mouse blastocysts are considered unethical and are thus prohibited. Therefore, the most stringent assay to assess the pluripotency of hiPS cells is injection under the skin of adult mice, where they spontaneously form differentiated benign tumors ('teratomas') [30]. These teratomas consist of derivatives of all three germ layers, e.g., neuroectoderm, endodermal gut epithelium, and mesodermal cartilage. As the teratoma assay is non-quantitative, time-consuming and animal-dependent alternative assays such as the analysis of global gene expression in undifferentiated hiPS cells in combination with directed in vitro differentiation into derivatives of the three germ layers have been proposed [30]. Finally, it should also be confirmed that hiPS cells have a normal karyotype.

4.6 Differentiation of hiPS Cells

A prerequisite to use hiPS cells for various applications is the efficient differentiation into the cell type(s) of interest. For example, hiPS cell-derived heart muscle cells (cardiomyocytes) can be used for testing potential cardiotoxic side effects of drugs or for modeling arrhythmias caused by a mutated gene encoding for a cardiac ion channel. Originally we differentiated human ES cells into cardiomyocytes by coculture with visceral-endoderm-like (END-2) cells [31]. Since then, significant progress has been made and now cytokines and small molecules are being used to induce a sequence of developmental events leading to cardiac differentiation: First the formation of mesoderm is induced by a combination of Activin A, bone morphogenic protein (BMP) 4, and a glycogen synthase kinase (GSK-3) small molecule inhibitor that activates the Wnt/β-catenin pathway. Further specification into cardiac progenitors is achieved by exposure to a small molecule Wnt/β-catenin inhibitor. About 1 week after the initiation of differentiation, spontaneously contracting cardiomyocytes can be observed [32]. The recent refinement of the method now enables the simultaneous generation of cardiomyocytes and endothelial cells [33]. Similar multi-step protocols mimicking early developmental processes have been developed for the differentiation into many other cell types, for example, pancreatic β-cells [34], skeletal muscle precursors [35], or retinal pigment epithelial (RPE) cells [36].

4.7 Application of hiPS Cells in Medical Research and Therapy

Due to their unique features, hiPS cells are currently being used in a number of applications in medical research and therapy (◘ Fig. 4.2).

4.7.1 Toxicology Testing

Doxorubicin was one of the first chemotherapeutic drugs and is still used for treatment in about 50% of the breast cancer patients. Its cardiotoxic side effects that lead to arrhythmia, cardiac infarction, and heart failure are well known. However, not all patients develop doxorubicin-induced side effects and at present it is impossible to predict who will be affected. Recently, Burridge et al. tested whether hiPS cell-derived cardiomyocytes were able to recapitulate the patient's doxorubicin susceptibility [37]. For this purpose, hiPS cells were generated from four healthy individuals and 8 doxorubicin-treated breast cancer patients, of which half experienced cardiotoxicity, whereas the other half was unaffected. Interestingly, hiPS cell-derived cardiomyocytes from doxorubicin-affected patients were also more susceptible to the drug in vitro: A higher degree of sarcomeric disarray was observed, arrhythmias occurred more frequently, and cell viability was significantly reduced [37]. These results nicely demonstrate the usefulness of hiPS cell derivatives for assays examining pharmacological toxicity.

4.7.2 Disease Modeling

Currently, hiPS cells are most widely used for developing models to study genetic diseases. The long-QT2 syndrome is a life-threatening cardiac disease resulting from a mutation in a cardiac ion channel gene. The defective potassium channel causes an abnormal (prolonged) repolarization of the heart after a heartbeat. Bellin et al. generated hiPS cells from a long-QT2 patient. As the genetic background influences the disease phenotype, the mutation was corrected in the long-QT hiPS cells in order to generate the matching control. In parallel, the long-QT2 mutation was introduced into a healthy hES cell line, which resulted in two isogenic pairs of mutant and normal cells. hiPS and hES cells were differentiated into cardiomyocytes and electrophysiological analysis confirmed the prolonged action potential in mutated cells when compared to controls [38].

Whereas the long-QT2 syndrome often affects young patients, other diseases have a late onset. For example, Parkinson's disease only becomes manifest in the sixth or seventh decade of life and is caused by loss of dopaminergic neurons. A major hurdle to efficiently use hiPS cell derivatives for disease modeling is the fact that differentiation of hiPS cells often results in immature cells. For example, hiPS cell-derived neurons resemble primary neurons from a fetal stage [39]. Accordingly, no disease-associated phenotype was observed in dopaminergic neurons differentiated from Parkinson's hiPS cells [40]. However, neurodegeneration eventually became apparent upon overexpression of a protein inducing cellular aging: Progerin-expressing dopaminergic neurons derived from Parkinson's hiPS cells had shorter dendrites and were more prone to apoptosis than their progerin-expressing healthy counterparts [40].

Whereas analysis of regular cultures of cardiomyocytes or neurons was sufficient to reveal the disease phenotype for long-QT2-syndrome or Parkinson's disease, respectively, modeling other diseases will likely require more complex in vitro assays. For example, hereditary hemorrhagic telangiectasia (HHT) is caused by defects in TGF-β signaling and leads to leaky blood vessels due to a disturbed interaction between inner endothelial cells and pericytes lining the outside of the blood vessel wall. A model mimicking the in vivo situation with three-dimensional luminal structures formed by both cell types has recently been developed [41]. Even more complex structures containing multiple cell types ('organoids') have been generated for various organs, e.g., the kidney [42] and will likely further improve disease modeling.

4.7.3 Drug Testing

Besides gaining further insight into the mechanisms of genetic diseases, an important goal for hiPS cell-based disease models is the identification and validation of specific drug candidates. Amyotrophic lateral sclerosis (ALS) is caused by the death of neurons controlling voluntary muscles. In 2014, the disease gained worldwide media attention by the 'ALS Ice Bucket Challenge'. In vitro hiPS cell-derived motoneurons recapitulate the disease phenotype and show hyperexcitability and reduced survival [43]. Interestingly, an already approved drug normally used for treatment of epilepsy, ezogabine was able to reduce neuronal excitability and improved cell survival [43, 44]. From this finding, it took less than 2 years to initiate a clinical trial to test ezogabine for treatment of ALS patients [45]. Ezogabine is a good example for identifying new targets for already existing drugs ('repurposing'), but hiPS cell-based disease models will be equally important for identification of new compounds.

4.7.4 Regenerative Medicine

Recently, the first clinical trials using differentiated cell types derived from hiPS cells have been started. The goal of such early-stage trials is to test the feasibility and the safety of potential cellular therapies. One disease which may be treated with hiPS cell derivatives in the future is age-related macular degeneration (AMD). Patients suffer from the degradation of pigmented epithelial (RPE) cells of the retina resulting in impaired vision. Mandai et al. generated hiPS cells from tissue material of two AMD patients [14]. hiPS cells generated from one AMD patient were differentiated into a RPE cell sheet which was subsequently transplanted into the eye. After 1 year of follow-up, the patient's vision had neither improved nor worsened. The graft appeared to have survived for an additional year. Importantly no adverse effects such as tumor formation were detected. For safety reasons, the second AMD patient did not undergo transplantation, as the DNA sequence of his hiPS cells contained small deletions in certain endogenous genes which might have altered their normal expression [14]. Future regenerative medicine based on hiPS cell derivatives may also include cardiac disease [46], diabetes, Parkinson's, and kidney disease.

In conclusion, disease models based on hiPS cell derivatives represent a significant improvement in understanding disease processes when compared to previously used animal models. In addition, hiPS cells reflect human genetic variants which influence the disease phenotype and are therefore particularly interesting for customized treatments ('personalized medicine'). hiPS cell-based models for drug testing hold a lot of promise, but need to be proven how well they can mimic the effect of a drug in vivo. The first clinical trials using hiPS cell-derived cell types are on their way. The inclusion of larger patient cohorts will prove whether such treatments are safe and efficient.

Take-Home Message

- hiPS cells can be generated from various somatic cell types. Minimally or non-invasive biopsies such as peripheral blood or urine are preferred.
- Delivery methods for the Yamanaka factors are preferentially nonintegrating, for example, RNA or SeV.
- By addition of cytokines and growth factors, hiPS cells can be differentiated into a multitude of cell types.
- Since hiPS cells capture the genotype of the donor, they are great tools for studying disease mechanisms and for drug testing. In addition, they are being used successfully for toxicology testing and developmental studies.
- hiPS cell derivatives hold great promise for future cell therapies. The first clinical trials with small patient cohorts aim at testing feasibility and safety.

Acknowledgments I am grateful to R. Schmidt-Ullrich, PhD (Max-Delbrück Center for Molecular Medicine, Berlin, Germany) and C. Mummery, PhD (LUMC, Leiden, The Netherlands) for critical reading of the manuscript.

References

1. Takahashi K, Tanabe K, Ohnuki M, Narita M, Ichisaka T, Tomoda K, et al. Induction of pluripotent stem cells from adult human fibroblasts by defined factors. Cell. 2007;131(5):861–72.
2. Takahashi K, Yamanaka S. Induction of pluripotent stem cells from mouse embryonic and adult fibroblast cultures by defined factors. Cell. 2006;126(4):663–76.
3. Gurdon JB. The developmental capacity of nuclei taken from intestinal epithelium cells of feeding tadpoles. J Embryol Exp Morphol. 1962;10:622–40.
4. Wilmut I, Schnieke AE, McWhir J, Kind AJ, Campbell KH. Viable offspring derived from fetal and adult mammalian cells. Nature. 1997;385(6619):810–3.
5. Davis RL, Weintraub H, Lassar AB. Expression of a single transfected cDNA converts fibroblasts to myoblasts. Cell. 1987;51(6):987–1000.
6. Xie H, Ye M, Feng R, Graf T. Stepwise reprogramming of B cells into macrophages. Cell. 2004;117(5):663–76.

7. Evans MJ, Kaufman MH. Establishment in culture of pluripotential cells from mouse embryos. Nature. 1981;292(5819):154–6.
8. Martin GR. Isolation of a pluripotent cell line from early mouse embryos cultured in medium conditioned by teratocarcinoma stem cells. Proc Natl Acad Sci U S A. 1981;78(12):7634–8.
9. Thomson JA, Itskovitz-Eldor J, Shapiro SS, Waknitz MA, Swiergiel JJ, Marshall VS, et al. Embryonic stem cell lines derived from human blastocysts. Science (New York, NY). 1998;282(5391):1145–7.
10. Yu J, Vodyanik MA, Smuga-Otto K, Antosiewicz-Bourget J, Frane JL, Tian S, et al. Induced pluripotent stem cell lines derived from human somatic cells. Science (New York, NY). 2007;318(5858):1917–20.
11. Koyanagi-Aoi M, Ohnuki M, Takahashi K, Okita K, Noma H, Sawamura Y, et al. Differentiation-defective phenotypes revealed by large-scale analyses of human pluripotent stem cells. Proc Natl Acad Sci U S A. 2013;110(51):20569–74.
12. Warlich E, Kuehle J, Cantz T, Brugman MH, Maetzig T, Galla M, et al. Lentiviral vector design and imaging approaches to visualize the early stages of cellular reprogramming. Mol Ther J Am Soc Gene Ther. 2011;19(4):782–9.
13. Okita K, Matsumura Y, Sato Y, Okada A, Morizane A, Okamoto S, et al. A more efficient method to generate integration-free human iPS cells. Nat Methods. 2011;8(5):409–12.
14. Mandai M, Kurimoto Y, Takahashi M. Autologous induced stem-cell-derived retinal cells for macular degeneration. N Engl J Med. 2017;377(8):792–3.
15. Nishimura K, Sano M, Ohtaka M, Furuta B, Umemura Y, Nakajima Y, et al. Development of defective and persistent Sendai virus vector: a unique gene delivery/expression system ideal for cell reprogramming. J Biol Chem. 2011;286(6):4760–71.
16. Yoshioka N, Gros E, Li HR, Kumar S, Deacon DC, Maron C, et al. Efficient generation of human iPSCs by a synthetic self-replicative RNA. Cell Stem Cell. 2013;13(2):246–54.
17. Dambrot C, van de Pas S, van Zijl L, Brandl B, Wang JW, Schalij MJ, et al. Polycistronic lentivirus induced pluripotent stem cells from skin biopsies after long term storage, blood outgrowth endothelial cells and cells from milk teeth. Differentiation Res Biol Divers. 2013;85(3):101–9.
18. Halaidych OV, Freund C, van den Hil F, Salvatori DCF, Riminucci M, Mummery CL, et al. Inflammatory responses and barrier function of endothelial cells derived from human induced pluripotent stem cells. Stem Cell Rep. 2018;10(5):1642–56.
19. Hou P, Li Y, Zhang X, Liu C, Guan J, Li H, et al. Pluripotent stem cells induced from mouse somatic cells by small-molecule compounds. Science (New York, NY). 2013;341(6146):651–4.
20. Ichida JK, Blanchard J, Lam K, Son EY, Chung JE, Egli D, et al. A small-molecule inhibitor of tgf-Beta signaling replaces sox2 in reprogramming by inducing nanog. Cell Stem Cell. 2009;5(5):491–503.
21. Weltner J, Balboa D, Katayama S, Bespalov M, Krjutskov K, Jouhilahti EM, et al. Human pluripotent reprogramming with CRISPR activators. Nat Commun. 2018;9(1):2643.
22. Kim JB, Greber B, Arauzo-Bravo MJ, Meyer J, Park KI, Zaehres H, et al. Direct reprogramming of human neural stem cells by OCT4. Nature. 2009;461(7264):649–3.
23. Zhou T, Benda C, Duzinger S, Huang Y, Li X, Li Y, et al. Generation of induced pluripotent stem cells from urine. J Am Soc Nephrol. 2011;22(7):1221–8.
24. Ohi Y, Qin H, Hong C, Blouin L, Polo JM, Guo T, et al. Incomplete DNA methylation underlies a transcriptional memory of somatic cells in human iPS cells. Nat Cell Biol. 2011;13(5):541–9.
25. Chen G, Gulbranson DR, Hou Z, Bolin JM, Ruotti V, Probasco MD, et al. Chemically defined conditions for human iPSC derivation and culture. Nat Methods. 2011;8(5):424–9.
26. Gagliano O, Luni C, Qin W, Bertin E, Torchio E, Galvanin S, et al. Microfluidic reprogramming to pluripotency of human somatic cells. Nat Protoc. 2019;14(3):722–37.
27. Polo JM, Anderssen E, Walsh RM, Schwarz BA, Nefzger CM, Lim SM, et al. A molecular roadmap of reprogramming somatic cells into iPS cells. Cell. 2012;151(7):1617–32.
28. Takahashi K, Yamanaka S. A decade of transcription factor-mediated reprogramming to pluripotency. Nat Rev Mol Cell Biol. 2016;17(3):183–93.
29. Zhao XY, Li W, Lv Z, Liu L, Tong M, Hai T, et al. iPS cells produce viable mice through tetraploid complementation. Nature. 2009;461(7260):86–90.
30. Bouma MJ, van Iterson M, Janssen B, Mummery CL, Salvatori DCF, Freund C. Differentiation-defective human induced pluripotent stem cells reveal strengths and limitations of the teratoma assay and in vitro pluripotency assays. Stem Cell Rep. 2017;8(5):1340–53.

31. Freund C, Ward-van Oostwaard D, Monshouwer-Kloots J, van den Brink S, van Rooijen M, Xu X, et al. Insulin redirects differentiation from cardiogenic mesoderm and endoderm to neuroectoderm in differentiating human embryonic stem cells. Stem Cells (Dayton, Ohio). 2008;26(3):724–33.
32. van den Berg CW, Elliott DA, Braam SR, Mummery CL, Davis RP. Differentiation of human pluripotent stem cells to cardiomyocytes under defined conditions. Methods Mol Biol (Clifton, NJ). 2016;1353:163–80.
33. Giacomelli E, Bellin M, Orlova VV, Mummery CL. Co-differentiation of human pluripotent stem cells-derived cardiomyocytes and endothelial cells from cardiac mesoderm provides a three-dimensional model of cardiac microtissue. Curr Protoc Hum Genet. 2017;95:21.9.1–2.
34. Rosado-Olivieri EA, Anderson K, Kenty JH, Melton DA. YAP inhibition enhances the differentiation of functional stem cell-derived insulin-producing beta cells. Nat Commun. 2019;10(1):1464.
35. Borchin B, Chen J, Barberi T. Derivation and FACS-mediated purification of PAX3+/PAX7+ skeletal muscle precursors from human pluripotent stem cells. Stem Cell Rep. 2013;1(6):620–31.
36. Maruotti J, Sripathi SR, Bharti K, Fuller J, Wahlin KJ, Ranganathan V, et al. Small-molecule-directed, efficient generation of retinal pigment epithelium from human pluripotent stem cells. Proc Natl Acad Sci U S A. 2015;112(35):10950–5.
37. Burridge PW, Li YF, Matsa E, Wu H, Ong SG, Sharma A, et al. Human induced pluripotent stem cell-derived cardiomyocytes recapitulate the predilection of breast cancer patients to doxorubicin-induced cardiotoxicity. Nat Med. 2016;22(5):547–56.
38. Bellin M, Casini S, Davis RP, D'Aniello C, Haas J, Ward-van Oostwaard D, et al. Isogenic human pluripotent stem cell pairs reveal the role of a KCNH2 mutation in long-QT syndrome. EMBO J. 2013;32(24):3161–75.
39. Mariani J, Simonini MV, Palejev D, Tomasini L, Coppola G, Szekely AM, et al. Modeling human cortical development in vitro using induced pluripotent stem cells. Proc Natl Acad Sci U S A. 2012;109(31):12770–5.
40. Miller JD, Ganat YM, Kishinevsky S, Bowman RL, Liu B, Tu EY, et al. Human iPSC-based modeling of late-onset disease via progerin-induced aging. Cell Stem Cell. 2013;13(6):691–705.
41. de Graaf MNS, Cochrane A, van den Hil FE, Buijsman W, van der Meer AD, van den Berg A, et al. Scalable microphysiological system to model three-dimensional blood vessels. APL Bioeng. 2019;3(2):026105.
42. van den Berg CW, Ritsma L, Avramut MC, Wiersma LE, van den Berg BM, Leuning DG, et al. Renal subcapsular transplantation of PSC-derived kidney organoids induces neo-vasculogenesis and significant glomerular and tubular maturation in vivo. Stem Cell Rep. 2018;10(3):751–65.
43. Wainger BJ, Kiskinis E, Mellin C, Wiskow O, Han SS, Sandoe J, et al. Intrinsic membrane hyperexcitability of amyotrophic lateral sclerosis patient-derived motor neurons. Cell Rep. 2014;7(1):1–11.
44. Kiskinis E, Sandoe J, Williams LA, Boulting GL, Moccia R, Wainger BJ, et al. Pathways disrupted in human ALS motor neurons identified through genetic correction of mutant SOD1. Cell Stem Cell. 2014;14(6):781–95.
45. McNeish J, Gardner JP, Wainger BJ, Woolf CJ, Eggan K. From dish to bedside: lessons learned while translating findings from a stem cell model of disease to a clinical trial. Cell Stem Cell. 2015;17(1):8–10.
46. Cyranoski D. 'Reprogrammed' stem cells approved to mend human hearts for the first time. Nature. 2018;557(7707):619–20.

Cellular Reprogramming and Aging

Sandrina Nóbrega-Pereira and Bruno Bernardes de Jesus

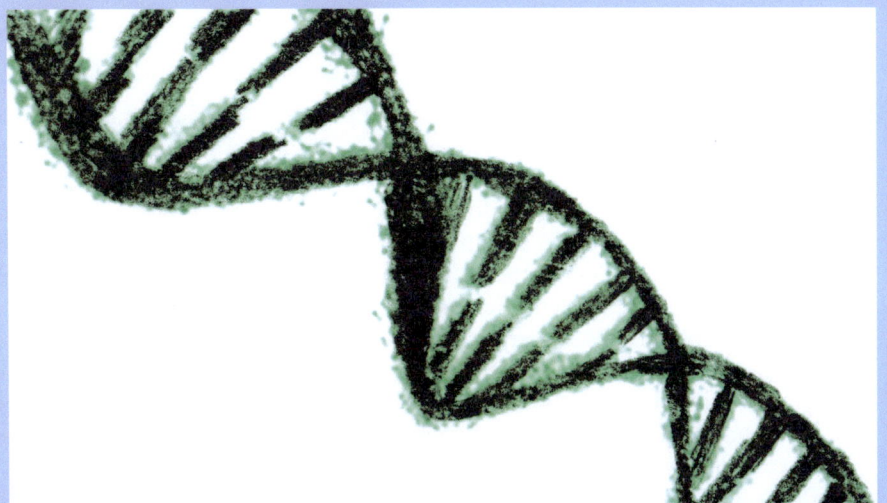

Contents

5.1 **Stem Cells in Health and Disease – 74**
5.1.1 Aging of Stem Cells – 75

5.2 **Cellular Reprogramming – 77**
5.2.1 Aging as a Barrier for Cellular Reprogramming – 78
5.2.2 Reprogramming In Vivo as an Antiaging Strategy – 84

References – 86

The original version of this chapter was revised by updating the affiliation of the second author. The correction to this chapter can be found at https://doi.org/10.1007/978-3-030-43939-2_14

© Springer Nature Switzerland AG 2020, corrected publication 2020
G. Rodrigues, B. A. J. Roelen (eds.), *Concepts and Applications of Stem Cell Biology*,
Learning Materials in Biosciences, https://doi.org/10.1007/978-3-030-43939-2_5

What Will You Learn in This Chapter?
In this chapter, we will discuss how aging impacts the dynamics of stem cells and affects cellular reprogramming. We will discuss certain pathways involved in the aging process, and how we can use those pathways to increase the efficiency of the reprogramming protocol. Furthermore, we will describe two protocols to establish primary cultures of a "young" cell line (mouse embryonic fibroblasts—MEFs) and an adult/old cell line (primary cultures of adult/old mouse ear fibroblasts). These cell lines are often the primary choice to study aging in vitro.

5.1 Stem Cells in Health and Disease

Stem cells are defined as cells that retain the potential to differentiate into several cell lineages and have a constant self-renewal capacity, although with a limited replicative capacity in vivo [1]. Three criteria should be met: (i) a stem cell should be able to self-renew, (ii) it should be unspecialized, meaning that it should not perform any function until required, and (iii) a stem cell should be able to specialize/differentiate into at least one cell type. One of the best-characterized pools are the stem cells belonging to the hematopoietic system (hematopoietic stem cells—HSC). These cells can differentiate in blood cells and are crucial for the normal functioning of the young and adult hematopoietic system. HSC follow a hierarchical cascade where multipotent stem cells give rise to fully committed lineages [1]. This traditional view has been challenged recently, and it has been hypothesized that stem cells, although residing at an organ-specific niche, can trans-differentiate to cell types, including cells from a different lineage. While the mechanism remains poorly understood and the field controversial, this could have implications for organ renewal [2].

Different tissues have variable rates of proliferation. In tissues with high proliferative capacity, stem cells were shown to generate large amounts of progeny, whereas in tissues with lower proliferative potential, it is believed that stem cells have less critical functions; recent evidence demonstrate, however, that almost every tissue (including heart and brain) retains fully functional niches of pluripotent cells which can replace damaged cells [3–8]. The safeguarding of these cell niches is crucial for the maintenance of body function, in particular, at older ages when tissues are more receptive to damaging signals. Indeed, mice with mutations directly affecting the pools of stem cells are characterized by an accelerated aging syndrome [9].

The function of stem cells seems to be modulated with age progression, displaying reduced developmental potency with division cycles. This could be correlated with the fact that, and despite being telomerase positive, stem cells experience a telomere loss with proliferative age [10]. Telomeres are repetitive DNA sequences at chromosome ends that are bound by a protective protein complex, which prevents them from eliciting a DNA-damage response (DDR). Longer telomeres are usually related to cells with division capacity, such as stem cells. The telomerase (a ribonucleic complex, which under certain circumstances may extend telomeres) activity in stem cells is not sufficient to fully support maintenance of telomere length, and this results in a loss of telomere repeats with cycling states, although the loss is lower compared with that of somatic cells. This decay in stem cell potential has been associated with age-dependent organ dysfunction [11, 12]. While the role of stem cells in organ renewal and tissue repair is a well-defined capacity inherent to this group of cells, and usually observed as a positive aptitude, stem cells were also shown to be involved in cancer susceptibility or progression [13, 14]. It is believed that tumors are composed of special groups of cells that fuel cancer formation and progression, being responsible for tumor survival to different therapies, either through activating DNA repair pathways or protection from apoptosis [15, 16]. The idea of a cancer initiating cell, however, could be restricted to certain tumor types. It has been documented that 1–4% of leukemic cells could form spleen colonies when transplanted in vivo [17–22]. On the other hand, around 25% of single melanoma cells isolated from patients could form tumors in mice [23]. Possibly, in particular conditions, stem cells could be at the basis of tumor initiation. Telomerase could be playing a role under these circumstances as well. While many tumors are telomerase positive, as previously mentioned, they present shorter telomeres compared to the surrounding tissues [24–29]. This observation indicates that cells with short telomeres accumulating mutations leading to an aberrant self-renewing capacity may develop into highly aggressive cancers. Although the tumors have the ability to activate other telomere lengthening pathways [such as alternative lengthening of telomeres (ALT) [30], therapies against telomerase are potentially anti-tumorigenic. Also, we cannot exclude that Wnt activation through telomerase [31, 32] (in a telomere-independent scenario) could play a role in self-renewal of stem cells under tumorigenic and non-tumorigenic conditions.

5.1.1 Aging of Stem Cells

Aging is a complex cellular and organismal process, driven by acquired and genetic factors [33, 34]. Aging is among the major known risk factors for most human diseases, including cancer [35–38]. Antiaging therapies, from which tissue regeneration is an emerging branch, reside on the concept of using our own material to regenerate age-related damaged tissues. Here, stem cells have a remarkable potential for counteracting several diseases. Stem cells have become a major scientific focus after the generation and culture of embryonic stem cells from human blastocysts (covered in ▶ Chap. 2), the discovery and identification of adult stem cells or, more recently, the advance in cellular reprogramming and the development of induced pluripotent stem cells (covered in ▶ Chap. 4).

Adult stem cells are found throughout the adult tissues and may function as self-renewing pools to replace dying cells throughout life [39]. Similar to differentiated cells, the pools of stem cells also age changing their properties with the biological clock [40, 41]. Their regenerative capacity declines with age and this directly impacts the healing of adult tissues. For instance, restoration of a fractured bone takes much longer in elderly than in young individuals [42–45].

Several pathways are believed to contribute directly or indirectly to the aging-associated stem cell dysfunction, including alterations at the microenvironment (such as changes at the hormonal, immunologic, and metabolic levels), mitochondrial dysfunction (respiratory chain dysfunction and accumulation of reactive oxygen species), epigenetic alterations (age-related changes on the epigenetics programming that compromise the activation of stem cells pluripotency network), and telomere shortening (described hereafter) [41, 46].

5.1.1.1 Telomerase and Telomere Dynamics in Stem Cells

As previously briefly stated, a telomere is a region of repetitive nucleotide sequences at the ends of the chromosomes. It protects chromosome ends of being recognized as double-strand breaks leading to nucleolytic degradation, unnecessary recombination, repair, or fusion [47]. There has been significant interest in the biology of telomeres during aging [34, 40, 48–50]. Although stem cells express telomerase they experience telomere shortening with age [51–53]. The use of gain of function and knockout models has contributed to unveil the role of telomerase in stem cell homeostasis [54], where, for instance, enforced telomerase reverse transcriptase (TERT) expression has been shown to increase the clonogenic capacity of hair follicle stem cells [53, 55]. Recently, it has been demonstrated that telomerase reactivation in an accelerated model of aging, involving an enhanced shortening of telomeres (and subsequently an increased percentage of short telomeres) through serial crosses of telomerase knockout mice, retains the capacity to mobilize stem cells [56]. Neural stem cells (NSCs) of aged mice seem to be reactive to a telomerase pulse, which impacted their proliferative capacity. This demonstrated that aged organisms can activate pools of stem cells that are dormant in normal conditions, relying on external factors to start dividing and repopulate tissues [56].

Another major advance in the understanding of the biology of stem cells and pluripotency was the contribution of the unexpected/pivotal studies of Yamanaka and colleagues [57]. The characterization/discovery that fully committed cells could return to an undifferentiated state (named reprogramming to induced pluripotent stem cells, iPSC, covered late during this chapter) through expression of four transcription factors, c-Myc, Oct4, Sox2, and Klf4 (several variations have been assigned to the original protocol, where c-Myc, for instance, was further confirmed as dispensable) [58], has opened new doors on cell differentiation, pluripotency, and, ultimately, regenerative medicine. Due to the fact that telomeres shorten with age, it was interesting to observe how telomeres react to a rollback of a committed condition (usually where short telomeres are present) to a pluripotent state (resembling a stem-like state with long telomeres). It was indeed confirmed that telomeres re-elongate during nuclear reprogramming, with respect to the telomere size presented in the parental cells of origin [59, 60]. The telomere-reprogramming efficiency was further shown to be related to the "age" of the parental cells; i.e., cells from a young donor give rise to a better telomere reprogramming compared with cells from an old donor. Also, it has been observed that this happens only in the presence of a functional telomerase complex, further demonstrating that telomere elongation is mediated through telomerase. The telomeres of iPSC not only present a reprogrammed telomere length, but are also characterized by chromatin markers found in embryonic stem cells, such as low trimethylation DNA density or the presence of telomeric repeat-containing RNA (TERRA) transcripts. Reprogrammed cells have shown to be good models for human diseases, allowing the avoidance of using human embryonic stem cells [61, 62]. At the same time, iPS cells can potentially be used therapeutically [63, 64], although safety issues related to viral induction and mutations rate have to be bypassed [65].

5.1.1.2 p53 Associated and Epigenetically Regulated Senescence

Whether Arf/p53 and Ink4/Rb play a role in stem cell deregulation has been recently addressed. p53 is a major cellular player involved in cell cycle arrest and oncogene-induced senescence. Mutations in p53 aid to bypass the senescent signaling and lead to an uncontrolled growth stimulus, one of the hallmarks of cancer [66]. The majority of malignant cancers have mutations in the p53 pathway. p53-mediated senescence is activated through DNA-damage-dependent signals. p21 is one of the most important targets of p53. Similar to a reduction of p53 levels, ablation of p21 prevents senescence and favors proliferation of cells [67–69]. Another player of the stress-induced senescence is p16INK4a (p16), which acts as a telomere length–sensor through CyclinD/Cdk4,6 inhibition, leading to the inhibition of Rb [66, 70, 71].

Lack of p53 in a mouse model leads to premature aging syndromes including decreased replacement of tissues through deregulated stem cell homeostasis [72]. These results show that deregulated p53 reduces life span by a decline in tissue stem cell regenerative function. On the other hand, an extra copy of p53 affects the preservation of the pools of stem cells of the brain and skin, resulting in better neuromuscular output and better skin regeneration properties [55, 73, 74]. Upregulation of p53 may result in slower proliferation contributing for long-term maintenance of stem cells.

With regard to p16, its role in aging is still debated. If, on the one hand, it negatively affects the preservation of stem cell in the brain, hematopoietic system, or pancreas [75–77], on the other hand, aged mice overexpressing p16 do not present any significant deleterious impact in neural stem cells (NSCs) [76, 77]. *Interestingly*, the constitutive expression of *TERT*, concomitant with the extra copies of *p16* and *p53* in mice, extends longevity, correlating with a delayed exhaustion of stem cells [55]. Similar results were obtained by expressing only TERT in aged mice [78, 80]. Nevertheless, senescent cells (usually characterized by higher expression of p16) are strictly related to the aging process. It was recently shown that clearance of senescent cells in aged mice leads to better tissue fitness, establishing a direct link between the accumulation of senescent cells and aging [79, 80].

5.2 Cellular Reprogramming

Several alternative in vitro methodologies have been introduced for the creation and expansion of embryonic-like stem cells, surpassing the need of primary culture of natural embryonic stem cells from human and mouse blastocysts. In particular, Yamanaka and colleagues found that expression of four different transcription factors in human and mouse differentiated cells was sufficient to revert the biological clock to a "stem-like" condition [57, 58]. The iPSC technology has many implications in the aging field, since many age-related diseases would benefit from regenerative therapies. Reprogrammed cells from old tissues may be a starting point to the development of organoids or functional tissues to replace the damaged ones.

Because of the regenerative potential of cellular reprogramming, it has been applied to the aging field [81, 82]. Interestingly, the induced pluripotent cells obtained during cellular reprogramming reset the stress- and senescence-associated epigenetic marks from the parental tissue [83–85]. Erasure of the aging marks is a crucial step during cellular and tissue regeneration strategies. The iPSC technology has been tested in the treatment of macular degeneration. The iPSCs were generated from skin fibroblasts obtained from two patients with advanced macular degeneration and were differentiated into retinal pigment epithelial (RPE) cells which were subsequently used to replace the damaged region [86, 87]. For clinical use, the reprogramming protocol needs to be optimized, and the role of iPSC in tissue regeneration is being widely studied [88, 89].

5.2.1 Aging as a Barrier for Cellular Reprogramming

Interestingly, one restriction for cellular reprogramming is the biological age of the starting material. Although it is now clear that cells from old patients can be reprogrammed, the frequency and efficiency of the reprograming process is greatly impaired by the age of the starting material. More studies will be needed to better understand how age impacts iPSC generation and quality. Compelling evidence shows, however, that biological aging is a barrier for cellular reprogramming in mice. Additionally, different tissues show different sensitivities to the age-dependent decline in the reprogramming process [90–93]. Several biochemical and physiological markers should be tested to survey the quality of the iPSC resulting from aged material. For instance, it will be important to characterize the epigenetic state of old cells before and after reprogramming in order to identify the age-dependent epigenetic changes that are reversible and those that endure as a "memory." Molecules that may rejuvenate the properties of aged cells to facilitate the reprogramming process, or that directly convert the characteristics of aged iPSC in iPSC with youth characteristics may open new avenues for the fields of aging and regenerative medicine.

Among the pathways involved in the age-related barriers for cellular reprogramming, cellular senescence may be one of the key mechanisms, at least in mice [59, 90, 94–99]. Senescent cells are characterized by an irreversible cell cycle arrest, higher expression of p16 and other cellular characteristics including changes in the chromatin condensation and secretory phenotype [100–102]. Cellular reprogramming is dependent on the capacity of cells to divide [103–105] being this loss probably the major barrier observed during senescence. Other barriers detected during aging that may be affecting cellular reprogramming are the changes affecting the mTOR (target of rapamycin) pathway. Genetically or pharmacological inhibition of the mTOR pathway promotes longevity from yeast to mammals [106] and improves the reprogramming efficiency of mouse embryonic fibroblasts (MEFs) several folds. TOR inhibitors may act by facilitating a mesenchymal-to-epithelial transition (MET) [107, 108], as cells of mesenchymal origin such as adult fibroblasts are thought to undergo MET during the reprogramming process into pluripotency [109, 110]. Indeed, expression of Zeb2 [111, 112], another EMT factor, is shown to increase with age and to be a barrier for cellular reprogramming [113]. Downregulation of Zeb2 in aged/old adult fibroblasts greatly impacts their efficiency to be reprogrammed [113, 114].

5.2.1.1 Primary Cultures of Mouse Embryonic Fibroblasts

Mouse embryonic fibroblasts (MEFs) are a primary source of cells commonly used in the laboratory for studying cell division, senescence, and immortalization [115, 116]. MEFs can be efficiently expanded in vitro until reaching a stress-associated senescence phenotype due to specific in vitro conditions, especially oxidative stress [117]. We describe hereafter a protocol for the isolation of MEFs from embryonic day 13.5–14.5 mouse embryos. The MEFs obtained are appropriate for use in biochemical and biological assays including genetic manipulations.

Step-by-Step Protocol

Materials and Reagents
1. *Pregnant mice female at 13–14 days postconception*
2. *Phosphate-buffered saline (PBS), without Ca2+ and Mg2+*
3. *DMEM containing 4.5 g/L D-glucose*
4. *Fetal bovine serum (FBS)*
5. *1× penicillin-streptomycin solution*
6. *L-glutamine*
7. *0.25% trypsin*
8. *70% ethanol*
9. *DMSO*

Equipment
- *Plastic dissecting board or sterile 100 mm petri dishes*
- *Sterile scissors*
- *Sterile forceps*
- *Sterile gauze*
- *Sterile scalpel blades*
- *Gloves*
- *Petri dishes for cell culture (100 mm and 60 mm)*
- *Pipettes, 5 and 10 ml*
- *Falcon tubes (50 ml)*
- *Cryovials*
- *37 °C 5% CO_2 tissue culture incubator*
- *Tissue culture hood*
- *Inverted microscope*

Procedure
A. Remove uterus at day 13.5 postconception.
   Note: Depending on the mouse strain, we can expect 6–10 embryos from each pregnant female. Euthanization procedure and removal of the uterus should be done according to the current and local laws, rules, regulations, and ethical guidelines. Briefly the procedure should be performed as stated in [118].
B. Transfer the uterus to a 50-ml falcon tube with PBS+1× penicillin-streptomycin pre-warmed at 37 °C
C. In the tissue culture hood transfer the uterus to a 100 mm petri dish with PBS+1× penicillin-streptomycin.
D. Separate the embryos by cutting the uterus in the regions between each embryo. The embryos usually pop out spontaneously or they may come out after pressing gently with forceps. If you find resistance to remove the embryos, carefully cut away the uterine tissue.
E. Transfer each embryo to a 60 mm petri dish with PBS+1× penicillin-streptomycin. Use the PBS to remove the blood from the embryo. Remove the head cutting

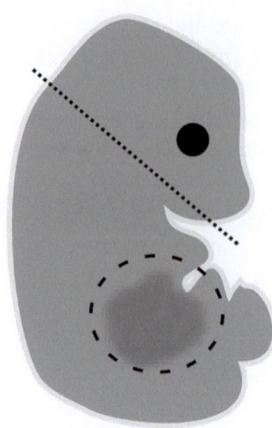

Fig. 5.1 Scheme of how to remove the head and red tissue (heart and liver) with a pair of fine forceps

Fig. 5.2 Scheme of a tilted dish, needed for an efficient trypsinization

below the eye with the scalpel blades and remove the red tissue (heart and liver) with a pair of fine forceps, as schematized in the *right figure* (Fig. 5.1).

Note: If genotyping is required, keep head in an Eppendorf-like tube and store at − 20 °C until use.

F. Transfer each embryo to a clean 60 mm petri dish with 1 ml of trypsin 0.25%. Chop up the embryo with a scalpel blade around 20 times; visible bits should remain (1–2 mm). Pipet up and down several times with a 10 ml pipet, and then place the dish tilted (as demonstrated in the *right picture* – Fig. 5.2) in the 37 °C tissue culture incubator for 15 min.

G. Remove the dish from the incubator and pipette the embryo pieces up and down several times with a 5 ml pipette to disperse the cells obtaining a cellular suspension. Return the dish to the incubator for another 15 min.

H. Add 4 ml complete medium [450 ml of DMEM + 10%FBS (50 ml) + 1× penicillin-streptomycin (5 ml 100× penicillin-streptomycin solution) + 1× L-Glutamine (5 ml 200 mM L-glutamine)] to each embryo in the 60 mm petri dish and pipet up and down several times. Let the cell suspension sit for about 5 min to allow larger embryo fragments to sink to the bottom of the tube. Transfer the supernatant to a 100 mm petri dish with 10 ml of MEF media.

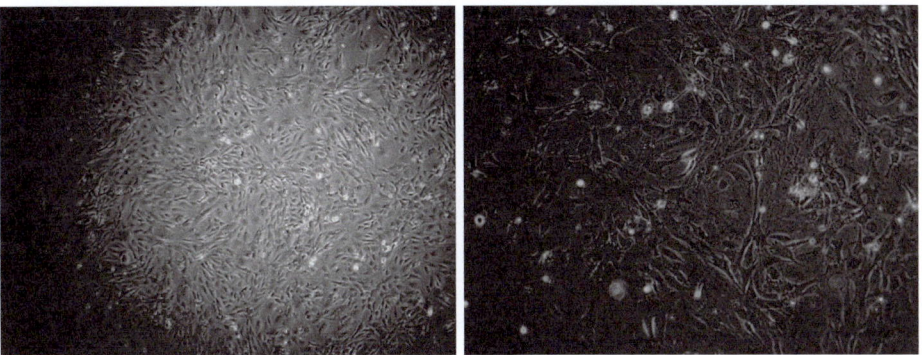

Fig. 5.3 Representative images of MEFs 3 days after the first passage at 4× objective (left) and 10× objective (right)

I. Change MEF media the next day and check the cells later in the day. If the culture is too confluent, split into 3 × 100 mm petri dishes. This is passage number 1.
J. When cells are confluent, harvest by trypsinization, count and resuspend the cell pellet in freezing medium (MEF media + 10%FBS + 10% DMSO). Aliquot in 3–4 × 10^6 cells per cryovial and freeze the cells using standard methods for mammalian cell cryopreservation.

Expected results In the images below (Fig. 5.3), we can observe representative images of MEFs 3 days after the first passage at 4× objective (left) and 10× objective (right). We can observe the homogeneity of the cultures.

5.2.1.2 Primary Cultures of Mouse Skin Fibroblasts

Primary cells are cells-derived directly from a particular tissue and cultured in vitro and closely resemble the physiological state and background of the tissue from which they originated. For this reason, primary cell cultures are a useful model for studying complex biological pathways, such as aging. Due to their nature, primary cells rapidly undergo senescence in culture, limiting their usage. Several types of primary cells are feasible for culture in vitro. Those include adult fibroblasts, neurons, epithelial cells, and bone marrow-derived dendritic cells, among others. Fibroblasts are often the cells of choice for reprogramming experiments, due to the facility to establish in vitro cultures, and to the minimal intervention required to obtain the primary biological source [119]. Additionally, fibroblasts do not require further purification steps prior to cell culture. In particular, for reprogramming experiments, fibroblasts can be efficiently transfected using biological, chemical, and physical protocols [120, 121].

We describe hereafter a protocol to establish adult fibroblast cultures from ears of mice of different ages [119].

Step-by-Step Protocol
Materials and Reagents
1. Mice
2. Phosphate-buffered saline (PBS), without Ca^{2+} and Mg^{2+}
3. DMEM containing 4.5 g/L D-glucose
4. Fetal bovine serum (FBS)
5. 1× penicillin-streptomycin solution
6. L-glutamine
7. 0.25% trypsin
8. 0.1% trypsin-EDTA
9. 70% ethanol
10. Alternative enzyme solutions follow the protocol described in [119]. Briefly prepare collagenase D solution in a 15 ml conical bottom tube by weighing 10 mg of collagenase D in 4 ml of complete medium. Prepare pronase solution weighing 10 mg of pronase in 5 µl of 1 M Tris buffer (pH 8.0). Add 1 µl of 0.5 M EDTA (pH 8.0), 494 µl of sterile water, and incubate the pronase solution at 37 °C in a water bath for 30 min. The collagenase D-pronase mix includes 250 µl of pronase solution with 4 ml of collagenase D solution (filter through a 0.2 µm pore into a sterile tube before use in the cell culture).

Equipment
- Plastic dissecting board or sterile 100 mm petri dishes
- Sterile scissors
- Sterile scalpel blades
- Gloves
- Petri dishes for cell culture (100 mm and 60 mm)
- Pipettes, 5 and 10 ml
- Falcon tubes (50 ml)
- 37 °C 5% CO_2 tissue culture incubator
- Tissue culture hood
- Inverted microscope

Procedure
Note: Mice should be housed in pathogen-free conditions in compliance with the current and local laws, rules, regulations, and ethical guidelines until euthanization. Euthanize mice according to the appropriate institutional guidelines. Tissue may be derived from several mouse strains, where C57BL/6 is probably the most commonly used. Additionally, tissue from living mice can be used resulting, for instance, from the leftovers of ear tagging. Noteworthy, the small pieces resulting from ear tagging give rise to limited amount of material and experiments should be planed accordingly taking this into consideration. After collection, ears can be stored in medium at room temperature or 4 °C for several days (up to 10 days). A simplified scheme of the workflow is depicted in the *upper picture* (◘ Fig. 5.4).

Fig. 5.4 Scheme of the workflow needed for the isolation of primary cultures of adult fibroblasts

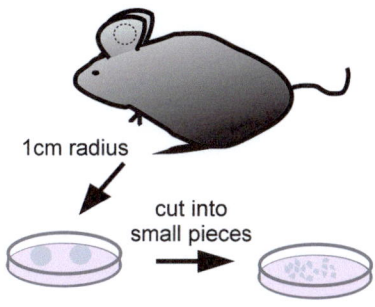

A. Start by placing the autoclaved surgical instruments (scissors and forceps) in the cell culture hood. Fill two 100 mm cell culture petri dishes with 10 ml of complete media.
B. Cut ears (pieces of 1 cm radius) of a mouse with scissors and incubate for 5 min in 40 ml 70% ethanol in a sterile 50 ml Falcon-like tube.
C. Air-dry the ear pieces for 5 min under the culture hood. Once dried, transfer each ear piece to a 100 mm cell culture petri dish containing 1 ml of trypsin 0.25% [alternatively, and for higher yields of tissue disaggregation, a solution of collagenase D-pronase may be used, see [119]].
D. Cut ears into small pieces using scissors or scalpel blades.
E. Transfer the cut tissues to a cell culture incubator at 37 °C for 20–30 minutes.
F. Remove the petri dish from the incubator, using the scalpel blades cut again the material into smaller pieces and return the dish to the incubator for another 20–30 min.
G. Add 10 ml complete medium [450 ml of DMEM + 10%FBS (50 ml) + 1 × penicillin-streptomycin (5 ml 100× penicillin-streptomycin solution) + 1× L-glutamine (5 ml 200 mM L-glutamine)] into the petri dish, and pipette all the media (including tissue) to a 50 ml Falcon-like tube.
H. Wash the dish with 10 ml complete medium and add the medium to the appropriate 50 ml Falcon-like tube.
I. Spin down the cell suspension for 5 min at ~500 × g at 4 °C using a refrigerated cell centrifuge adapted for conical tubes.
J. Remove supernatant, add 15 ml complete medium to the cell pellet and resuspend the cells in a 100 mm cell culture petri dish. Culture overnight in a 37 °C 5% CO_2 tissue culture incubator.
K. On the next day remove supernatant and resuspend cells in 15 ml complete medium.
L. Change media each third day until you start to see tissue attaching to the plate and fibroblasts growing.
M. When culture reaches around 70% confluence, remove the medium and wash the cells with 5 ml sterile 1× phosphate-buffered saline (PBS).
N. Remove PBS and add 3 ml sterile 1× trypsin-EDTA solution to the cells. Incubate the cells for 5–10 min at 37 °C in a humidified 5% CO_2 incubator. After, add 6 ml complete medium to the cells and pipette up and down to release the fibroblasts. Transfer the cell suspension to a conical bottom tube and spin the tube for 5 min at ~500 × g and 4 °C using a refrigerated cell centrifuge. Resuspend the cell pellet in 15 ml complete medium and seed in a new 100 mm petri dish.

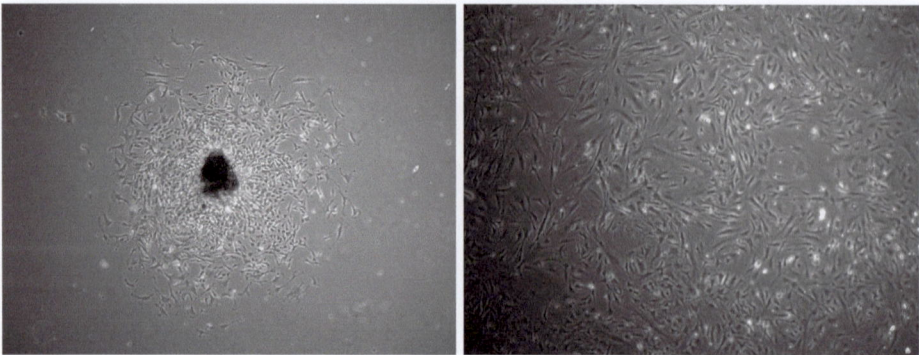

Fig. 5.5 Primary culture of adult/old skin fibroblasts. Usually, we can start to see tissue attaching after 3–10 days post-plating (left image, 4× magnification). When numerous fibroblasts are being released from the tissue, or the plate is confluent you can add trypsin and pass to a new 100 mm cell culture petri dish. After passage 1, the primary culture of adult fibroblasts should be homogeneous (right image, 10× magnification)

O. In case keratinocytes are detected in the culture, proceed with a short trypsinization with 0.1% trypsin/EDTA, as previously described [122, 123], in order to establish second and third passage pure fibroblasts cultures.

Expected results In the images below (Fig. 5.5), we can observe different steps of the primary culture of adult/old skin fibroblasts. Usually, you can start to see tissue attaching after 3–10 days post-plating the tissue (left image, 4× magnification). When numerous fibroblasts are being released from the tissue, or the plate is confluent you can add trypsin and pass to a new 100 mm cell culture petri dish. After passage 1 the primary culture of adult fibroblasts should be homogeneous (right image, 10× magnification). Of note, primary cultures of adult/old fibroblasts can endure a very limited number of passages in vitro reaching senescence very quickly.

5.2.2 Reprogramming In Vivo as an Antiaging Strategy

One of the major breakthroughs of cellular reprogramming was the understanding that differentiated and mature cells have the plasticity to return to a pluripotent condition [124, 125]. Nevertheless, and despite the great advances on the understanding of cellular plasticity, the seminal works on cellular reprogramming were achieved under in vitro cell culture conditions [124, 125]. Although in principle adult tissues and the surrounding microenvironment pose a barrier to cellular reprogramming, examples in mice demonstrate the capacity of several tissues to trans-differentiate from one cell type into another [126, 127]. Following this observation various groups used reprogrammable mice in which the "cocktail" of reprogramming factors (OKSM) was expressed in vivo under an inducible promoter. This allows the time and tissue-specific activation of the reprogramming factors [128]. The studies have shown that forced expression of the OKSM factors is possible in vivo leading to a

In vivo direct reprogramming

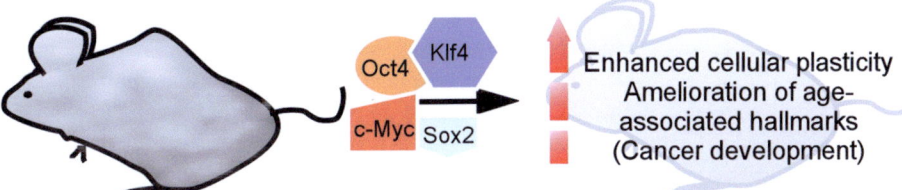

Fig. 5.6 In vivo short-term activation of the reprogramming factors is enough to ameliorate the age-related phenotypes of progeria mice and naturally aged mice, impacting on mice health span, enhancing, for instance, the glucose tolerance and the muscle regenerative capacity

myriad of effects spanning from DNA damage to inflammation, senescence, cellular reprogramming, and formation of teratomas [88, 129]. Intriguingly, the major outcome during in vivo activation of the reprogramming factors was the appearance of senescence and pluripotency, two complex and somehow opposing pathways [130–133]. Interestingly, senescent cells were aiding the reprogramming efficiency of adjacent cells. Mechanistically, senescent cells are characterized by the release of several factors to the neighboring tissues (named senescence-associated secretory phenotype—SASP) [100]. In particular, one of the components of the SASP named IL6 (usually produced during senescence and inflammatory responses) was shown to facilitate in vivo reprogramming probably by promoting damaged-induced tissue regeneration [130]. These results clearly demonstrate the limitations of in vitro cell culture systems in the assessment of complex pathways such as cell plasticity. Still, based on the in vitro results demonstrating how cellular reprogramming may be a useful tool for the generation of different tissues; researchers investigate whether OSKM activation in vivo may affect the age-related phenotypes [89]. Indeed, in vivo short-term activation (2 days) of the reprogramming factors was enough to ameliorate the age-related phenotypes of progeria mice. Additionally, induction of OSKM in naturally aged mice also impacts health span, enhancing, for instance, the glucose tolerance and the muscle regenerative capacity (summarized in the picture above – Fig. 5.6). While these results are promising, it should be emphasized that extended activation (more than 7 days) of the OSKM factors in vivo leads primarily to the formation of teratomas containing iPSC [128, 134]. Indeed, the reprogramming process toward pluripotency greatly resembles the tumorigenic process at the level of activated pathways, metabolic reprogramming and cellular characteristics. Activation of the OSKM factors during 7 days, at least, caused cancers in multiple organs [128], resembling the first steps of cellular reprogramming, being characterized by a loss of cell identity and acquisition of pluripotency-associated genes. The close gap between aging-ameliorative effects and cancer phenotypes represents a central barrier for the use of reprogramming in regenerative medicine. Indeed, some reports found the existence of late tumors in therapies aiming at in vivo regeneration through the use of stem cells [135–137]. Overall, further studies are needed to better distinguish the dual role of in vivo reprogramming strategies during organismal rejuvenation or cancer progression.

> **Take-Home Message**
>
> Stem cells and iPSC are the major players in regenerative medicine. Here, we explored the current knowledge on the potential roles of stem cells in tissue replacement and, in particular, on how aging impacts on stem cells dynamics and in the capacity to reprogram adult cells to pluripotency. Antiaging strategies, in particular approaches exploring age-related pathways, greatly impact on the plasticity of mature cells and are envisioned as the future of tissue regenerative medicine. These results are exciting; however, additional efforts will be necessary to translate these findings into actual patient-specific therapies.

References

1. Weissman IL. Stem cells: units of development, units of regeneration, and units in evolution. Cell. 2000;100(1):157–68.
2. Korbling M, Estrov Z. Adult stem cells for tissue repair - a new therapeutic concept? N Engl J Med. 2003;349(6):570–82.
3. Slack JM. Stem cells in epithelial tissues. Science. 2000;287(5457):1431–3.
4. McKay R. Stem cells in the central nervous system. Science. 1997;276(5309):66–71.
5. Charge SB, Rudnicki MA. Cellular and molecular regulation of muscle regeneration. Physiol Rev. 2004;84(1):209–38.
6. da Silva ML, Chagastelles PC, Nardi NB. Mesenchymal stem cells reside in virtually all post-natal organs and tissues. J Cell Sci. 2006;119(Pt 11):2204–13.
7. Doetsch F, Caille I, Lim DA, Garcia-Verdugo JM, Alvarez-Buylla A. Subventricular zone astrocytes are neural stem cells in the adult mammalian brain. Cell. 1999;97(6):703–16.
8. Smart N, Bollini S, Dube KN, Vieira JM, Zhou B, Davidson S, et al. De novo cardiomyocytes from within the activated adult heart after injury. Nature. 2011;474:640.
9. Ruzankina Y, Pinzon-Guzman C, Asare A, Ong T, Pontano L, Cotsarelis G, et al. Deletion of the developmentally essential gene ATR in adult mice leads to age-related phenotypes and stem cell loss. Cell Stem Cell. 2007;1(1):113–26.
10. Flores I, Cayuela ML, Blasco MA. Effects of telomerase and telomere length on epidermal stem cell behavior. Science. 2005;309(5738):1253–6.
11. Rossi DJ, Bryder D, Seita J, Nussenzweig A, Hoeijmakers J, Weissman IL. Deficiencies in DNA damage repair limit the function of haematopoietic stem cells with age. Nature. 2007;447(7145):725–9.
12. Schlessinger D, Van Zant G. Does functional depletion of stem cells drive aging? Mech Ageing Dev. 2001;122(14):1537–53.
13. Jordan CT, Guzman ML, Noble M. Cancer stem cells. N Engl J Med. 2006;355(12):1253–61.
14. Reya T, Morrison SJ, Clarke MF, Weissman IL. Stem cells, cancer, and cancer stem cells. Nature. 2001;414(6859):105–11.
15. Singh SK, Clarke ID, Terasaki M, Bonn VE, Hawkins C, Squire J, et al. Identification of a cancer stem cell in human brain tumors. Cancer Res. 2003;63(18):5821–8.
16. Wang J, Sakariassen PO, Tsinkalovsky O, Immervoll H, Boe SO, Svendsen A, et al. CD133 negative glioma cells form tumors in nude rats and give rise to CD133 positive cells. Int J Cancer. 2008;122(4):761–8.
17. Mushinski JF, Koziol JA, Marini M. Cluster analysis of aminoacyl-tRNAs from mouse plasmacytomas correlates chromatographic profiles with myeloma protein similarity, clonal origin of tumour lines, and the neoplastic nature of the tissues. J Theor Biol. 1980;85(3):507–21.
18. Huntly BJ, Gilliland DG. Leukaemia stem cells and the evolution of cancer-stem-cell research. Nat Rev Cancer. 2005;5(4):311–21.
19. Kamel-Reid S, Letarte M, Sirard C, Doedens M, Grunberger T, Fulop G, et al. A model of human acute lymphoblastic leukemia in immune-deficient SCID mice. Science. 1989;246(4937):1597–600.

20. Lapidot T, Sirard C, Vormoor J, Murdoch B, Hoang T, Caceres-Cortes J, et al. A cell initiating human acute myeloid leukaemia after transplantation into SCID mice. Nature. 1994;367(6464):645–8.
21. Sirard C, Lapidot T, Vormoor J, Cashman JD, Doedens M, Murdoch B, et al. Normal and leukemic SCID-repopulating cells (SRC) coexist in the bone marrow and peripheral blood from CML patients in chronic phase, whereas leukemic SRC are detected in blast crisis. Blood. 1996;87(4):1539–48.
22. Bonnet D, Dick JE. Human acute myeloid leukemia is organized as a hierarchy that originates from a primitive hematopoietic cell. Nat Med. 1997;3(7):730–7.
23. Quintana E, Shackleton M, Sabel MS, Fullen DR, Johnson TM, Morrison SJ. Efficient tumour formation by single human melanoma cells. Nature. 2008;456(7222):593–8.
24. Engelhardt M, Drullinsky P, Guillem J, Moore MA. Telomerase and telomere length in the development and progression of premalignant lesions to colorectal cancer. Clin Cancer Res. 1997;3(11):1931–41.
25. Engelhardt M, Albanell J, Drullinsky P, Han W, Guillem J, Scher HI, et al. Relative contribution of normal and neoplastic cells determines telomerase activity and telomere length in primary cancers of the prostate, colon, and sarcoma. Clin Cancer Res. 1997;3(10):1849–57.
26. Odagiri E, Kanada N, Jibiki K, Demura R, Aikawa E, Demura H. Reduction of telomeric length and c-erbB-2 gene amplification in human breast cancer, fibroadenoma, and gynecomastia. Relationship to histologic grade and clinical parameters. Cancer. 1994;73(12):2978–84.
27. Meeker AK, Argani P. Telomere shortening occurs early during breast tumorigenesis: a cause of chromosome destabilization underlying malignant transformation? J Mammary Gland Biol Neoplasia. 2004;9(3):285–96.
28. Meeker AK, Hicks JL, Iacobuzio-Donahue CA, Montgomery EA, Westra WH, Chan TY, et al. Telomere length abnormalities occur early in the initiation of epithelial carcinogenesis. Clin Cancer Res. 2004;10(10):3317–26.
29. van Heek NT, Meeker AK, Kern SE, Yeo CJ, Lillemoe KD, Cameron JL, et al. Telomere shortening is nearly universal in pancreatic intraepithelial neoplasia. Am J Pathol. 2002;161(5):1541–7.
30. Bryan TM, Englezou A, Dalla-Pozza L, Dunham MA, Reddel RR. Evidence for an alternative mechanism for maintaining telomere length in human tumors and tumor-derived cell lines. Nat Med. 1997;3(11):1271–4.
31. Choi J, Southworth LK, Sarin KY, Venteicher AS, Ma W, Chang W, et al. TERT promotes epithelial proliferation through transcriptional control of a Myc- and Wnt-related developmental program. PLoS Genet. 2008;4(1):e10.
32. Park JI, Venteicher AS, Hong JY, Choi J, Jun S, Shkreli M, et al. Telomerase modulates Wnt signalling by association with target gene chromatin. Nature. 2009;460(7251):66–72.
33. Ahmed AS, Sheng MH, Wasnik S, Baylink DJ, Lau KW. Effect of aging on stem cells. World J Exp Med. 2017;7(1):1–10.
34. Sharpless NE, DePinho RA. How stem cells age and why this makes us grow old. Nat Rev Mol Cell Biol. 2007;8(9):703–13.
35. Dillin A, Gottschling DE, Nystrom T. The good and the bad of being connected: the integrons of aging. Curr Opin Cell Biol. 2014;26:107–12.
36. Shane Anderson A, Loeser RF. Why is osteoarthritis an age-related disease? Best Pract Res Clin Rheumatol. 2010;24(1):15–26.
37. DeLong MR, Huang KT, Gallis J, Lokhnygina Y, Parente B, Hickey P, et al. Effect of advancing age on outcomes of deep brain stimulation for Parkinson disease. JAMA Neurol. 2014;71(10):1290–5.
38. Niccoli T, Partridge L. Ageing as a risk factor for disease. Curr Biol. 2012;22(17):R741–52.
39. Boyette LB, Tuan RS. Adult stem cells and diseases of aging. J Clin Med. 2014;3(1):88–134.
40. Schultz MB, Sinclair DA. When stem cells grow old: phenotypes and mechanisms of stem cell aging. Development. 2016;143(1):3–14.
41. Oh J, Lee YD, Wagers AJ. Stem cell aging: mechanisms, regulators and therapeutic opportunities. Nat Med. 2014;20(8):870–80.
42. Ho AD, Wagner W, Mahlknecht U. Stem cells and ageing. The potential of stem cells to overcome age-related deteriorations of the body in regenerative medicine. EMBO Rep. 2005;6 Spec No:S35-8.
43. Sousounis K, Baddour JA, Tsonis PA. Aging and regeneration in vertebrates. Curr Top Dev Biol. 2014;108:217–46.
44. Paxson JA, Gruntman A, Parkin CD, Mazan MR, Davis A, Ingenito EP, et al. Age-dependent decline in mouse lung regeneration with loss of lung fibroblast clonogenicity and increased myofibroblastic differentiation. PLoS One. 2011;6(8):e23232.

45. Keller K, Engelhardt M. Strength and muscle mass loss with aging process. Age and strength loss. Muscles Ligaments Tendons J. 2013;3(4):346–50.
46. Sharpless NE, Schatten G. Stem cell aging. J Gerontol A Biol Sci Med Sci. 2009;64(2):202–4.
47. Shammas MA. Telomeres, lifestyle, cancer, and aging. Curr Opin Clin Nutr Metab Care. 2011;14(1):28–34.
48. Bernardes de Jesus B, Vera E, Schneeberger K, Tejera AM, Ayuso E, Bosch F, et al. Telomerase gene therapy in adult and old mice delays aging and increases longevity without increasing cancer. EMBO Mol Med. 2012;4(8):691–704.
49. Bernardes de Jesus B, Blasco MA. Telomerase at the intersection of cancer and aging. Trends Genet. 2013;29(9):513–20.
50. Bernardes de Jesus B, Blasco MA. Aging by telomere loss can be reversed. Cell Stem Cell. 2011;8(1):3–4.
51. Bonab MM, Alimoghaddam K, Talebian F, Ghaffari SH, Ghavamzadeh A, Nikbin B. Aging of mesenchymal stem cell in vitro. BMC Cell Biol. 2006;7:14.
52. Ferron SR, Marques-Torrejon MA, Mira H, Flores I, Taylor K, Blasco MA, et al. Telomere shortening in neural stem cells disrupts neuronal differentiation and neuritogenesis. J Neurosci. 2009;29(46):14394–407.
53. Flores I, Canela A, Vera E, Tejera A, Cotsarelis G, Blasco MA. The longest telomeres: a general signature of adult stem cell compartments. Genes Dev. 2008;22(5):654–67.
54. Hiyama E, Hiyama K. Telomere and telomerase in stem cells. Br J Cancer. 2007;96(7):1020–4.
55. Tomas-Loba A, Flores I, Fernandez-Marcos PJ, Cayuela ML, Maraver A, Tejera A, et al. Telomerase reverse transcriptase delays aging in cancer-resistant mice. Cell. 2008;135(4):609–22.
56. Jaskelioff M, Muller FL, Paik JH, Thomas E, Jiang S, Adams AC, et al. Telomerase reactivation reverses tissue degeneration in aged telomerase-deficient mice. Nature. 2011;469(7328):102–6.
57. Takahashi K, Yamanaka S. Induction of pluripotent stem cells from mouse embryonic and adult fibroblast cultures by defined factors. Cell. 2006;126(4):663–76.
58. Yamanaka S. A fresh look at iPS cells. Cell. 2009;137(1):13–7.
59. Marion RM, Strati K, Li H, Murga M, Blanco R, Ortega S, et al. A p53-mediated DNA damage response limits reprogramming to ensure iPS cell genomic integrity. Nature. 2009;460(7259):1149–53.
60. Marion RM, Strati K, Li H, Tejera A, Schoeftner S, Ortega S, et al. Telomeres acquire embryonic stem cell characteristics in induced pluripotent stem cells. Cell Stem Cell. 2009;4(2):141–54.
61. Lee G, Papapetrou EP, Kim H, Chambers SM, Tomishima MJ, Fasano CA, et al. Modelling pathogenesis and treatment of familial dysautonomia using patient-specific iPSCs. Nature. 2009;461(7262):402–6.
62. Batista LF, Pech MF, Zhong FL, Nguyen HN, Xie KT, Zaug AJ, et al. Telomere shortening and loss of self-renewal in dyskeratosis congenita induced pluripotent stem cells. Nature. 2011;474(7351):399–402.
63. Hanna J, Wernig M, Markoulaki S, Sun CW, Meissner A, Cassady JP, et al. Treatment of sickle cell anemia mouse model with iPS cells generated from autologous skin. Science. 2007;318(5858):1920–3.
64. Xu D, Alipio Z, Fink LM, Adcock DM, Yang J, Ward DC, et al. Phenotypic correction of murine hemophilia A using an iPS cell-based therapy. Proc Natl Acad Sci U S A. 2009;106(3):808–13.
65. Sun N, Longaker MT, Wu JC. Human iPS cell-based therapy: considerations before clinical applications. Cell Cycle. 2010;9(5):880–5.
66. Collado M, Serrano M. The power and the promise of oncogene-induced senescence markers. Nat Rev Cancer. 2006;6(6):472–6.
67. Brown JP, Wei W, Sedivy JM. Bypass of senescence after disruption of p21CIP1/WAF1 gene in normal diploid human fibroblasts. Science. 1997;277(5327):831–4.
68. Beausejour CM, Krtolica A, Galimi F, Narita M, Lowe SW, Yaswen P, et al. Reversal of human cellular senescence: roles of the p53 and p16 pathways. EMBO J. 2003;22(16):4212–22.
69. Gire V, Roux P, Wynford-Thomas D, Brondello JM, Dulic V. DNA damage checkpoint kinase Chk2 triggers replicative senescence. EMBO J. 2004;23(13):2554–63.
70. Collado M, Blasco MA, Serrano M. Cellular senescence in cancer and aging. Cell. 2007;130(2):223–33.
71. Serrano M, Lee H, Chin L, Cordon-Cardo C, Beach D, DePinho RA. Role of the INK4a locus in tumor suppression and cell mortality. Cell. 1996;85(1):27–37.
72. Dumble M, Moore L, Chambers SM, Geiger H, Van Zant G, Goodell MA, et al. The impact of altered p53 dosage on hematopoietic stem cell dynamics during aging. Blood. 2007;109(4):1736–42.

73. Carrasco-Garcia E, Arrizabalaga O, Serrano M, Lovell-Badge R, Matheu A. Increased gene dosage of Ink4/Arf and p53 delays age-associated central nervous system functional decline. Aging Cell. 2015;14(4):710–4.
74. Matheu A, Maraver A, Klatt P, Flores I, Garcia-Cao I, Borras C, et al. Delayed ageing through damage protection by the Arf/p53 pathway. Nature. 2007;448(7151):375–9.
75. Janzen V, Forkert R, Fleming HE, Saito Y, Waring MT, Dombkowski DM, et al. Stem-cell ageing modified by the cyclin-dependent kinase inhibitor p16INK4a. Nature. 2006;443(7110):421–6.
76. Krishnamurthy J, Ramsey MR, Ligon KL, Torrice C, Koh A, Bonner-Weir S, et al. p16INK4a induces an age-dependent decline in islet regenerative potential. Nature. 2006;443(7110):453–7.
77. Molofsky AV, Slutsky SG, Joseph NM, He S, Pardal R, Krishnamurthy J, et al. Increasing p16INK4a expression decreases forebrain progenitors and neurogenesis during ageing. Nature. 2006;443(7110):448–52.
78. de Jesus BB, Blasco MA. Potential of telomerase activation in extending health span and longevity. Curr Opin Cell Biol. 2012;24(6):739–43.
79. Baker DJ, Childs BG, Durik M, Wijers ME, Sieben CJ, Zhong J, et al. Naturally occurring p16(Ink4a)-positive cells shorten healthy lifespan. Nature. 2016;530(7589):184–9.
80. Baker DJ, Wijshake T, Tchkonia T, LeBrasseur NK, Childs BG, van de Sluis B, et al. Clearance of p16Ink4a-positive senescent cells delays ageing-associated disorders. Nature. 2011;479(7372):232–6.
81. Soria-Valles C, Lopez-Otin C. iPSCs: on the road to reprogramming aging. Trends Mol Med. 2016;22(8):713–24.
82. Ocampo A, Reddy P, Belmonte JCI. Anti-aging strategies based on cellular reprogramming. Trends Mol Med. 2016;22(8):725–38.
83. Lapasset L, Milhavet O, Prieur A, Besnard E, Babled A, Ait-Hamou N, et al. Rejuvenating senescent and centenarian human cells by reprogramming through the pluripotent state. Genes Dev. 2011;25(21):2248–53.
84. Liu GH, Barkho BZ, Ruiz S, Diep D, Qu J, Yang SL, et al. Recapitulation of premature ageing with iPSCs from Hutchinson-Gilford progeria syndrome. Nature. 2011;472(7342):221–5.
85. Zhang J, Lian Q, Zhu G, Zhou F, Sui L, Tan C, et al. A human iPSC model of Hutchinson Gilford Progeria reveals vascular smooth muscle and mesenchymal stem cell defects. Cell Stem Cell. 2011;8(1):31–45.
86. Schwartz SD, Hubschman JP, Heilwell G, Franco-Cardenas V, Pan CK, Ostrick RM, et al. Embryonic stem cell trials for macular degeneration: a preliminary report. Lancet. 2012;379(9817):713–20.
87. Mandai M, Kurimoto Y, Takahashi M. Autologous induced stem-cell-derived retinal cells for macular degeneration. N Engl J Med. 2017;377(8):792–3.
88. Taguchi J, Yamada Y. In vivo reprogramming for tissue regeneration and organismal rejuvenation. Curr Opin Genet Dev. 2017;46:132–40.
89. Ocampo A, Reddy P, Martinez-Redondo P, Platero-Luengo A, Hatanaka F, Hishida T, et al. In vivo amelioration of age-associated hallmarks by partial reprogramming. Cell. 2016;167(7):1719–33 e12.
90. Li H, Collado M, Villasante A, Strati K, Ortega S, Canamero M, et al. The Ink4/Arf locus is a barrier for iPS cell reprogramming. Nature. 2009;460(7259):1136–9.
91. Kim K, Doi A, Wen B, Ng K, Zhao R, Cahan P, et al. Epigenetic memory in induced pluripotent stem cells. Nature. 2010;467(7313):285–90.
92. Wang B, Miyagoe-Suzuki Y, Yada E, Ito N, Nishiyama T, Nakamura M, et al. Reprogramming efficiency and quality of induced Pluripotent Stem Cells (iPSCs) generated from muscle-derived fibroblasts of mdx mice at different ages. PLoS Curr. 2011;3:RRN1274.
93. Cheng Z, Ito S, Nishio N, Xiao H, Zhang R, Suzuki H, et al. Establishment of induced pluripotent stem cells from aged mice using bone marrow-derived myeloid cells. J Mol Cell Biol. 2011;3(2):91–8.
94. Banito A, Rashid ST, Acosta JC, Li S, Pereira CF, Geti I, et al. Senescence impairs successful reprogramming to pluripotent stem cells. Genes Dev. 2009;23(18):2134–9.
95. Tat PA, Sumer H, Pralong D, Verma PJ. The efficiency of cell fusion-based reprogramming is affected by the somatic cell type and the in vitro age of somatic cells. Cell Reprogram. 2011;13(4):331–44.
96. Utikal J, Polo JM, Stadtfeld M, Maherali N, Kulalert W, Walsh RM, et al. Immortalization eliminates a roadblock during cellular reprogramming into iPS cells. Nature. 2009;460(7259):1145–8.
97. Hong H, Takahashi K, Ichisaka T, Aoi T, Kanagawa O, Nakagawa M, et al. Suppression of induced pluripotent stem cell generation by the p53-p21 pathway. Nature. 2009;460(7259):1132–5.

98. Kawamura T, Suzuki J, Wang YV, Menendez S, Morera LB, Raya A, et al. Linking the p53 tumour suppressor pathway to somatic cell reprogramming. Nature. 2009;460(7259):1140–4.
99. Zhao Y, Yin X, Qin H, Zhu F, Liu H, Yang W, et al. Two supporting factors greatly improve the efficiency of human iPSC generation. Cell Stem Cell. 2008;3(5):475–9.
100. de Jesus BB, Blasco MA. Assessing cell and organ senescence biomarkers. Circ Res. 2012;111(1):97–109.
101. Kuilman T, Michaloglou C, Mooi WJ, Peeper DS. The essence of senescence. Genes Dev. 2010;24(22):2463–79.
102. Campisi J, d'Adda di Fagagna F. Cellular senescence: when bad things happen to good cells. Nat Rev Mol Cell Biol. 2007;8(9):729–40.
103. Hanna J, Saha K, Pando B, van Zon J, Lengner CJ, Creyghton MP, et al. Direct cell reprogramming is a stochastic process amenable to acceleration. Nature. 2009;462(7273):595–601.
104. Hanna JH, Saha K, Jaenisch R. Pluripotency and cellular reprogramming: facts, hypotheses, unresolved issues. Cell. 2010;143(4):508–25.
105. Lynch CJ, Bernad R, Calvo I, Nobrega-Pereira S, Ruiz S, Ibarz N, et al. The RNA Polymerase II Factor RPAP1 Is Critical for Mediator-Driven Transcription and Cell Identity. Cell Rep. 2018;22(2):396–410.
106. Kapahi P, Zid B. TOR pathway: linking nutrient sensing to life span. Sci Aging Knowl Environ. 2004;2004(36):PE34.
107. Chen T, Shen L, Yu J, Wan H, Guo A, Chen J, et al. Rapamycin and other longevity-promoting compounds enhance the generation of mouse induced pluripotent stem cells. Aging Cell. 2011;10(5):908–11.
108. Santos F, Moreira C, Nobrega-Pereira S, Bernardes de Jesus B. New insights into the role of epithelial(-)mesenchymal transition during aging. Int J Mol Sci. 2019;20(4):891.
109. Li R, Liang J, Ni S, Zhou T, Qing X, Li H, et al. A mesenchymal-to-epithelial transition initiates and is required for the nuclear reprogramming of mouse fibroblasts. Cell Stem Cell. 2010;7(1):51–63.
110. Samavarchi-Tehrani P, Golipour A, David L, Sung HK, Beyer TA, Datti A, et al. Functional genomics reveals a BMP-driven mesenchymal-to-epithelial transition in the initiation of somatic cell reprogramming. Cell Stem Cell. 2010;7(1):64–77.
111. Beltran M, Puig I, Pena C, Garcia JM, Alvarez AB, Pena R, et al. A natural antisense transcript regulates Zeb2/Sip1 gene expression during Snail1-induced epithelial-mesenchymal transition. Genes Dev. 2008;22(6):756–69.
112. Wang G, Guo X, Hong W, Liu Q, Wei T, Lu C, et al. Critical regulation of miR-200/ZEB2 pathway in Oct4/Sox2-induced mesenchymal-to-epithelial transition and induced pluripotent stem cell generation. Proc Natl Acad Sci U S A. 2013;110(8):2858–63.
113. Bernardes de Jesus B, Marinho SP, Barros S, Sousa-Franco A, Alves-Vale C, Carvalho T, et al. Silencing of the lncRNA Zeb2-NAT facilitates reprogramming of aged fibroblasts and safeguards stem cell pluripotency. Nat Commun. 2018;9(1):94.
114. Sousa-Franco A, Rebelo K, da Rocha ST, Bernardes de Jesus B. LncRNAs regulating stemness in aging. Aging Cell. 2019;18(1):e12870.
115. Hahn WC, Weinberg RA. Modelling the molecular circuitry of cancer. Nat Rev Cancer. 2002;2(5):331–41.
116. Zuckerman V, Wolyniec K, Sionov RV, Haupt S, Haupt Y. Tumour suppression by p53: the importance of apoptosis and cellular senescence. J Pathol. 2009;219(1):3–15.
117. Parrinello S, Samper E, Krtolica A, Goldstein J, Melov S, Campisi J. Oxygen sensitivity severely limits the replicative lifespan of murine fibroblasts. Nat Cell Biol. 2003;5(8):741–7.
118. Durkin ME, Qian X, Popescu NC, Lowy DR. Isolation of mouse embryo fibroblasts. Bio Protoc. 2013;3(18):e908.
119. Khan M, Gasser S. Generating primary fibroblast cultures from mouse ear and tail tissues. J Vis Exp. 2016;(107):53565. https://doi.org/10.3791/53565.
120. Lim J, Dobson J. Improved transfection of HUVEC and MEF cells using DNA complexes with magnetic nanoparticles in an oscillating field. J Genet. 2012;91(2):223–7.
121. Li M, Jayandharan GR, Li B, Ling C, Ma W, Srivastava A, et al. High-efficiency transduction of fibroblasts and mesenchymal stem cells by tyrosine-mutant AAV2 vectors for their potential use in cellular therapy. Hum Gene Ther. 2010;21(11):1527–43.
122. Hentzer B, Kobayasi T. Separation of human epidermal cells from fibroblasts in primary skin culture. Arch Dermatol Forsch. 1975;252(1):39–46.

123. Siengdee P, Klinhom S, Thitaram C, Nganvongpanit K. Isolation and culture of primary adult skin fibroblasts from the Asian elephant (Elephas maximus). Peer J. 2018;6:e4302.
124. Takahashi K, Yamanaka S. Induced pluripotent stem cells in medicine and biology. Development. 2013;140(12):2457–61.
125. Yamanaka S. Induced pluripotent stem cells: past, present, and future. Cell Stem Cell. 2012;10(6):678–84.
126. Merrell AJ, Stanger BZ. Adult cell plasticity in vivo: de-differentiation and transdifferentiation are back in style. Nat Rev Mol Cell Biol. 2016;17(7):413–25.
127. Fu L, Zhu X, Yi F, Liu GH, Izpisua Belmonte JC. Regenerative medicine: transdifferentiation in vivo. Cell Res. 2014;24(2):141–2.
128. Abad M, Mosteiro L, Pantoja C, Canamero M, Rayon T, Ors I, et al. Reprogramming in vivo produces teratomas and iPS cells with totipotency features. Nature. 2013;502(7471):340–5.
129. Srivastava D, DeWitt N. In vivo cellular reprogramming: the next generation. Cell. 2016;166(6):1386–96.
130. Mosteiro L, Pantoja C, de Martino A, Serrano M. Senescence promotes in vivo reprogramming through p16(INK)(4a) and IL-6. Aging Cell. 2018;17(2):e12711.
131. Cazin C, Chiche A, Li H. Evaluation of injury-induced senescence and in vivo reprogramming in the skeletal muscle. J Vis Exp. 2017;(128):56201.
132. Chiche A, Le Roux I, von Joest M, Sakai H, Aguin SB, Cazin C, et al. Injury-induced senescence enables in vivo reprogramming in skeletal muscle. Cell Stem Cell. 2017;20(3):407–14 e4.
133. Mosteiro L, Pantoja C, Alcazar N, Marion RM, Chondronasiou D, Rovira M, et al. Tissue damage and senescence provide critical signals for cellular reprogramming in vivo. Science. 2016;354(6315).
134. Shibata H, Komura S, Yamada Y, Sankoda N, Tanaka A, Ukai T, et al. In vivo reprogramming drives Kras-induced cancer development. Nat Commun. 2018;9(1):2081.
135. Folkerth RD, Durso R. Survival and proliferation of nonneural tissues, with obstruction of cerebral ventricles, in a parkinsonian patient treated with fetal allografts. Neurology. 1996;46(5):1219–25.
136. Amariglio N, Hirshberg A, Scheithauer BW, Cohen Y, Loewenthal R, Trakhtenbrot L, et al. Donor-derived brain tumor following neural stem cell transplantation in an ataxia telangiectasia patient. PLoS Med. 2009;6(2):e1000029.
137. Dlouhy BJ, Awe O, Rao RC, Kirby PA, Hitchon PW. Autograft-derived spinal cord mass following olfactory mucosal cell transplantation in a spinal cord injury patient: case report. J Neurosurg Spine. 2014;21(4):618–22.

Cloning

Bernard A. J. Roelen

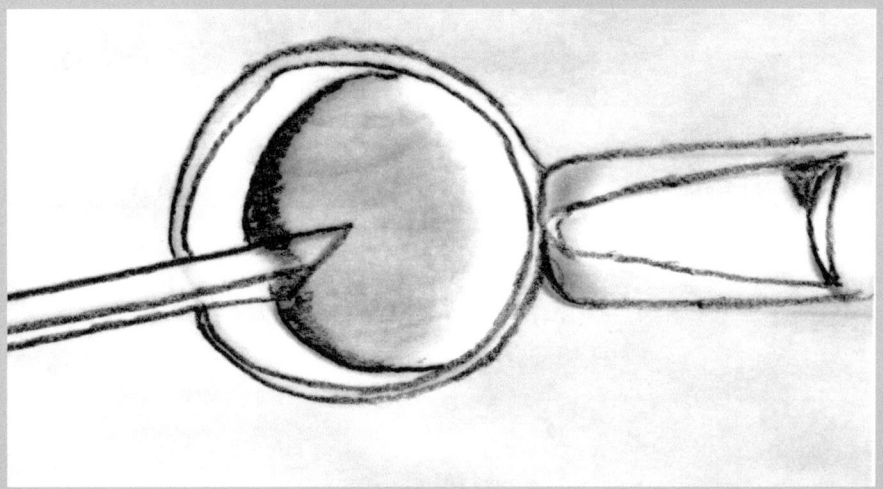

Contents

6.1 What Is Cloning? – 95

6.2 Regeneration – 96

6.3 Somatic Cell Nuclear Transfer – 98
6.3.1 Frog Cloning – 98
6.3.2 Dolly – 99

6.4 Epigenetics – 100

6.5 Ways in Which Cloning Might Be Useful – 102
6.5.1 Cloning and Age – 102
6.5.2 Cloning to Understand Cell Biology – 104
6.5.3 Pet Cloning – 104

© Springer Nature Switzerland AG 2020
G. Rodrigues, B. A. J. Roelen (eds.), *Concepts and Applications of Stem Cell Biology*,
Learning Materials in Biosciences, https://doi.org/10.1007/978-3-030-43939-2_6

6.5.4	Cloning of Commercially Valuable Animals – 105
6.5.5	Cloning Endangered and Extinct Animal Species – 105
6.5.6	Cloning of Equids – 107
6.5.7	Generation of Transgenic Animals – 107
6.6	**Human Cloning – 108**
	References – 110

Cloning

What Will You Will Learn in This Chapter?

Cloning is the generation of a genetically identical organism from an existing organism. The best known example of cloning is taking plant cuttings, but also in various animals; genetically identical clones can be formed from cells or tissues, for example in some types of flatworms. Differentiation of cells in more complex animals leads to loss of developmental capacity and is caused by epigenetic events: methylation of cytosines in CpG dinucleotides and histone tail modifications. Differentiated cells can be reprogrammed to a totipotent state when introduced into an enucleated oocyte, a process known as somatic cell nuclear transfer. Although inefficient, an embryo developed with this method can give rise to offspring. The sheep Dolly was the first clone born after somatic cell nuclear transfer using an adult cell as donor. Since the birth of Dolly, several animal species have been cloned, including pets and commercially valuable animals. In therapeutic cloning, cells of a patient are injected into a human oocyte to form a blastocyst of which embryonic stem cells are generated rather than transplantation of the embryo to the uterus. These patient-specific pluripotent cells may be used for drug screening, study of the disease, or regenerative medicine.

6.1 What Is Cloning?

Before every cell division, the genomic DNA is precisely copied so that the progeny cells contain the same biological information. This implies that in a living organism almost all cells have the capacity to form all cells and tissues of that organism. Indeed it has been demonstrated that in plants individual root cells have the capacity to grow out to a complete plant with roots, leaves, and flowers [1]. The newly formed plant would be genetically identical to the plant from which the individual cell was taken; they would be clones (◘ Fig. 6.1).Taking cuttings to make more of your favorite plants, albeit not from individual cells, is a way of cloning: the asexual generation of genetically identical organisms that descend from a single cell or organism.

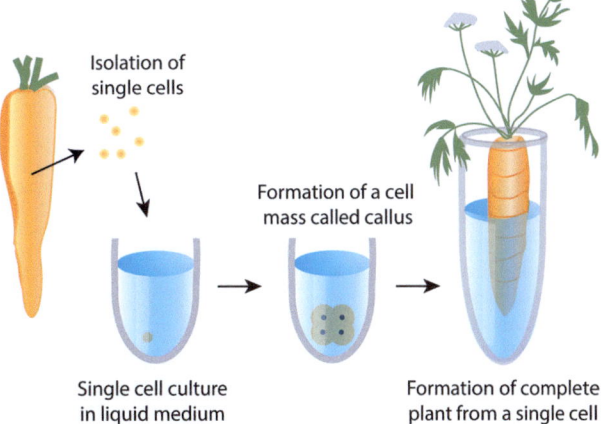

◘ **Fig. 6.1** Plant cloning. Schematic representation of the cloning of a carrot plant from a single cell. Parts of a carrot root are isolated, fractionated to single cells, and culture in a suitable medium. The cells dedifferentiate to form a clump of totipotent cells called callus. The callus can develop to a complete plant with differentiated structures such as roots, stem, leaves, and flowers

6.2 Regeneration

Not all asexual reproduction is cloning. Particularly in more complex bisexual organisms, including snakes, lizards, sharks, and domestic fowl, asexual reproduction occurs by oocytes that develop without being fertilized. In 2006, parthenogenetic offspring was reported from two independent female Komodo dragons (*Varanus komodeonsis*) that were kept individually in separate zoos (◘ Fig. 6.2). These animals each produced clutches of more than 20 eggs from which about one third hatched. The offspring were all male, and although the offspring from one mother were homozygous, they were not clones. The exclusive male offspring can be explained by the sex chromosomes in these lizards, where females are heterozygous for the sex chromosomes Z and W, while males are homozygous ZZ. Since the parthenogenetic mechanism can only produce homozygotes and WW is not viable, the only viable parthenotes will be ZZ males [2]. Similar observations have been done in other species and even in wild non-captive snakes. This could be a functional strategy to maintain a population when females become isolated from males. The exact mechanism of the parthenogenesis is probably different in various vertebrates and can include fusion of cells before or after meiosis [3].

Less complex organisms demonstrate a regenerative ability similar to that of plants. A classic example is the freshwater polyp *Hydra*, belonging to the Phylum Cnidaria, that can regenerate a complete organism from small fragments and dissociated cells [4, 5]. Another well-known example for studying regeneration is the group of freshwater planarians, types of flatworms in the Phylum Platyhelminthes and the Class Turbellaria. These are relatively simple animals but more complex than *Hydra* as they are composed of three germ layers: ectodermal nerve cells, endodermal intestinal cells, and mesodermal muscle. Planarians have formidable regenerative capacities; when the animal is cut in half between the head and tail, the tail-part will form a new head and the head part will form a new tail. Similarly, when a planarian is cut lengthwise, each half will form a new half. These newly formed animals are all clones of the original dissected animal. Although many planarians are hermaphrodites and can reproduce sexually, there are also triploid populations that reproduce asexually

◘ Fig. 6.2 Komodo dragon. Photograph of a preserved Komodo dragon (*Varanus komodoensis*) at a museum display. This large lizard's natural habitat is in the Indonesian archipelago. Examples exist of female Komodo dragons living in captivity that laid unfertilized eggs of which only male animals hatched. Whether this parthenogenesis occurs in the wild or only during captivity is not known

by cloning. They do so by simply tearing themselves into a head and a tailpiece that each produces a new worm. Intriguingly the process of fission is socially controlled by a mechanism in the brain and fission is promoted after decapitation of the worm (◘ Fig. 6.3).

It has been identified that the tissue regeneration, either when externally disrupted or after fission, is mediated via pluripotent neoblasts that accumulate at the wound edge and can give rise to fully patterned body parts [6]. Not all planarians are equal; however, there are various planaria species that can form a new tail from a head but lack the capacity to form a new head from a tail. Intriguingly, in these regeneration-non-competent species, downregulating Wnt/β-catenin signaling could lead to head regeneration from regeneration-deficient tails [7–9]. The question of how the neoblasts are instructed to form the correct tissue types of appropriate sizes is yet to be resolved.

Various vertebrate species such as the South African clawed frog *Xenopus laevis* and the axolotl *Ambystoma mexicanum* have similar, although more limited, regeneration potential (◘ Fig. 6.4). *Xenopus* tadpoles can regenerate their tails including nerves and muscles, *Xenopus* froglets can partly regenerate their limb to form a carti-

◘ **Fig. 6.3** Planaria. Freshwater planarians of the class of Turbellaria have a remarkable regenerative capacity. In this particular species, rudimentary eyes are visible and the ear-like auricles that function as sensory organs

◘ **Fig. 6.4** Axolotl. The axolotl (*Ambystoma mexicanum*) can regenerate an entire limb including digits

laginous protrusion, while the axolotl is well-known for its capacity to grow a new completely functional limb after injury [10]. For the axolotl, it has been established that amputation induces the formation of a blastema with a heterogeneous set of progenitor cells that are not pluripotent but instead have a restricted developmental potential [11]. Programmed cell death or apoptosis, plays a key role in tissue regeneration of both planarians and amphibians as it controls the exact numbers of cells needed to form the various tissues [12].

6.3 Somatic Cell Nuclear Transfer

The regeneration of appendages in amphibians indicate that, at least some, vertebrate cells maintain or regain the capacity to differentiate into derivatives of the three germ layers.

6.3.1 Frog Cloning

In the 1950s, Briggs and King tested whether nuclei of differentiated cells were intrinsically changed by transplanting nuclei of various stages of embryos into enucleated eggs. For their experiments, they made use of eggs and embryos of the American leopard frog *Rana pipiens*. These nuclear transfer experiments were partially successful and Briggs and King obtained feeding stage larva from blastula nuclei, but not from nuclei of later stages. It was concluded that although nuclei of early blastula stage embryos are pluripotent, nuclei from neurula stage embryos onwards are restricted in developmental capacity [13, 14]. Similar experiments were made by Gurdon, only he used *Xenopus* eggs and nuclei for the transplantation experiments. By that time *Xenopus* was widely distributed in European and American labs for its use as a pregnancy test since female adult *Xenopus* frogs injected with urine of a pregnant woman started to lay eggs the following day, the so-called Hogben test [15, 16]. *Xenopus* had several advantages above *Rana* and was becoming the preferred model organism: *Xenopus* is a wholly aquatic animal making it easier to maintain in a laboratory, it can produce eggs year-round and has a relatively short life cycle. Most importantly for nuclear transfer experiments, different *Xenopus* strains could be identified by the number of nucleoli, enabling to distinguish between donor and recipient cells. Nuclei from the intestinal epithelium of *Xenopus* feeding tadpoles were transplanted to eggs from which the chromosomes had been destroyed after ultraviolet light exposure. These transplant embryos could develop to swimming tadpoles and even become adult fertile frogs, although in a low percentage [17, 18]. Similar experiments were performed using cells from adult frogs as nuclear donor, but although the formed embryos did develop to healthy appearing tadpoles, these never transformed into adult frogs [19]. From these data, it was concluded that, at least in amphibians, embryonic nuclei are unrestricted in their developmental potential but that the hereditary material in the nucleus does not remain intact during differentiation.

6.3.2 Dolly

Experiments similar to those performed in frogs were performed with mouse embryonic cells as donor and fertilized and enucleated one-cell zygotes as recipients. This led to controversial results; while one research group claimed the birth of healthy mice from transplanted nuclei of embryonic cells [20], another group claimed that it is biologically impossible to clone mammals by somatic cell nuclear transfer [21].

The birth of the Scottish cloned sheep Dolly in 1996 [22], therefore, came for many as a complete surprise, but there had been several tell-tale signs (◘ Fig. 6.5).

Already in 1986, it was published that sheep can be produced by transplantation of nuclei of the blastomeres from eight cell embryos to enucleated unfertilized eggs [23]. With the development of pluripotent embryonic stem cell lines, combined with homologous recombination and chimaera formation it had become possible to generate mice with targeted mutations [24–26]. That, together with the availability of inbred strains limited the interest in the cloning technology for mouse embryologists. For other animal species including farm animals, embryonic stem cell lines were not available. Therefore the theoretical possibility of replicating those animals with favorable genetic characteristics, or rapidly enhancing those characteristics using cloning, spurred the research on cloning particularly in farm animals. Before the birth of Dolly, live calves and sheep had already been born by transfer of nuclei from cultured embryonic cells [27, 28]. It was Keith Campbell, then at the Roslin Institute in Scotland, who realized that the cell cycle stage of the donor cell and the recipient oocyte should match [29]. In subsequent somatic cell nuclear transfer experiments the donor cells were therefore induced to a state of quiescence by serum starvation (◘ Fig. 6.6). Eventually, this led to the birth of the sheep Dolly using cultured udder cells from a 6-year-old ewe as a donor [22].

◘ **Fig. 6.5** Dolly. The most famous sheep in the world Dolly has been stuffed after her death in 2003 and is now at display at the National Museum of Scotland in Edinburgh

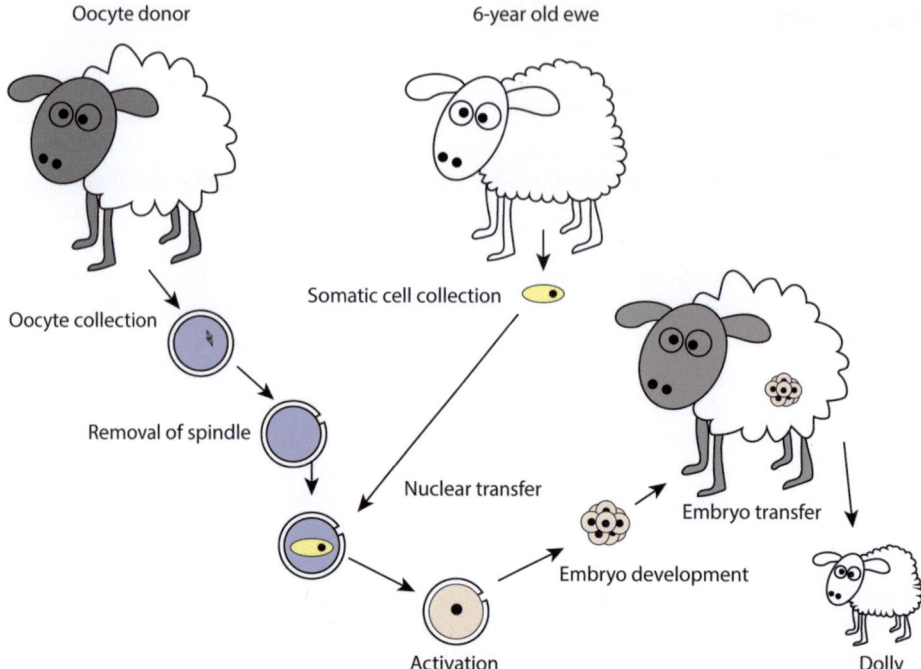

Fig. 6.6 How to clone a sheep? Schematic representation of the somatic cell nuclear transfer experiments that led to the birth of Dolly. Oocytes were collected from Scottish blackface sheep. The genomic DNA was removed by aspiration of the meiotic spindle. Somatic fibroblasts were isolated from the mammary gland of a Finn Dorset ewe and cultured in vitro for a maximum of six passages before being microinjected into the enucleated oocyte. Electrical pulses induced fusion of the donor cell and the enucleated oocyte. Embryos developed to morula and blastocyst stages were transferred to Scottish Blackface recipient ewes. Of the total 434 fused oocytes, 247 embryos were formed which led to 29 blastocysts that were transferred to 13 recipient ewes. One animal was pregnant after embryo transfer and gave birth to Dolly

6.4 Epigenetics

The successful nuclear reprogramming of an adult cell and the birth of an intact organism from that nucleus unequivocally demonstrated that even in cells of complex organisms the DNA sequence remains intact during differentiation. In the 1940s, Conrad Waddington coined the term 'epigenetics' for complex developmental processes that take place in a one-way direction during differentiation that gradually become more severe. Waddington envisioned cellular differentiation as a marble that rolls down a hill that is pulled by gravity and steered by ridges and canals. On top of the hill, the marble can roll to many directions, but as the ball rolls further down the hill the alternative routes become more and more limited. This route is clearly unidirectional as gravity and ridges prevent the marble from changing the path [30].

The contemporary view of epigenetics is that of hereditable changes in gene function or activity that are not due to changes in the DNA sequence. The most well-known epigenetic modifications are methylation of DNA at position 5 of cysteine in CG dinucleotides (◘ Fig. 6.7), and histone tail modifications. DNA within the

◻ **Fig. 6.7** DNA methylation. Skeletal formula of a cytosine (red) nucleotide followed by a guanine (blue) nucleotide separated by a phosphate group of the DNA backbone, a so-called CpG dinucleotide (top). The cytosine in a CpG dinucleotide can be methylated at the 5-position (green dotted circle) to form methylcytosine (bottom). Methylation of CpG sites at promotor regions leads to gene silencing

nucleus is not bare but wrapped around protein complexes to form repeating units known as nucleosomes.

The core proteins that constitute the nucleosome occur as octamers containing 2 copies each of histones H2A, H2B, H3, and H4. Transcriptional activity from the chromatin can be regulated by covalent reversible modifications of the amino acid tails protruding from the central histone proteins. Biochemical modification of the histone N-terminal amino acid sequences can change the accessibility of the DNA for transcription or by promoting the association of chromatin-binding proteins. The best known modifications in this respect are methylation of lysine and arginine residues, ubiquitination of lysine residues, and acetylation of lysine residues. To add more complexity to this system, lysine residues can be mono-, di-, or trimethylated that can have different roles in gene regulation [31]. DNA methylation represses transcription, while histone tail modifications such as acetylation and methylation can be repressive or activating, depending on the type of modification (◻ Fig. 6.8).

Nowadays, there is some debate on how strict heritability fits within the definition of epigenetics since, for instance, histone modifications are reversible and not all stable during cell divisions [32]. Also, the birth of Dolly the sheep unequivocally demonstrated that epigenetic changes are reversible, albeit rather inefficiently. In this respect, the epigenetic landscape resembles a pinball machine: without action the pinball rolls down the table but player-controlled flippers can get the ball back to the top. No matter how good you are at pinball, it seems virtually impossible to control the ball, quite similar to cloning by somatic cell nuclear transfer.

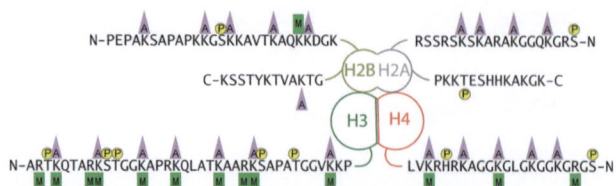

Fig. 6.8 Histone tail modifications. Schematic representation of various reversible modifications that can occur on the amino acids that compose the N- and C-tails of the core histone proteins and that are involved in chromatin reorganization. Methylation is indicated with a green rectangle, acetylation with a purple triangle, and phosphorylation with a yellow circle. Note that residues can carry multiple modifications, for instance, H3K27 can carry three methylation marks. Other modifications that can occur, such as ubiquitination, sumoylation and propionylation, are not indicated in this scheme

6.5 Ways in Which Cloning Might Be Useful

The publication of 'Dolly' evoked many emotions and sparked many questions from scientists, students, and the press. What is the real 'age' of Dolly? Why sheep? Will this lead to human cloning? Should these types of experiments be allowed?

6.5.1 Cloning and Age

Some of these questions have been answered in the years since the publication of the 'Dolly' paper, others are still left unanswered. The birth of a healthy living animal after somatic cell nuclear transfer demonstrated that nuclei from adult differentiated cells can be fully reprogrammed to a totipotent state. The procedure was, and still is, highly inefficient, however. In the case of Dolly, 277 oocytes were injected with a somatic cell leading to the birth of one healthy lamb [22]. In the years, since Dolly, quite a few animal species have been cloned (Table 6.1), but the efficiency has remained low.

One way of determining the 'age' of a cell is by examining the repetitive DNA sequences that decorate ends of chromosomes, the so-called telomeres. It is generally believed that in somatic cells telomeres shorten with each cell division in a way that telomeres are shorter in aged individuals [33] (for further information see ▶ Chap. 5). When Dolly was one-year old, her telomeres indeed were significantly smaller than those from age-matched control sheep, which could mean that the telomeres could prematurely reach a critical size [34]. Telomere sizes of cloned cattle and pigs, however, were normal or even slightly increased, suggesting that telomere lengths are restored after nuclear transfer [35, 36]. When mice were repeatedly recloned, telomere size did not differ compared with controls, not even after 23 generations of cloning [37], strengthening the hypothesis that genomic reprogramming leads to a resetting of telomere size and argues against premature aging of cloned animals. Indeed the life span of cloned and repeatedly recloned mice is similar to that of naturally conceived mice [37].

The world famous sheep, Dolly was rather corpulent during her life which may have been caused by the excess food she received to perform well for the press. She was a celebrity after all. Perhaps because of the weight problem she developed arthritis, but arthritis is also a sign of aging. On Valentine's day 2003, when Dolly was

◼ **Table 6.1** List of mammalian species cloned by somatic cell nuclear transfer that led to live births

Common name	Scientific name	Reference
African wild cat	*Felis silvestris lybica*	[60]
Bactrian camel	*Camelus bactrianus*	[61]
Banteng	*Bos javanicus*	[78]
Brown rat	*Rattus norvegicus*	[62]
Cattle	*Bos taurus*	[63]
Coyote	*Canis latrans*	[64]
Cynomolgus monkey[a] (Crab eating macaque)	*Macaca fascicularis*	[57]
Domestic cat	*Felis silvestris catus*	[40]
Dog	*Canis lupus familiaris*	[41]
Dromedary	*Camelus dromedaries*	[65]
European mouflon	*Ovis orientalis musimon*	[66]
Ferret	*Mustela putorius furo*	[67]
Gaur[b]	*Bos gaurus*	[68]
Goat	*Capra aegagrus hircus*	[69]
Horse	*Equus caballus*	[47]
House mouse	*Mus musculus*	[70]
Mule	*Equus asinus* × *Equus caballus*	[46]
Pig	*Sus scrofa*	[71, 72]
Pyrenean ibex[c]	*Capra pyrenaica pyrenaica*	[43]
Rabbit	*Oryctolagus cuniculus*	[73]
Red Deer	*Cervus elaphus*	[74]
Sand cat	*Felis margarita*	[75]
Sheep	*Ovis aries*	[22]
Swamp buffalo	*Bubalus bubalis*	[76]
Wolf	*Canis lupus*	[77]

[a]From embryonic cells only
[b]Died 2 days after birth
[c]Died minutes after birth

6.5 years old (she was born 5 July 1996), she was euthanized after veterinarians confirmed that she suffered from a contagious lung cancer caused by a virus. There is no indication that this disease was related to cloning, and in fact several other normal sheep at the Roslin institute had gone down with it. Importantly, Dolly herself had given birth to 6 healthy lambs through natural matings as a demonstration of her vitality. Even after her death, Dolly remains in the limelight and is exhibited in the National Museum of Scotland in Edinburgh.

Cloning by somatic cell nuclear transfer can have several applications. It can teach us about cell function and differentiation. It can be used to generate commercially or emotionally valuable animals; it can be used to reproduce animals that are almost extinct or even already extinct; and it can possibly be used to clone humans either for reproduction or for therapeutics.

6.5.2 Cloning to Understand Cell Biology

A great deal of research has been devoted to making the somatic cell nuclear transfer procedure more efficient and to decipher how many healthy cloned animals are similar to naturally born animals. By the use of various chemical histone modifiers, cloning has been made more efficiently, albeit only marginally [38]. In addition, procedures have been described that facilitate nuclear transfer without the use of expensive micromanipulators, the so-called handmade cloning. In this process, rather than injecting somatic cells into oocytes, the zona pellucida is first removed from the oocyte after which the cell is denuded by hand with a disposable blade. A somatic cell is subsequently combined with the cytoplast by electrofusion [39]. Another way of making cloning more efficient by somatic cell nuclear transfer could be by identification of the most suitable oocytes. Using in vitro fertilization, the efficiency of oocytes that can develop to blastocyst is 30–50% in cattle and globally the success rate of human IVF is between 20% and 40%, in terms of babies born. This indicates that there is a large variability in the developmental capacity in oocytes. Understanding what causes this variability could increase the efficiency of IVF and also cloning by somatic cell nuclear transfer.

6.5.3 Pet Cloning

Shortly after the cloning of a sheep, it was suggested that cloning could be used to commercially 'restore' pets that are critically ill or had already deceased. Similarly, cloning was suggested as a way to reproduce dogs or cats that, for instance, have great intelligence and a gentle temperament but that are unable to reproduce naturally because they have been neutered. Indeed both cats [40] and dogs [41] have been cloned from adult cells. Cloning of dogs poses an additional complication since dog oocytes are ovulated at the first meiotic prophase and mature in the oviduct for another 48–72 hours before reaching the metaphase II stage. Therefore, oocytes have to be retrieved by flushing oviducts by laparotomy. Interestingly, for the first successful cloning of a cat, a cumulus cell from a calico cat was used. In cats, coat color is partly caused by genes on the X-chromosomes, and since in queens one of the X-chromosomes in every cell is randomly inactivated, coat patches originated from different cells with either the paternal or maternal X inactivated can have different colors in calico cats. Copy Cat, as is the name of the first cloned cat, was phenotypically not an exact copy of her twin mother because of the random events associated with the coat color. She was predominantly white with tabby grey patches while her donor/twin sister Rainbow had orange, white, and black fur patches. Scientifically very interesting, but for those interested in creating an exact copy of their beloved pet animal, it is less appealing. Similarly, in cloned dogs, the spot pattern may vary. The possible differences in appearance, together with the high costs of cloning, are among the factors by which commercial pet cloning never became a success, although in South Korea you can still have your favorite dog cloned commercially for around 100,000 euros. Applications of dog cloning as a help to study disease are likely to be more supported by the society, for instance cloning transgenic dogs that show hallmarks of Alzheimer's disease.

6.5.4 Cloning of Commercially Valuable Animals

Cloning has also been associated with the food industry, particularly for meat and milk production. Elite animals can be reproduced by cloning as sires for pigs, beef and dairy cattle. This would seem to be economically more valuable than using the meat or milk from cloned animals directly. Either way, eventually the meat or milk from cloned animals, directly or after several generations, subsequently enters the food chain. In the USA, already in 2008, the US Food and Drug administration approved the use of milk and meat from the offspring of pig, cattle, and goat clones. Interestingly, the use of milk and meat from other animal species, including sheep was not recommended due to lack of sufficient information regarding safety. Since cloned food cannot be recognized as such, no special labeling is required. In 2015, the European parliament has banned cloned meat and milk of all farm animals from the market. The European Commission decided that labeling of meat from offspring of clones is unrealistic and it is therefore inevitable that semen or embryos from cloned animals or their descendants will enter the agricultural market in Europe as well.

6.5.5 Cloning Endangered and Extinct Animal Species

In January 2001, a Gaur (*Bos gaurus*), a wild ox from Southeast Asia that faces extinction, was born from a regular cow (*Bos taurus*). The Gaur was cloned using Gaur skin cells and enucleated oocytes from a cow. Not only was this the first cloned Gaur, but it was also the first cloned animal born from interspecies somatic cell nuclear transfer where the species of donor cell and recipient egg and carrying mother are different [42]. Although the Guar calf, Noah, was initially healthy she died within two days after birth from scours, a disease characterized by diarrhea. More embryos were produced and transplanted to foster cow mothers but the majority never implanted into the cow's womb and those that did experience spontaneous abortions, except for Noah.

Not only endangered animals but also animal species or at least a subspecies that are extinct have been cloned. The Pyrenaan ibex *Capra pyrenaica* is a subspecies of the Spanish ibex of which in the late 1990s reportedly only one animal, a female, was

Fig. 6.9 Thylacine. Taxidermy specimen of a thylacine (Tasmanian tiger) at a museum. It seems unlikely that DNA from taxidermy specimens is sufficiently intact to allow cloning, but fetal animals preserved in ethanol may be useful

left. Since no males were known to exist, the subspecies was doomed to extinction. Fortunately, the last remaining animal could be captured and a skin sample secured. The Pyrenean ibex became extinct when in 2000 the last remaining specimen died in 2000 by a falling tree that crashed her skull. Cells from the skin sample were used in an attempt to clone the animal. Since obviously no oocytes were available, oocytes from domestic goats were used for nuclear transfer and Spanish ibex and hybrids of Spanish ibex with domestic goats were used as a surrogate mother. One hybrid goat pregnancy continued and a morphologically normal animal was born by caesarian section. Unfortunately, the animal died within minutes due to respiratory stress resulting in a second extinction of the subspecies [43].

Bringing back extinct animals by ways of cloning has been repeatedly discussed in popular press and the topic is one of the favorites of the entertainment industry. One necessity for cloning extinct animals would be the availability of cells, and therefore two animal species top the list of animals to be cloned: the Woolly mammoth (*Mammuthus primigenius*) and the Thylacine or Tasmanian tiger (*Thylacinus cynocephalus*) (Fig. 6.9).

Cells of the Woolly mammoth can be obtained from specimens that have been fairly conserved in the American and Russian permafrost. An important question is whether the DNA of tissues that have been frozen without cryoprotectants would be intact enough for the generation of viable offspring after nuclear transfer. Viable mice have been born from an animal that had been frozen for 16 years without any cryoprotection, thereby demonstrating the feasibility of cloning from frozen bodies [44]. An additional problem in cloning extinct animals is the oocyte for reprogramming and embryo formation and a surrogate that could accommodate the cloned embryo. In the case of the Wooly mammoth, the Asian elephant (*Elephas maximus*) would be a logical choice based on evolutionary relationship and size. With the inefficiency of cloning it can be argued, however, whether sufficient numbers of oocytes and surrogate elephant cows are available or even present. The question remains as to why we would want to de-extinct animals. It has been proposed that it is our obligation, as many animals became extinct because of humans. Others argue that it would be better to invest time and money in trying to prevent extinction of plants and animal species that are now critically endangered.

Recently a large part of the Thylacine genome has been sequenced from a pouch young specimen [45]. Whether the DNA is of sufficient quality for nuclear reprogramming is not known. Another difficulty with this animal is the oocyte and surrogate mother. The closest living relative of the Thylacine seems to be the Numbat (*Myrmecobius fasciatus*) but this animal by itself is endangered.

6.5.6 Cloning of Equids

It may be worthwhile to clone valuable animals that cannot reproduce in the normal fashion. The first cloned equid was a mule, a hybrid from the breeding of a male donkey (*Equus asinus*) with a horse (*Equus caballus*) mare [46]. Mules are by definition sterile but can give offspring via cloning using horse oocytes and a mare as recipient. Mules can be commercially valuable when used for sports. Indeed, the first mule clone, Idaho Gem, was cloned from a fetus that could have been a race winner and has already won races himself.

For valuable mares that have been successful in sports but are too old for breeding, or champion geldings, cloning might be an interesting option. Shortly after the birth of the first cloned mule, the first cloned horse was born. This animal was not only the first of its species, but was also the first animal that was carried by her 'twin' sister: adult cell donor and surrogate mother were the same animal [47]. This demonstrated that recognition of the embryo by the mother and maintenance of gestation is not dependent on immunological recognition. Whether horse cloning will have a future is partly dependent on the breeding associations, as many of these associations do not allow admittance of cloned horses and participation to some equestrian sports requires listing in a breed registry. Horse cloning could be valuable on the other hand for the preservation of genetic lines [48]. Irrespectively, horse cloning can teach us many aspects of equine peri-implantation development and maternal-fetal interactions.

6.5.7 Generation of Transgenic Animals

For the generation of transgenic or knockin/knockout animals, the use of embryonic stem cell lines targeted using homologous recombination in combination with chimera formation has been very successful indeed [49]. For mammalian species other than rodents or primates, however, it has been demonstrated to be extremely difficult to derive and maintain pluripotent stem cell lines [50]. Transgenic farm animals can be made by microinjection of DNA into zygotes optionally in combination with, for instance, CRISPR/Cas9 technology. Selection of animals with the correct transgenes and subsequent production of F1 animals from founders is, however, a time-consuming process in animals with long generation intervals. When the aim is, for instance, to produce a pharmaceutically active human protein in the milk of a cow, the process can take years. Once such an animal has been generated, cloning by somatic cell nuclear transfer would be a relatively efficient strategy to enhance the numbers of animals [51].

6.6 Human Cloning

The creation of Dolly ignited many scientific and ethical discussions on the possibilities of human cloning and their consequences. Animal cloning is already inefficient, unreliable, and risky, so what about human cloning? First of all, why cloning humans? To clone humans for reproductive purposes, for instance, as an alternative for subfertile couples, seems unrealistic. The predictable inefficiency resulting in the large numbers of human oocytes and surrogate mothers needed and the expected occurrences of spontaneous abortions exclude human cloning as a way of reproduction. Besides reproductive cloning, in combination with embryonic stem cell culture, human cloning could be useful in (regenerative) medicine. Instead of transferring a cloned human embryo to a womb, embryonic stem cells can be generated from a cloned embryo. When adult cells of a patient are used for cloning, the clone-derived pluripotent stem cells would carry the patient's genotype and could be used for autologous transplantation without being rejected. Maybe even more important, the patient-specific pluripotent cells could be used to study disease progression and to test drug efficacy for personalized medicine.

Due to the inefficiency of cloning by somatic cell nuclear transfer, large numbers of oocytes are needed for cloning. Human oocytes are scarce, however, and retrieval is not without risks for women as they can develop ovarian hyperstimulation syndrome [52]. Most women who qualify for hyperstimulation are those who do so for immediate in vitro fertilization procedures or for egg freezing and in vitro fertilization at a later stage. Surplus eggs of such procedures could be used after informed consent of the women. Alternatively, eggs may be donated altruistically or after commercial payment, depending on the country's legislation.

Although human oocytes can reprogram somatic cells to a pluripotent state, removal of the oocyte's genetic material (metaphase II spindle) led to developmental arrest at the early morula stage following nuclear transfer [53]. Apparently, in human oocytes, critical factors for development are physically associated with the meiotic spindle apparatus. Removal of these factors leads to spontaneous exit from meiosis, thereby disturbing reprogramming and development. Enucleation and somatic cell fusion in the presence of caffeine, functioning as a phosphatase inhibitor, protects the oocyte from premature meiosis exit. This technique has enabled the generation of patient-specific human embryonic stem cell lines [54] (◘ Fig. 6.10). Patient-specific pluripotent stem cell lines can, however, also be produced using induced pluripotent stem (iPS) cell technology and with fewer ethical and legal barriers [55] (for further information, read ▶ Chap. 4). It has been suggested, however, that human ES cells derived after somatic cell nuclear transfer are more faithfully reprogrammed and contain less genomic errors than iPS cells [56].

A slightly different approach was adapted for the first successful reproductive cloning of the crabeating macaque *Macaca fascicularis*, a primate. Oocytes from these animals were retrieved by laparoscopy after ovarian superovulation. The meiotic spindle of the oocytes was visualized and removed using a spindle imaging microscopic system. Critical for the procedure was the treatment of the enucleated cells with the histone deacytelase inhibitor trichostatin A. In addition, after the nuclear transfer procedure, the embryos were injected with human *KDM4D* mRNA, coding for a histone demethylase with the incentive of opening up the chromatin to facilitate nuclear reprogramming. Similar to the cloning of other mammals, the procedure was highly inefficient; injection of 127 oocytes with somatic cells from a 61-day-old aborted fetal monkey led to the birth of two healthy individuals from 79 transferred embryos. Interestingly, no live births were obtained when cells from adult animals were used [57].

Human-cloned embryos have been generated to study whether a disease caused by a genetic mutation can be corrected with CRISPR/ Cas9 technology. For this, skin cells from a patient suffering from the genetic disease β-Thalassemia were used to produce cloned zygotes. The patient was homozygous for the disease and thus carried two abnormal copies of the *HBB* gene. Since the zygotes were genetically identical to the patient, they carried the same homozygous mutation, and base-editing technology with a modified CRISPR/ Cas9 protocol was used to correct the disease. Sequencing of the blastomeres after embryo culture revealed that indeed the gene was corrected in 8 out of 20 embryos in at least one copy, albeit with a high degree of mosaicism, meaning that in most blastomeres the gene was still defective [58]. Cloning was used in this study as a tool to generate sufficient embryos with a specific mutation to test a gene-editing system. Both cloning and gene editing of human embryos are controversial and not allowed in many countries.

When it comes to reproductive cloning, monozygotic twins can help us understand what it means to have a genetic 'copy' or how it may limit, or enhance, selfness. Monozygotic twins are by definition of the same age and share the same mitochondrial DNA, while clones from somatic cell nuclear transfer would be of dissimilar age and have different mitochondrial DNA. Not surprisingly, monozygotic twins are less likely to object to human cloning [59].

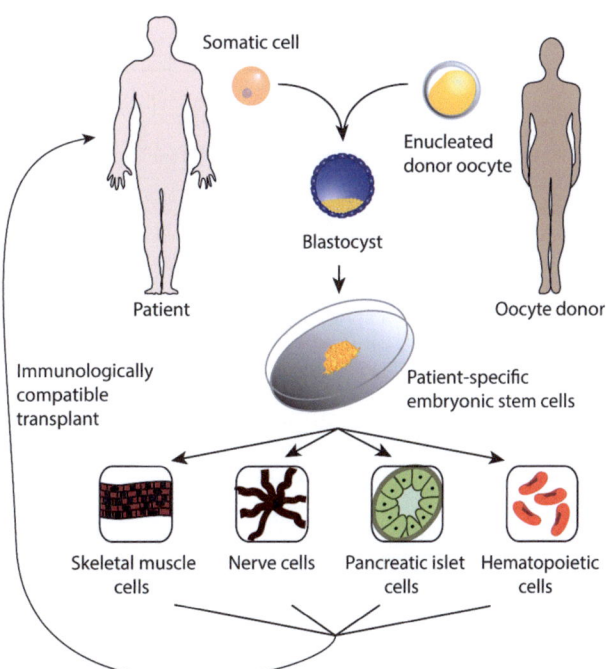

◻ **Fig. 6.10** Therapeutic cloning. Schematic representation of therapeutic human cloning. Theoretically, a somatic cell from a patient can be injected into an enucleated oocyte which can result in a blastocyst stage embryo that is genetically identical to the patient (a clone). Instead of transfer of the embryo to a uterus, embryonic stem cells can be derived from the inner cell mass of the blastocyst. Theoretically pluripotent stem cells can be used to generate tissues that can be transplanted to the patient. The tissues would not be immune rejected by the patient

> **Take-Home Message**
>
> *This chapter covers the following topics:*
> - A clone is a genetically identical organism that descends from a single cell or organism.
> - Cloning is the generation of clones.
> - Various examples are provided of asexual reproduction in vertebrates.
> - Experiments with frogs demonstrated that the nucleus of a differentiated embryonic cell can be reprogrammed by the cytoplasm of an oocyte. The method is known as somatic cell nuclear transfer.
> - Dolly, the sheep, was the first animal cloned from a single cell of an adult animal by somatic cell nuclear transfer.
> - Cloning is an inefficient procedure.
> - Various mammalian species have been cloned.
> - During differentiation, the genetic code stays intact but access to the DNA changes: epigenetics.
> - The most important epigenetic changes are methylation of cytosines in CG dinucleotides and histone tail modifications.
> - Cloning by somatic cell nuclear transfer has been used to clone pets and economically valuable animals and is considered as a possibility to repopulate endangered species.
> - Human cloning is a method of generating patient-specific pluripotent stem cells.
> - Cloning is not allowed in most countries.

References

1. Steward FC. Totipotency, variation and clonal development of cultured cells. Endeavour. 1970;29(108):117–24.
2. Watts PC, Buley KR, Sanderson S, Boardman W, Ciofi C, Gibson R. Parthenogenesis in Komodo dragons. Nature. 2006;444(7122):1021–2.
3. Booth W, Smith CF, Eskridge PH, Hoss SK, Mendelson JR 3rd, Schuett GW. Facultative parthenogenesis discovered in wild vertebrates. Biol Lett. 2012;8(6):983–5.
4. Gierer A, Berking S, Bode H, David CN, Flick K, Hansmann G, et al. Regeneration of hydra from reaggregated cells. Nat New Biol. 1972;239(91):98–101.
5. Gierer A. The Hydra model - a model for what? Int J Dev Biol. 2012;56(6–8):437–45.
6. Wagner DE, Wang IE, Reddien PW. Clonogenic neoblasts are pluripotent adult stem cells that underlie planarian regeneration. Science. 2011;332(6031):811–6.
7. Liu SY, Selck C, Friedrich B, Lutz R, Vila-Farre M, Dahl A, et al. Reactivating head regrowth in a regeneration-deficient planarian species. Nature. 2013;500(7460):81–4.
8. Sikes JM, Newmark PA. Restoration of anterior regeneration in a planarian with limited regenerative ability. Nature. 2013;500(7460):77–80.
9. Umesono Y, Tasaki J, Nishimura Y, Hrouda M, Kawaguchi E, Yazawa S, et al. The molecular logic for planarian regeneration along the anterior-posterior axis. Nature. 2013;500(7460):73–6.
10. Suzuki M, Yakushiji N, Nakada Y, Satoh A, Ide H, Tamura K. Limb regeneration in Xenopus laevis froglet. Sci World J. 2006;6(Suppl 1):26–37.
11. Kragl M, Knapp D, Nacu E, Khattak S, Maden M, Epperlein HH, et al. Cells keep a memory of their tissue origin during axolotl limb regeneration. Nature. 2009;460(7251):60–5.
12. Tseng AS, Adams DS, Qiu D, Koustubhan P, Levin M. Apoptosis is required during early stages of tail regeneration in Xenopus laevis. Dev Biol. 2007;301(1):62–9.

13. Briggs R, King TJ. Transplantation of living nuclei from blastula cells into enucleated frogs' eggs. Proc Natl Acad Sci U S A. 1952;38(5):455–63.
14. King TJ, Briggs R. Changes in the nuclei of differentiating gastrula cells, as demonstrated by nuclear transplantation. Proc Natl Acad Sci U S A. 1955;41(5):321–5.
15. Hogben L. History of the Hogben test. Br Med J. 1946;2:554.
16. Gurdon JB, Hopwood N. The introduction of Xenopus laevis into developmental biology: of empire, pregnancy testing and ribosomal genes. Int J Dev Biol. 2000;44(1):43–50.
17. Gurdon JB. The developmental capacity of nuclei taken from intestinal epithelium cells of feeding tadpoles. J Embryol Exp Morphol. 1962;10:622–40.
18. Gurdon JB, Uehlinger V. "Fertile" intestine nuclei. Nature. 1966;210(5042):1240–1.
19. Laskey RA, Gurdon JB. Genetic content of adult somatic cells tested by nuclear transplantation from cultured cells. Nature. 1970;228(5278):1332–4.
20. Illmensee K, Hoppe PC. Nuclear transplantation in Mus musculus: developmental potential of nuclei from preimplantation embryos. Cell. 1981;23(1):9–18.
21. McGrath J, Solter D. Inability of mouse blastomere nuclei transferred to enucleated zygotes to support development in vitro. Science. 1984;226(4680):1317–9.
22. Wilmut I, Schnieke AE, McWhir J, Kind AJ, Campbell KH. Viable offspring derived from fetal and adult mammalian cells. Nature. 1997;385(6619):810–3.
23. Willadsen SM. Nuclear transplantation in sheep embryos. Nature. 1986;320(6057):63–5.
24. Evans MJ, Kaufman MH. Establishment in culture of pluripotential cells from mouse embryos. Nature. 1981;292(5819):154–6.
25. Koller BH, Hagemann LJ, Doetschman T, Hagaman JR, Huang S, Williams PJ, et al. Germ-line transmission of a planned alteration made in a hypoxanthine phosphoribosyltransferase gene by homologous recombination in embryonic stem cells. Proc Natl Acad Sci U S A. 1989;86(22): 8927–31.
26. Capecchi MR. Altering the genome by homologous recombination. Science. 1989;244(4910):1288–92.
27. Sims M, First NL. Production of calves by transfer of nuclei from cultured inner cell mass cells. Proc Natl Acad Sci U S A. 1994;91(13):6143–7.
28. Campbell KH, McWhir J, Ritchie WA, Wilmut I. Sheep cloned by nuclear transfer from a cultured cell line. Nature. 1996;380(6569):64–6.
29. Campbell KH, Ritchie WA, Wilmut I. Nuclear-cytoplasmic interactions during the first cell cycle of nuclear transfer reconstructed bovine embryos: implications for deoxyribonucleic acid replication and development. Biol Reprod. 1993;49(5):933–42.
30. Waddinngton CH. The strategy of genes: a discussion of some aspects of theoretical biology. London: George Allen and Unwin; 1957.
31. Berger SL. The complex language of chromatin regulation during transcription. Nature. 2007;447(7143):407–12.
32. Bird A. Perceptions of epigenetics. Nature. 2007;447(7143):396–8.
33. Sahin E, Depinho RA. Linking functional decline of telomeres, mitochondria and stem cells during ageing. Nature. 2010;464(7288):520–8.
34. Shiels PG, Kind AJ, Campbell KH, Waddington D, Wilmut I, Colman A, et al. Analysis of telomere lengths in cloned sheep. Nature. 1999;399(6734):316–7.
35. Lanza RP, Cibelli JB, Blackwell C, Cristofalo VJ, Francis MK, Baerlocher GM, et al. Extension of cell life-span and telomere length in animals cloned from senescent somatic cells. Science. 2000;288(5466):665–9.
36. Jiang L, Carter DB, Xu J, Yang X, Prather RS, Tian XC. Telomere lengths in cloned transgenic pigs. Biol Reprod. 2004;70(6):1589–93.
37. Wakayama S, Kohda T, Obokata H, Tokoro M, Li C, Terashita Y, et al. Successful serial recloning in the mouse over multiple generations. Cell Stem Cell. 2013;12(3):293–7.
38. Vajta G. Cloning: a sleeping beauty awaiting the kiss? Cell Reprogram. 2018;20(3):145–56.
39. Vajta G. Handmade cloning: the future way of nuclear transfer? Trends Biotechnol. 2007;25(6): 250–3.
40. Shin T, Kraemer D, Pryor J, Liu L, Rugila J, Howe L, et al. A cat cloned by nuclear transplantation. Nature. 2002;415(6874):859.
41. Lee BC, Kim MK, Jang G, Oh HJ, Yuda F, Kim HJ, et al. Dogs cloned from adult somatic cells. Nature. 2005;436(7051):641.
42. Vogel G. Endangered species. Cloned gaur a short-lived success. Science. 2001;291(5503):409.

43. Folch J, Cocero MJ, Chesne P, Alabart JL, Dominguez V, Cognie Y, et al. First birth of an animal from an extinct subspecies (Capra pyrenaica pyrenaica) by cloning. Theriogenology. 2009;71(6):1026–34.
44. Wakayama S, Ohta H, Hikichi T, Mizutani E, Iwaki T, Kanagawa O, et al. Production of healthy cloned mice from bodies frozen at −20 degrees C for 16 years. Proc Natl Acad Sci U S A. 2008;105(45):17318–22.
45. Feigin CY, Newton AH, Doronina L, Schmitz J, Hipsley CA, Mitchell KJ, et al. Genome of the Tasmanian tiger provides insights into the evolution and demography of an extinct marsupial carnivore. Nat Ecol Evol. 2018;2(1):182–92.
46. Woods GL, White KL, Vanderwall DK, Li GP, Aston KI, Bunch TD, et al. A mule cloned from fetal cells by nuclear transfer. Science. 2003;301(5636):1063.
47. Galli C, Lagutina I, Crotti G, Colleoni S, Turini P, Ponderato N, et al. Pregnancy: a cloned horse born to its dam twin. Nature. 2003;424(6949):635.
48. Church SL. Nuclear transfer saddles up. Nat Biotechnol. 2006;24(6):605–7.
49. Mak TW. Gene targeting in embryonic stem cells scores a knockout in Stockholm. Cell. 2007;131(6):1027–31.
50. Ezashi T, Yuan Y, Roberts RM. Pluripotent stem cells from domesticated mammals. Ann Rev Anim Biosci. 2016;4:223–53.
51. Niemann H, Lucas-Hahn A. Somatic cell nuclear transfer cloning: practical applications and current legislation. Reprod Domest Anim. 2012;47(Suppl 5):2–10.
52. Beeson D, Lippman A. Egg harvesting for stem cell research: medical risks and ethical problems. Reprod Biomed Online. 2006;13(4):573–9.
53. Noggle S, Fung HL, Gore A, Martinez H, Satriani KC, Prosser R, et al. Human oocytes reprogram somatic cells to a pluripotent state. Nature. 2011;478(7367):70–5.
54. Tachibana M, Amato P, Sparman M, Gutierrez NM, Tippner-Hedges R, Ma H, et al. Human embryonic stem cells derived by somatic cell nuclear transfer. Cell. 2013;153(6):1228–38.
55. Mummery CL, Roelen BA. Stem cells: cloning human embryos. Nature. 2013;498(7453):174–5.
56. Ma H, Morey R, O'Neil RC, He Y, Daughtry B, Schultz MD, et al. Abnormalities in human pluripotent cells due to reprogramming mechanisms. Nature. 2014;511(7508):177–83.
57. Liu Z, Cai Y, Wang Y, Nie Y, Zhang C, Xu Y, et al. Cloning of macaque monkeys by somatic cell nuclear transfer. Cell. 2018;172(4):881–7 e7.
58. Liang P, Ding C, Sun H, Xie X, Xu Y, Zhang X, et al. Correction of beta-thalassemia mutant by base editor in human embryos. Protein Cell. 2017;8(11):811–22.
59. Prainsack B, Cherkas LF, Spector TD. Attitudes towards human reproductive cloning, assisted reproduction and gene selection: a survey of 4600 British twins. Hum Reprod. 2007;22(8):2302–8.
60. Gomez MC, Pope CE, Giraldo A, Lyons LA, Harris RF, King AL, et al. Birth of African wildcat cloned kittens born from domestic cats. Cloning Stem Cells. 2004;6(3):247–58.
61. Wani NA, Vettical BS, Hong SB. First cloned Bactrian camel (Camelus bactrianus) calf produced by interspecies somatic cell nuclear transfer: a step towards preserving the critically endangered wild Bactrian camels. PLoS One. 2017;12(5):e0177800.
62. Zhou Q, Renard JP, Le Friec G, Brochard V, Beaujean N, Cherifi Y, et al. Generation of fertile cloned rats by regulating oocyte activation. Science. 2003;302(5648):1179.
63. Kato Y, Tani T, Sotomaru Y, Kurokawa K, Kato J, Doguchi H, et al. Eight calves cloned from somatic cells of a single adult. Science. 1998;282(5396):2095–8.
64. Hwang I, Jeong YW, Kim JJ, Lee HJ, Kang M, Park KB, et al. Successful cloning of coyotes through interspecies somatic cell nuclear transfer using domestic dog oocytes. Reprod Fertil Dev. 2013;25(8):1142–8.
65. Wani NA, Wernery U, Hassan FA, Wernery R, Skidmore JA. Production of the first cloned camel by somatic cell nuclear transfer. Biol Reprod. 2010;82(2):373–9.
66. Loi P, Ptak G, Barboni B, Fulka J Jr, Cappai P, Clinton M. Genetic rescue of an endangered mammal by cross-species nuclear transfer using post-mortem somatic cells. Nat Biotechnol. 2001;19(10):962–4.
67. Li Z, Sun X, Chen J, Liu X, Wisely SM, Zhou Q, et al. Cloned ferrets produced by somatic cell nuclear transfer. Dev Biol. 2006;293(2):439–48.
68. Lanza RP, Cibelli JB, Diaz F, Moraes CT, Farin PW, Farin CE, et al. Cloning of an endangered species (Bos gaurus) using interspecies nuclear transfer. Cloning. 2000;2(2):79–90.

69. Baguisi A, Behboodi E, Melican DT, Pollock JS, Destrempes MM, Cammuso C, et al. Production of goats by somatic cell nuclear transfer. Nat Biotechnol. 1999;17(5):456–61.
70. Wakayama T, Perry AC, Zuccotti M, Johnson KR, Yanagimachi R. Full-term development of mice from enucleated oocytes injected with cumulus cell nuclei. Nature. 1998;394(6691):369–74.
71. Onishi A, Iwamoto M, Akita T, Mikawa S, Takeda K, Awata T, et al. Pig cloning by microinjection of fetal fibroblast nuclei. Science. 2000;289(5482):1188–90.
72. Polejaeva IA, Chen SH, Vaught TD, Page RL, Mullins J, Ball S, et al. Cloned pigs produced by nuclear transfer from adult somatic cells. Nature. 2000;407(6800):86–90.
73. Chesne P, Adenot PG, Viglietta C, Baratte M, Boulanger L, Renard JP. Cloned rabbits produced by nuclear transfer from adult somatic cells. Nat Biotechnol. 2002;20(4):366–9.
74. Berg DK, Li C, Asher G, Wells DN, Oback B. Red deer cloned from antler stem cells and their differentiated progeny. Biol Reprod. 2007;77(3):384–94.
75. Gomez MC, Pope CE, Kutner RH, Ricks DM, Lyons LA, Ruhe M, et al. Nuclear transfer of sand cat cells into enucleated domestic cat oocytes is affected by cryopreservation of donor cells. Cloning Stem Cells. 2008;10(4):469–83.
76. Shi D, Lu F, Wei Y, Cui K, Yang S, Wei J, et al. Buffalos (Bubalus bubalis) cloned by nuclear transfer of somatic cells. Biol Reprod. 2007;77(2):285–91.
77. Kim MK, Jang G, Oh HJ, Yuda F, Kim HJ, Hwang WS, et al. Endangered wolves cloned from adult somatic cells. Cloning Stem Cells. 2007;9(1):130–7.
78. Jansen DL, Edwards ML, Koster JA, Lanza RP, Ryder OA. Postnatal management of cryptorchid Banteng calves cloned by nuclear transfer utilizing frozen fibroblast cultures and enucleated cow ova. Reprod Fertil Dev. 2004;16:206.

Stem Cells in Plant Development

Beatriz Gonçalves

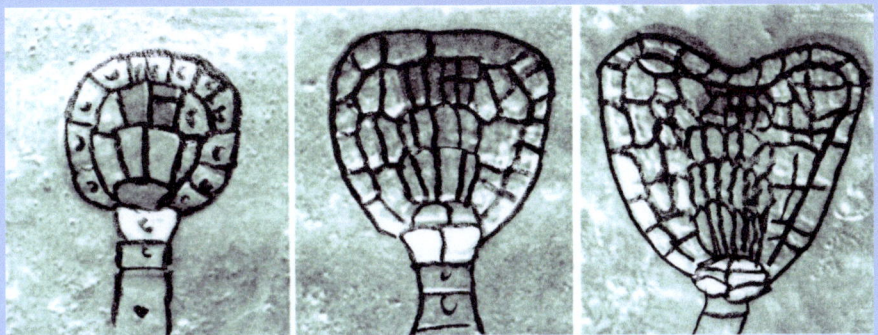

Contents

7.1 Overview of Plant Development and Stem Cells – 116

7.2 Primary Meristems Organization – 117
7.2.1 Organization of the Shoot Apical Meristem – 118
7.2.2 Organization of the Root Apical Meristem – 119

7.3 Stem Cell Initiation – 119
7.3.1 Shoot Apical Meristem Initiation – 121
7.3.2 Root Apical Meristem Initiation – 122

7.4 Stem Cell Maintenance – 123
7.4.1 Stem Cell Maintenance in the Shoot Apical Meristem – 123
7.4.2 Stem Cell Maintenance in the Root Apical Meristem – 124
7.4.3 The Role of Hormones in Stem Cell Maintenance in Plant Meristems – 125
7.4.4 Conservation of Meristem Maintenance Mechanisms in Plants – 126

References – 127

© Springer Nature Switzerland AG 2020
G. Rodrigues, B. A. J. Roelen (eds.), *Concepts and Applications of Stem Cell Biology*,
Learning Materials in Biosciences, https://doi.org/10.1007/978-3-030-43939-2_7

What Will You Learn in This Chapter?

The aim of this chapter is to give you a basic understanding of how stem cells function in plant development. Plants and animals lead quite different lives. Plants, for one, do not often move and cannot escape adverse environmental conditions. The most important consequence of their sessile mode of life is that plant development is not restricted to embryogenesis. Instead, it continues throughout the plant's life, as it continues to grow and produce new organs. We will briefly discuss the implications of plant immobility on their developmental strategies, and how plant stem cell activity contributes to their indeterminate growth mode. Next, we will discuss the concept of plant meristems. These are highly specialized tissues that contain the plant stem cells and control both their maintenance and the production of new organs and tissues. Two main meristems are responsible for most of the growth in plants, the shoot and root apical meristems, and will be the focus of this chapter. We begin by reviewing the organization of apical meristems. Next, we discuss their embryonic origin, and we explore in finer detail the key signalling pathways involved in both the specification and maintenance of stem cell identity and activity. We will highlight the mechanisms that underlie the coordination of cell proliferation and differentiation. Hopefully, the concepts exposed in this chapter will provide you with a base from which to further explore stem cell activity and maintenance in plants, but also with the tools to draw interesting comparisons between animal and plant stem cells.

7.1 Overview of Plant Development and Stem Cells

As you have learned in previous chapters, stem cells are undifferentiated cells with an unlimited capacity to self-renew and generate the different cell types that compose an organism. In animals, they act during embryogenesis to produce all the tissues required for organogenesis, but also during adult life, repairing and replenishing certain adult tissues. Plant stem cells are also initiated in the embryo, however, unlike in animal development, they remain inactive throughout embryogenesis. In plants, unlike animals, the product of embryogenesis in plants is not a complete body that will grow into maturity, but a minimal body plan with a main axis and two poles, the shoot and the root. Upon germination, after embryonic development ceases, stem cells are activated to divide and form the mature plant body, with organs like the leaves and flowers. Hence, plant development differs from animal development in that it is not restricted to the embryonic stages, but it occurs throughout a plant's life with the continuous production of new organs such as leaves, roots and flowers.

Plants are sessile organisms for the most part. This means they cannot escape unsuitable abiotic conditions or biotic threats. Instead, they must constantly adapt, shaping their body plans and architecture to the surrounding environment. The postembryonic development strategy is the answer of plants to the constraints imposed by their sessile lifestyles. It provides plants with an enormous capacity to adapt by generating new structures or replacing damaged ones. It also underlies their indeterminate growth and record-breaking longevity of some plant species.

Plant stem cells, therefore, arise in a variety of contexts and are maintained throughout the plant's life. Developmental stem cells are contained in specialized tissues called meristems. Plants have two main meristems initiated during embryogenesis and localized at the two tips of the main body axis, called the apical meristems. The shoot apical meristem (SAM) localized at the apex of the main stem produces all aerial organs, such as leaves, flowers and primary vascular tissues. The root apical meristem (RAM) localized at the tip of the main root axis gives rise to the network of underground tissues. Together, the SAM and RAM contribute to growth along the primary plant body plan or axis. In some species, another type of meristem, called cambium, produces the secondary vascular tissues, xylem and phloem, contributing to growth along the lateral axis. Outside of the primary meristems and cambium, plant stem cells can also be regenerated by de-differentiation of some tissues to generate new axes of growth in the shoot and root, or in response to wounding.

The capacity to produce unlimited new organs and repair wounded ones over the whole life of the organism poses the problem of maintaining and regenerating stem cells over a long span of time and in multiple cellular contexts. In plants, this is possible due to a great capacity to de-differentiate and change cell fate depending on context and positional cues. Despite the diversity of contexts in which plant stem cells are active there is an overarching theme to the way they are organized into meristems, their specification and the maintenance of a stable pool of undifferentiated and proliferating cells. Whether in the shoots, roots or regenerating organs, a conserved set of transcriptional and hormonal signalling pathways interact to regulate meristem function.

Note on Tissue Coordination in Plants Plant cells are surrounded by cell walls that constrain their movement, fixing them in the position they arise via division. An emerging property of this constraint is that cell–cell communication plays an important role in cell fate specification. This is in contrast with animal development where lineage-specific fate is a major stem cell specification mechanism and cells can migrate to their final niche. The importance of cell–cell communication in plants is highlighted by the many non-cell autonomous actors in cell fate specification. In addition, coordination of developmental and cellular events at a tissue and organ level requires mid to longer range signals. In plants, hormones (phytohormones) often play this role, bridging the cellular and tissue scales.

In this chapter, we will focus on the mechanisms of stem cell initiation and maintenance that have been highlighted in the plant model species *Arabidopsis thaliana*. Elegant work in other species such as tomato, petunia, rice and maize has brought to light deeper conservation of these mechanisms at an evolutionary scale.

7.2 Primary Meristems Organization

As mentioned briefly, plant stem cells arise in different contexts. Here and in the following sections, we will concentrate on the two main stem cell populations which are localized at strategic places in the plant called the apical meristems (◘ Fig. 7.1). The meristem is a specialized tissue that coordinates cell proliferation and organ or tissue differentiation. It contains the stem cells proper, additional proliferating cells and the mitotically less active niche cells that induce and maintain stem cell identity in a non-cell autonomous way.

The terms 'meristematic cells' and 'stem cells' are sometimes used interchangeably. While meristematic cells may include the broader range of cell types in a meristem, here we will use 'stem cells' to designate specifically the cells that remain undifferentiated and proliferating at the centre of the meristem.

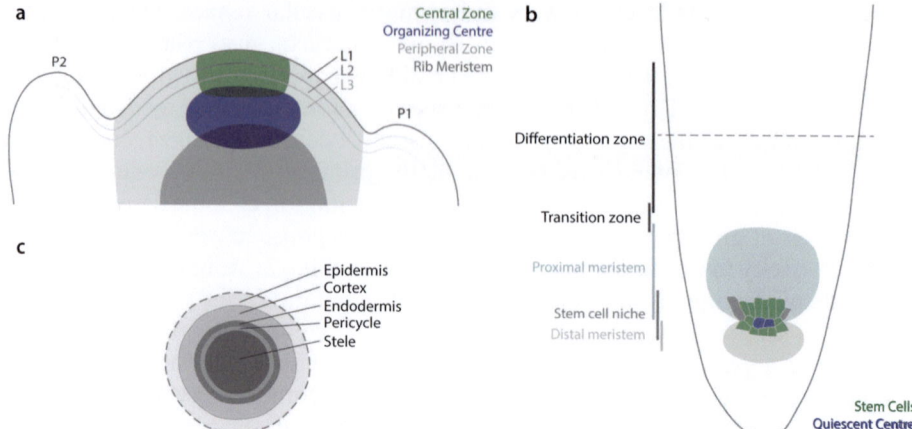

Fig. 7.1 Schematic representation of *Arabidopsis thaliana* apical meristems. **a** Depiction of the shoot apical meristem. The central bulge contains the meristem proper which can be divided into the central zone (CZ, green), organizing centre (OC, blue), peripheral zone (PZ, light grey) and rib meristem (RM, darker grey). The three distinguishable cell layers are indicated (L1–L3), as well as two developing organ primordia (P1, P2) which are separated from the meristem by a dip indicating the boundary zone. **b** Depiction of the root apical meristem and radial organization of root tissues. The meristem is divided into stem cell niche, distal and proximal meristems. The stem cell niche is further decomposed into the quiescent centre (blue) and stem cells or initials (green). At its proximal end, the meristem is flanked by the transition zone. In the differentiation zone, the cell files acquire their mature identities and the concentric organization of tissues can be seen. **c** Radial organization of root tissues at the level of the dashed line in **c**. From the outside, the root is organized into epidermis, cortex, endodermis, pericycle and the stele where vascular tissues differentiate. More detailed views of **a** and **b** can be found in Figs. 7.3 and 7.4, respectively

7.2.1 Organization of the Shoot Apical Meristem

The shoot apical meristem contributes the cells that will produce shoot growth along the main body axis, but also all cells in aerial lateral organs such as leaves, lateral shoots (via axillary meristems) and flowers (via inflorescence and floral meristems).

From an anatomical point of view, the shoot apical meristem can be described by two well-defined cell layers, called the L1 and L2, covering a third cell layer, L3, and a deeper mass of less organized inner cells (Fig. 7.1a). The L1 gives rise to the epidermis in one continuous layer of cells across all aerial tissues. The L2 and L3 produce the ground internal tissues and the primary vascular tissue.

Conceptually, the SAM can be further organized into four zones that have been identified based on the properties of the cells within, or the expression of certain molecular markers [1, 2]. At the centre of the apex and spanning all three layers, the *central zone* (CZ) contains the pool of undifferentiated stem cells, as defined by the expression of plant stem cell marker, *CLAVATA3*. Below the CZ sits the *organizing centre* (OC) which transmits positional cues to the cells above it to induce and maintain their stem cell identity. Surrounding the OC and CZ, and creating a transition zone between the meristem and organ initiation, is the *peripheral zone* (PZ). Underlying these zones, deeper within the apex, is the rib zone or *rib meristem* (RM) where proliferating cells contribute to main shoot growth.

The stem cells contained in the CZ divide slowly. As they divide, some of the daughter cells stay in the CZ replenishing the stem cell pool while others are pushed into the peripheral zone. In the PZ, cells divide more quickly amplifying the initial population. After a few rounds of division, cells reach the periphery of the meristem where they are incorporated into organ initiation as organ primordia [2]. As we will see in future sections, the integrity of the meristem depends on a strong coordination of all these cellular events.

7.2.2 Organization of the Root Apical Meristem

The root apical meristem contains the pool of stem cells that give rise to the underground tissues. Along with the SAM, it contributes to the main growth axis of the plant. Additional growth axes also occur in underground tissues and can be formed by root branching. This, however, is not a direct consequence of RAM activity but of de-differentiation of pericycle cells in the mature root.

Although analogous structures can be found in the SAM and RAM, these are organized in different ways. Localized at the tip of the primary root, the RAM is composed of the stem cell niche, the distal meristem and the proximal meristem (◘ Fig. 7.1b). The *stem cell niche* comprises the *quiescent centre* (QC) and the surrounding *stem cells or initials* [3]. Like the OC in the SAM, the mitotically inactive QC provides the positional cues that specify stem cell identity and suppress differentiation [4]. However, while in the SAM stem cells make up a mass localized above the OC, in the root, stem cells are found in a single layer directly around the QC, thus ensuring that all stem cells maintain cell to cell contact with the instructing niche.

Root stem cells are also called initials as they are the first, or initial, cells in the files that give rise to the different root tissues in a stereotypical root organization. The initials below the QC belong to the *distal meristem* region and will produce the columella and lateral root cap cells. The initials to the sides and above the QC will give rise to the epidermis, cortex, endodermis, pericycle and vascular tissues and belong to the *proximal meristem* zone. The proximal meristem connects to the *transition zone* just above it, where cells proliferate and after a few rounds of cell division exit the meristem entering the *differentiation zone* where they elongate and acquire a fate.

7.3 Stem Cell Initiation

The shoot and root apical meristems, containing the two main populations of plant stem cells, are initiated during embryogenesis, as the embryo is patterned into the apical and basal poles, by the action of interacting hormone signalling pathways and transcriptional networks.

Although the first signs of the shoot and root apical meristems and can be seen at mid-embryo stages, the precise series of cell division patterns that eventually gives rise to these cell populations can be traced to the first asymmetric division of the single-cell zygote (◘ Fig. 7.2) (reviewed in [5, 6]). This first mitosis splits the zygote into a small apical cell that will give rise to most embryo tissues, and a large basal cell that will produce the extra-embryonic suspensor and the basal-most embryonic

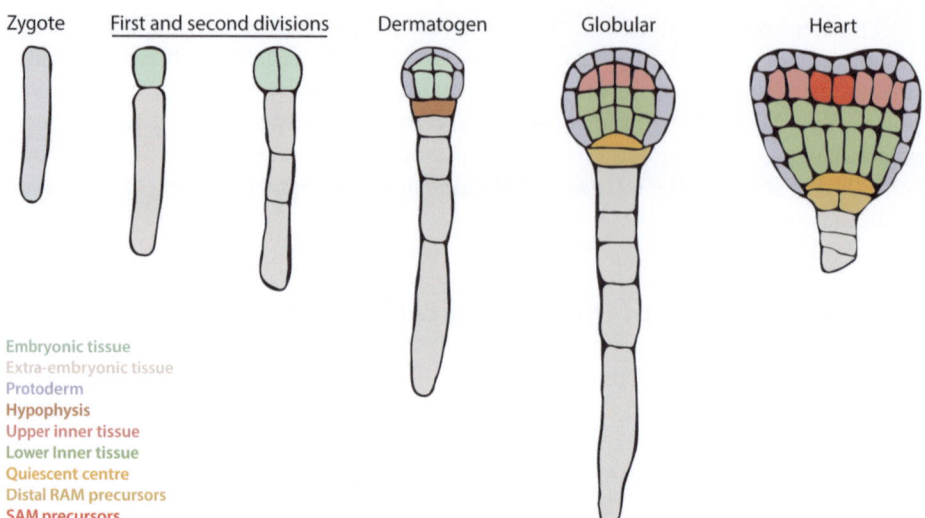

Fig. 7.2 Schematic representation of key stages in *Arabidopsis thaliana* embryo development. All stages are represented in longitudinal section. The first division produces the small apical cell (light green) and the larger basal cell (light grey). Both apical and basal cells go on to divide. The apical cell produces most of the embryonic tissues. By the dermatogen stage, the apical region splits into an outer layer, called the protoderm (light purple), and an inner mass of cells. The inner cells further divide into upper tissue (pink) and lower tissue (green). By early heart stage, the two cotyledon bulges can be seen in the apical region of the embryo, and between them the precursors of the SAM start to express meristem markers. The basal cell produces the extra-embryonic tissue called suspensor, as well as the basal-most embryo tissues. At the dermatogen stage, the top cell of the suspensor divides asymmetrically producing the hypophysis (brown). This cell further divides to give rise to the quiescent centre precursor (yellow) and the precursors to the distal RAM

tissues, including the root stem cell niche. As the apical cell divides, the embryo transitions to the dermatogen stage where the future L1 (protoderm) and the inner tissues become individualized. At the globular stage, the inner tissues further subdivide. The lower inner tissues acquire vascular precursor identity while the upper inner tissues will contribute to the shoot meristem niche which starts to be visible by the heart stage embryo, between the two cotyledon bulges. Meanwhile, at the dermatogen stage, the uppermost cell of the suspensor called the hypophysis divides asymmetrically forming the precursors to the root quiescent centre and the distal meristem cells.

The patterning of the embryo into apical and basal regions involves several transcriptional and signalling pathways. Among those, the phytohormones auxin and cytokinin (CK) play crucial roles in the specification of the shoot and root identities. This was elegantly demonstrated by Skoog and Miller [7] who experimented with different ratios of auxin and CK in growth media. They showed that a high CK/auxin ratio promotes shoot formation and a low ratio promotes root differentiation. More recent experiments confirmed the role of auxin and CK in embryo patterning into apical and basal poles [8] and detailed how their signalling pathways interact with transcription factors to initiate meristems. Below we will explore the mechanisms and molecular pathways involved in the specification of stem cell identity at the apical and basal poles.

7.3.1 Shoot Apical Meristem Initiation

Our understanding of shoot meristem maintenance predates the efforts to unveil the mechanisms of its origin. Hence, candidates for the initiation of shoot meristem stem cells during embryogenesis have been borrowed from the extensively researched field of mature SAM organization.

At the centre of mature SAM maintenance and activity is a feedback loop between shoot stem cell marker *CLAVATA3* (*CLV3*) and stem cell-inducing factor *WUSCHEL* (*WUS*) (◘ Fig. 7.3 detailed in ▶ Sect. 7.4.1). In the embryo, *WUS* expression in the presumptive OC precedes the first signs of shoot apical meristem initiation and the concomitant expression of *CLV3* [9, 10]. However, despite sustaining *CLV3* expression in the mature plant WUS is not required for stem cell initiation in the embryo [11]. Instead, the WUSCHEL-RELATED HOMEOBOX (WOX) transcription factors in the WOX2 module (comprised of WOX2 and the closely related WOX1, WOX3 and WOX5) contribute to shoot stem cell initiation and embryo patterning by promoting the expression of *CLV3* and shoot patterning genes of the class III HD-ZIP (Homeodomain Leucine Zipper) family of transcription factors [12]. The WOX2-module roles in embryo patterning and shoot meristem specification are partially achieved by regulating the auxin/cytokinin balance in this tissue, promoting cytokinin signalling and repressing auxin transport in the presumptive shoot meristem [12, 13]. How cytokinin signalling helps promote shoot meristem identity is still largely unknown. However, seedlings with impaired cytokinin signalling in *ARABIDOPSIS RESPONSE REGULATOR* (*ARR*) overexpressing lines have arrested meristem phenotypes, highlighting its importance in meristem development [14].

Another early regulator of shoot meristem activity is the class II KNOX (KNOTTED-LIKE HOMEOBOX) homeodomain transcription factor *SHOOT MERISTEMLESS* (STM). Loss of function *stm* mutants lack meristem formation [15]. Weak *stm* alleles, however, form a meristem that arrests shortly after initiation

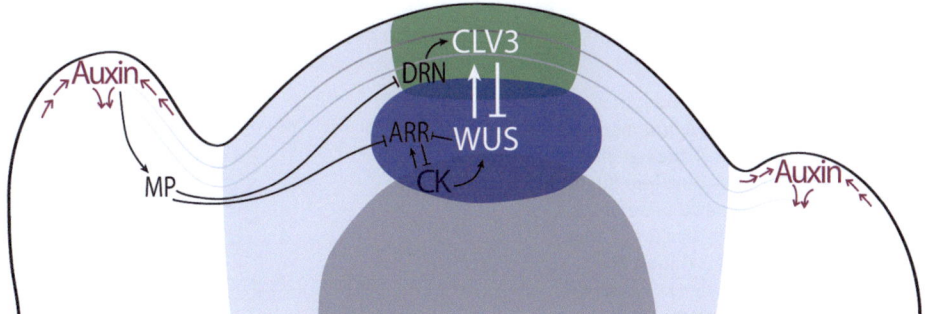

◘ Fig. 7.3 Schematic representation of the *Arabidopsis thaliana* shoot apical meristem. The CZ (green) contains the CLAVATA3 expressing stem cells. CLV3 moves to the OC below it (blue) and inhibits the WUSCHEL transcription factor. WUS in turn moves to the cells above it, promoting stem cell identity maintenance and activity via the expression of *CLV3*. The two side bulges represent organ primordia, where cells enter differentiation programs. Phytohormone signalling and transcriptional network-mediated cross-talk between the meristem and the organ primordia ensures that a stable pool of stem cells is maintained, and meristem size remains constant. ARR, ARABIDOPSIS RESPONSE REGULATOR. CK, cytokinin. DRN, DORNROSCHEN. MP, MONOPTEROS. See main text for details

[16] suggesting that STM may not be required for stem cell specification in the embryo but rather to maintain meristematic activity levels at later stages. Accordingly, *CLV3* expression is not completely lost in *stm* mutants [11].

7.3.2 Root Apical Meristem Initiation

The root apical meristem is first evident at the mid-globular stage with the asymmetric division of the hypophysis which produces QC precursor cells, expressing the future QC marker *WOX5* [17]. Upstream of *WOX5* expression, an intricate network of transcription factors and auxin-dependent signalling, specifies the root stem cell niche.

WOX5 expression is mediated by auxin signalling via the AUXIN RESPONSE FACTORS, ARF10 and ARF16 [18]. Asymmetric distribution of auxin efflux carrier PIN-FORMED1 (PIN1) creates an auxin flow that results in a response maximum in the uppermost cell of the suspensor as detected by the auxin response reporter DR5 [8]. Concomitantly, the auxin response factor MONOPTEROS (MP or ARF5) mediates auxin signalling in the basal embryo region, inducing the expression of *PLETHORA* (*PLT*) genes [19]. *PLT* genes encode APETALA2-type transcription factors that feedback on PIN1 distribution to reinforce the auxin distribution pattern [20]. In the incipient embryonic root region, PLT transcription factors specify root stem cell identity [19] and pattern the meristem zones through protein gradients that inhibit differentiation and promote cell proliferation [21].

In a parallel pathway, the GRAS-type transcription factors SHORTROOT (SHR) and SCARECROW (SCR) help pattern the root radial organization specifying QC centre activity and stem cell identity [22, 23]. SHR and SCR induce expression of *WOX5* in the QC [18] which may partially account for their roles in the specification of both QC activity and, non-cell autonomously, stem cell identity in the surrounding cells (◘ Fig. 7.4) [23].

A third set of genes may play a role in root meristem specification by regulating early embryo patterning. Studies of *POLTERGEIST* (*POL*) and *POLTERGEIST-LIKE1* (*PLL1*) double mutants show a role in early embryo cell division and apical-

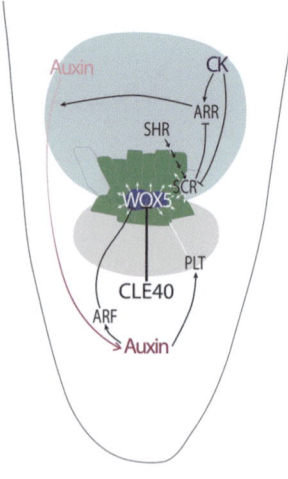

◘ **Fig. 7.4** Schematic representation of *Arabidopsis thaliana* root apical meristem (RAM) depicting some of the signalling pathways involved in its function. Stem cells (green) are maintained undifferentiated and proliferating by positional cues from the quiescent centre (blue). The stem cell niche receives signals from both proximal and distal meristem regions (grey areas), and differentiating tissues, and feeds back to them to coordinate proliferation and differentiation. ARF, AUXIN RESPONSE FACTOR. ARR, ARABIDOPSIS RESPONSE REGULATOR. CK, cytokinin. PLT, PLETHORA. SCR, SCARECROW. SHR, SHORTROOT. See main text for details

basal axis patterning [24]. POL and PLL1 activity is required for *WOX5* expression and loss of meristem in *pol pll1* mutants is accompanied by the loss of *SCR* and *SHR* expression but not *PLT* [24].

7.4 Stem Cell Maintenance

Although stem cells are constantly dividing to produce new organs and tissues, meristems remain at a near-constant size throughout plant development. To achieve meristem maintenance, plants must coordinate cell proliferation in the meristem niche with cell differentiation outside of it. By coordinating these two events, plants ensure a balance between the cells that remain in the meristem niche and those that leave to join organ and tissue formation.

Two main pathways are known to regulate the balance between stem cell maintenance and differentiation across meristems in plants. Central to stem cell specification and maintenance is a transcriptional and signalling feedback loop between members of the CLE (CLAVATA3/EMBRYO SURROUNDING REGION-RELATED) and WOX families. Coordinating stem cell proliferation at the centre of the meristem and cell differentiation at its periphery involves the action of phytohormones auxin and cytokinin. Crosstalk between transcriptional regulators and hormone signalling pathways integrates all the information into one coherent network.

7.4.1 Stem Cell Maintenance in the Shoot Apical Meristem

As mentioned above, the CLE–WOX feedback loop, represented here by *CLV3* and *WUS*, is a key part of the regulatory mechanisms involved in the maintenance of a stable pool of stem cells in the shoot apical meristem (Fig. 7.3). Mutations in *WUS* and *CLV3* signalling demonstrate how this dynamic regulatory loop maintains a stable pool of proliferating stem cells. *wus* mutants have a strong reduction or complete loss of the stem cell pool, illustrating WUS requirement for the maintenance of stem cell identity [25]. *clv3* mutants have an enlarged stem cell pool, suggesting that CLV3 regulates stem cell proliferation promoting differentiation [26]. *WUS* and *CLV3* have non-overlapping expression domains but their activities affect the expression of one another which led to the proposal of a non-cell autonomous feedback mechanism [9]. The details of this loop have been worked out over the last two decades and are briefly detailed below.

WUS encodes a homeodomain transcription factor that is expressed in the OC [9] and migrates to the CZ above it via cell–cell connections called plasmodesmata [27, 28]. In the CZ, the non-cell autonomous WUS signal induces the expression of *CLV3* and promotes the maintenance of an undifferentiated proliferative state [9, 29].

Under the influence of WUS, *CLV3* is exclusively expressed in the CZ where it produces a short mobile peptide that acts non-cell autonomously, diffusing away from the stem cells and into the OC below [10]. In the OC, the CLV3 peptide is perceived by a combination of receptor-like kinases triggering a signalling cascade that results in the repression of WUS [29–31]. Thus, with two elements, WUS promotion of CLV3 and CLV3 repression of WUS, the CLV3-WUS loop provides a self-regulatory mechanism for meristem size maintenance.

The signalling cascade triggered by CLV3 signalling has not been fully clarified but several of the leucine-rich-repeat (LRR) receptors and receptor-like kinases (RLK) involved are known. Among the better characterized receptors involved in CLV3 signalling is CLV1, a LRR-RLK [10, 32, 33] and the LRR/pseudo-kinase dimer CLV2/CORYNE (CLV2/CRN) [34–36]. Other members of the CLV1 family of RLK such as BARELY ANY MERISTEM1-3 (BAM1-3) and the more distantly related RECEPTOR-LIKE PROTEIN KINASE 2 (RPK2) also play a role in meristem activity and size maintenance [37–39].

One reason for our incomplete understanding of this intricate network of receptors is that the relationships between the different components are complex. For example, CLV1 and BAM1 directly bind CLV3 but the CLV2/CRN dimer and RPK2, although capable of integrating CLV3 signalling, do not [40, 41]. In addition, genetic redundancy and feedback mechanisms that add to the robustness of this system can also mask the functions of individual actors. For instance, CLV3 binding to CLV1 regulates this receptor availability at the membrane inducing its trafficking and degradation [33], while CLV1 activity negatively regulates *BAM* expression [42, 43].

The signalling cascade downstream of CLV3 that results in the reduction of WUS also remains poorly understood with only a few identified elements, like the protein phosphatases POL and PLL1. While POL and PLL1 are required for *WUS* expression, CLV1 signalling inhibits POL and PLL1 activity, thus providing one explanation for how CLV signalling might regulate WUS [44]. Additional factors contributing to the activity of WUS in meristem maintenance and activity are the KNOX transcription factor STM and the GRAS transcription factors HAIRY MERISTEM (HAM). STM contributes to meristem maintenance by suppressing differentiation and inducing cell proliferation [45]. Both WUS and STM can ectopically induce meristematic identity, including cell proliferation and eventual organ initiation [46]. While WUS and STM have similar roles in suppressing cell differentiation, STM has a broader effect maintaining an undifferentiated state of cells in the peripheral zone as well as the central zone [47]. HAM transcription factors act in parallel with WUS and STM to promote the maintenance of undifferentiated cells in the CZ [48, 49]. Despite having an effect in stem cell maintenance in the CZ, HAM transcription factors are expressed in the deeper layers of the meristem, likely acting in a non-cell autonomous way like WUS [49]. Indeed, WUS and HAM share a number of targets with roles in cell division and stem cell identity maintenance [50].

The events downstream of WUS activity in the OC and CZ are not entirely elucidated. Nevertheless, two genome-wide studies of transcriptional targets provide insight into the mechanisms of WUS regulatory activity. Notably, they show that WUS works as a repressor, limiting the expression of differentiation factors like *ASYMMETRIC LEAVES 2* and *KANADI*, but also promoting the expression of shoot stem cell specification factors such as *TOPLESS* [51, 52].

7.4.2 Stem Cell Maintenance in the Root Apical Meristem

Stem cell maintenance in the RAM reiterates the basic elements of the CLE–WOX feedback loop found in the shoot, while also echoing some of the stem cell initiation mechanisms reviewed in ▶ Sect. 7.3.2.

In the root, CLE–WOX clades are represented by *WOX5* and *CLE40* (◘ Fig. 7.4). Like WUS, WOX5 marks the niche that induces stem cells, which in the root consists of the mitotically inactive quiescent centre. Similarly, WOX5 acts non-cell autonomously moving to the adjacent cells to promote the maintenance of an undifferentiated state [18]. On the other side of the loop, the CLE-related mobile peptide CLE40 is produced in differentiated root cap cells. CLE40 is perceived by CLV1 and ARABIDOPSIS CRINKLY 4 (ACR4) in both the initials and the differentiated cells in the distal meristem, restricting *WOX5* expression to the QC [53]. The CLV2/CRN complex is also expressed in the root and mediates CLE40 signalling in the proximal meristem [36]. Likewise, BAM1 and RPK2 mediate CLE signalling in the root to regulate cell proliferation [54].

One regulator of *WOX5* expression is REPRESSOR OF WUSCHEL 1 (ROW1), a plant homeodomain protein involved in transcriptional regulation via chromatin modification [55]. ROW1 was first identified in the shoot apical meristem where it is required to repress WUS [56]. In the root, ROW1 restricts *WOX5* expression to the QC, repressing it in the proximal meristem [55].

In parallel to the CLE40-WOX5 loop, root stem cells are maintained by the PLT and SHR/SCR pathways initiated during embryonic development. The auxin response maximum at the root tip induces the expression of *PLT* in the QC [20]. PLT proteins diffuse via cell–cell connections and form a gradient that tails towards the transition zone and patterns the meristem in a dose-dependent manner [57]. High PLT levels maintain stem cell activity while lower levels allow cell differentiation [21]. The two pathways are linked by WOX5, which upregulates *PLT* expression in distal stem cell activity maintenance linking the two pathways [58].

Recapitulating its role in the embryo, in the mature root, *SHR* is expressed in the stele and moves to the QC inducing the expression of *SCR* [22, 59]. SCR in turn prevents SHR from moving to additional cell layers by sequestering it in the endodermis initials [60]. SHR and SCR maintain stem cell activity non-cell autonomously, partly by regulating cell-cycle genes [23, 61]. Downstream of SHR/SCR, the zinc-finger proteins MAGPIE and JACKDAW help maintain *SHR* expression domains and preserve asymmetric cell division in the stem cell niche [62].

7.4.3 The Role of Hormones in Stem Cell Maintenance in Plant Meristems

In the SAM, the CLV3-WUS network interacts with cytokinin signalling to regulate meristem activity. Cytokinin acts at the centre of the meristem inhibiting cell differentiation, and promoting cell proliferation and meristematic maintenance of stem cells via WUS activation [63, 64]. WUS in turn represses the expression of *ARR* family members [14]. ARR proteins are negative regulators of cytokinin signalling, thus producing a positive feedback on *WUS*' own expression. The combinatorial effects of cytokinin and CLV3-WUS signalling in the apex are sufficient to explain the WUS domain position in the shoot apical meristem in a minimal computational model [65]. Auxin acts oppositely at the periphery of the meristem, inducing organ primordia initiation and cell differentiation [66]. Like the cross-talk in the CZ and OC mediated by CLV3 and WUS, the central zone and peripheral zone hormone signalling pathways also feedback on each other. Notably, MP-mediated auxin signalling represses ARR members, promoting cytokinin response and *WUS* expression [64]. In addition, MP-mediated auxin signalling directly represses the expression of APETALA2 type factor *DORNROSCHEN (DRN)* [67]. *DRN* is expressed in the central zone where it promotes *CLV3* expression and stem cell maintenance [67, 68].

In the RAM, the auxin/cytokinin balance also plays a role in maintaining a balance between cell division and cell differentiation. However, in the RAM their roles are reversed, with auxin inducing stem cell positioning and cytokinin promoting cell differentiation [69]. Auxin polar transport and biosynthesis in the root creates a response maximum in the distal meristem and a gradient that tails towards the differentiation zone. The auxin response maximum is required for QC maintenance [20], while ARF-mediated signalling in the distal meristem promotes differentiation of root tip cells and limits *WOX5* expression to the QC [58]. The auxin gradient helps guide root patterning [70] with the auxin minimum defining the boundary between proliferation and differentiation zones [71]. Cytokinin accumulation in the transition zone promotes cell differentiation [72] and inhibits stem cell identity genes *SCR* and *WOX5* [73]. Cytokinin and auxin interact in a context-dependent manner to achieve a dual balance. In the proximal meristem, SCR inhibits cytokinin signalling factor ARR1. ARR1 in turn stimulates auxin biosynthesis in the stem cell niche, tipping the balance towards auxin activity and stem cell maintenance [74]. In the transition zone, however, *ARR1* expression in turn represses auxin accumulation via the PIN1 transporter, tipping the balance towards cytokinin [74].

Other phytohormone groups like gibberellins and brassinosteroids have also been shown to promote root meristem activity and root growth by regulating cell proliferation in the stem cell niche, but their mechanisms in this context are still poorly known [75, 76].

7.4.4 Conservation of Meristem Maintenance Mechanisms in Plants

There are a number of elements that suggest the fundamental mechanisms that regulate stem cell activity in plants are conserved. In addition to being present in both apical meristems, the CLE and WOX family members share deeper functional homologies. WUS and WOX5, for example, can replace each other's functions in shoot and root stem cell maintenance [18], while CLE40 can substitute for CLV3 in the shoot meristem and CLV3 can promote distal meristem differentiation in the root [53, 77]. Furthermore, the same overlapping network of receptor-like kinases is active in CLV signalling in both the shoot and the root meristems [36, 42]. The CLE–WOX module is further conserved in the maintenance and differentiation of the vascular meristem (cambium). Like WUS, WOX4 maintains cell proliferation in the cambium. However, unlike CLV3 and CLE40, CLE41/44 signalling induces the expression of *WOX4* rather than repressing it [78]. In addition, CLE41 and CLE44 signal through a novel LRR-RLK called PHLOEM INTERCALATED WITH XYLEM (PXY) [79]. Like other LRR-RLKs, PXY mediates *WOX4* expression through an as yet unknown mechanism [79].

The same two hormonal pathways are also involved in the specification and maintenance of meristem activity in different contexts with interesting implications. While auxin and cytokinin signalling have opposing effects in the meristem context, their roles are reversed from one meristem to the other. Although not yet understood, this reversal of roles could be part of a long-distance communication mechanism between the two main meristems. Meristem activity in root and shoot could thus be coordinated in response to overall auxin/cytokinin levels. Such a mechanism could allow a balanced growth of aerial and underground tissues and a quick integrated response to environmental changes.

> **Take-Home Message**
> - Plant development occurs mostly post-embryonically in an indeterminate fashion, with nearly constant growth and production of new organs.
> - The indeterminate growth mode of plants is possible due to the continuous activity of plant stem cells which remain active throughout the plant's life.
> - Plant stem cells are organized into specialized cell niches called the meristems.
> - Meristems are responsible for maintaining a stable pool of stem cells, coordinating cell proliferation and cell differentiation.
> - Two key signalling pathways are involved in meristem patterning and activity, including the specification of stem cell identity and the promotion of cell proliferation.
> - The feedback loop between CLE and WOX family members ensures communication between stem cells and the organizing or quiescent centre, creating a self-regulatory mechanism that maintains a stable pool of stem cells.
> - The phytohormones auxin and cytokinin keep a balance between cell proliferation in the centre of the meristems and cell differentiation at the periphery.

Acknowledgments This summary of literature on plant meristems was compiled with the best of intentions in 2019. Apologies to all colleagues whose work was not included due to time and space constraints.

References

1. Clark SE. Organ formation at the vegetative shoot meristem. Plant Cell. 1997;9(7):1067–76.
2. Laufs P, Grandjean O, Jonak C, Kiêu K, Traas J. Cellular parameters of the shoot apical meristem in Arabidopsis. Plant Cell. 1998;10(August):1375–90.
3. Dolan L, Janmaat K, Willemsen V, Linstead P, Poethig S, Roberts K, et al. Cellular organisation of the Arabidopsis thaliana root. Development. 1993;119(1):71–84.
4. van den Berg C, Willemsen V, Hendriks G, Weisbeek P, Scheres B. Short-range control of cell differentiation in the Arabidopsis root meristem. Nature. 1997;390(6657):287–9.
5. Boscá S. Embryonic development in Arabidopsis thaliana: from the zygote division to the shoot meristem. Front Plant Sci. 2011;2(December):1–6.
6. ten Hove CA, Lu K-J, Weijers D. Building a plant: cell fate specification in the early Arabidopsis embryo. Development. 2015;142(3):420–30.
7. Skoog F, Miller CO. Chemical regulation of growth and organ formation in plant tissues cultured in vitro. Symp Soc Exp Biol England. 1957;11:118–30.
8. Friml J, Vieten A, Sauer M, Weijers D, Schwarz H, Hamann T, et al. Efflux-dependent auxin gradients establish the apical-basal axis of Arabidopsis. Nature. 2003;426(6963):147–53.

9. Mayer KFX, Schoof H, Haecker A, Lenhard M, Jürgens G, Laux T. Role of WUSCHEL in regulating stem cell fate in the Arabidopsis shoot meristem. Cell. 1998;95(6):805–15.
10. Fletcher JC, Brand U, Running MP, Simon R, Meyerowitz EM. Signaling of cell fate decisions by CLAVATA3 in Arabidopsis shoot meristems. Science. 1999;19(5409):1911–4.
11. Brand U, Grunewald M, Hobe M, Simon R. Regulation of CLV3 expression by two Homeobox genes in Arabidopsis. Plant Physiol. 2002;129(2):565–75.
12. Zhang Z, Tucker E, Hermann M, Laux T. A molecular framework for the embryonic initiation of shoot meristem stem cells. Dev Cell. 2017;40(3):264–277.e4.
13. Breuninger H, Rikirsch E, Hermann M, Ueda M, Laux T. Differential expression of WOX genes mediates apical-basal Axis formation in the Arabidopsis embryo. Dev Cell. 2008;14(6):867–76.
14. Leibfried A, To JPC, Busch W, Stehling S, Kehle A, Demar M, et al. WUSCHEL controls meristem function by direct regulation of cytokinin-inducible response regulators. Nature. 2005;438(7071):1172–5.
15. Barton MK, Poethig RS. Formation of the shoot apical meristem in Arabidopsis thaliana: an analysis of development in the wild type and in the shoot meristemless mutant. Development. 1993;119:823–31.
16. Endrizzi K, Moussian B, Haecker A, Levin JZ, Laux T. The SHOOT MERISTEMLESS gene is required for maintenance of undifferentiated cells in Arabidopsis shoot and floral meristems and acts at a different regulatory level than the meristem genes WUSCHEL and ZWILLE. Plant J. 1996;10(6):967–79.
17. Haecker A, Groß-Hardt R, Geiges B, Sarkar A, Breuninger H, Herrmann M, et al. Expression dynamics of WOX genes mark cell fate decisions during early embryonic patterning in Arabidopsis thaliana. Development. 2004;131(3):657–68.
18. Sarkar AK, Luijten M, Miyashima S, Lenhard M, Hashimoto T, Nakajima K, et al. Conserved factors regulate signalling in Arabidopsis thaliana shoot and root stem cell organizers. Nature. 2007;446(7137):811–4.
19. Aida M, Beis D, Heidstra R, Willemsen V, Blilou I, Galinha C, et al. The PLETHORA genes mediate patterning of the Arabidopsis root stem cell niche. Cell. 2004;119(1):119–20.
20. Blilou I, Xu J, Wildwater M, Willemsen V, Paponov I, Frimi J, et al. The PIN auxin efflux facilitator network controls growth and patterning in Arabidopsis roots. Nature. 2005;433(7021):39–44.
21. Galinha C, Hofhuis H, Luijten M, Willemsen V, Blilou I, Heidstra R, et al. PLETHORA proteins as dose-dependent master regulators of Arabidopsis root development. Nature. 2007;449(7165):1053–7.
22. Helariutta Y, Fukaki H, Wysocka-Diller J, Nakajima K, Jung J, Sena G, et al. The SHORT-ROOT gene controls radial patterning of the Arabidopsis root through radial signaling. Cell. 2000;101(5):555–67.
23. Sabatini S, Heidstra R, Wildwater M, Scheres B. SCARECROW is involved in positioning the stem cell niche in the Arabidopsis root meristem. Genes Dev. 2002;17(3):354–8.
24. Song SK, Hofhuis H, Lee MM, Clark SE. Key divisions in the early Arabidopsis embryo require POL and PLL1 phosphatases to establish the root stem cell organizer and vascular axis. Dev Cell. 2008;15(1):98–109.
25. Laux T, Mayer KF, Berger J, Jürgens G. The WUSCHEL gene is required for shoot and floral meristem integrity in Arabidopsis. Development. 1996;122(1):87–96.
26. Clark SE, Running MP, Meyerowitz EM. CLAVATA3 is a specific regulator of shoot and floral meristem development affecting the same processes as CLAVATA1. Development. 1995;121(May):2057–67.
27. Yadav RK, Perales M, Gruel J, Girke T, Jonsson H, Reddy GV. WUSCHEL protein movement mediates stem cell homeostasis in the Arabidopsis shoot apex. Genes Dev. 2011;25(19):2025–30.
28. Daum G, Medzihradszky A, Suzaki T, Lohmann JU. A mechanistic framework for noncell autonomous stem cell induction in *Arabidopsis*. Proc Natl Acad Sci. 2014;111(40):14619–24.
29. Schoof H, Lenhard M, Haecker A, Mayer KFX, Jürgens G, Laux T. The stem cell population of Arabidopsis shoot meristems is maintained by a regulatory loop between the CLAVATA and WUSCHEL genes. Cell. 2000;100(6):635–44.
30. Müller R, Borghi L, Kwiatkowska D, Laufs P, Simon R. Dynamic and compensatory responses of Arabidopsis shoot and floral meristems to CLV3 signaling. Plant Cell. 2006;18(5):1188–98.
31. Brand U, Fletcher JC, Hobe M, Meyerowitz EM, Simon R. Dependence of stem cell fate in Arabidopsis on a feedback loop regulated by CLV3 activity. Science. 2000;289(5479):617–9.

32. Clark SE, Williams RW, Meyerowitz EM. The CLAVATA1 gene encodes a putative receptor kinase that controls shoot and floral meristem size in Arabidopsis. Cell. 1997;89(4):575–85.
33. Nimchuk ZL, Tarr PT, Ohno C, Qu X, Meyerowitz EM. Plant stem cell signaling involves ligand-dependent trafficking of the CLAVATA1 receptor kinase. Curr Biol. 2011;21(5):345–52.
34. Müller R, Bleckmann A, Simon R. The receptor kinase CORYNE of Arabidopsis transmits the stem cell-limiting signal CLAVATA3 independently of CLAVATA1. Plant Cell. 2008;20(4):934–46.
35. Bleckmann A, Weidtkamp-Peters S, Seidel CAM, Simon R. Stem cell signaling in Arabidopsis requires CRN to localize CLV2 to the plasma membrane. Plant Physiol. 2010;152(1):166–76.
36. Somssich M, Bleckmann A, Simon R. Shared and distinct functions of the pseudokinase CORYNE (CRN) in shoot and root stem cell maintenance of Arabidopsis. J Exp Bot. 2016;67(16):4901–15.
37. DeYoung BJ, Clark SE. BAM receptors regulate stem cell specification and organ development through complex interactions with CLAVATA signaling. Genetics. 2008;180(2):895–904.
38. DeYoung BJ, Bickle KL, Schrage KJ, Muskett P, Patel K, Clark SE. The CLAVATA1-related BAM1, BAM2 and BAM3 receptor kinase-like proteins are required for meristem function in Arabidopsis. Plant J. 2006;45(1):1–16.
39. Kinoshita A, Betsuyaku S, Osakabe Y, Mizuno S, Nagawa S, Stahl Y, et al. RPK2 is an essential receptor-like kinase that transmits the CLV3 signal in Arabidopsis. Development. 2010;137(24):4327.
40. Shinohara H, Matsubayashi Y. Reevaluation of the CLV3-receptor interaction in the shoot apical meristem: dissection of the CLV3 signaling pathway from a direct ligand-binding point of view. Plant J. 2015;82(2):328–36.
41. Guo Y, Han L, Hymes M, Denver R, Clark SE. CLAVATA2 forms a distinct CLE-binding receptor complex regulating Arabidopsis stem cell specification. Plant J. 2010;63(6):889–900.
42. Nimchuk ZL, Zhou Y, Tarr PT, Peterson BA, Meyerowitz EM. Plant stem cell maintenance by transcriptional cross-regulation of related receptor kinases. Development. 2015;142(6):1043–9.
43. Nimchuk ZL. CLAVATA1 controls distinct signaling outputs that buffer shoot stem cell proliferation through a two-step transcriptional compensation loop. PLoS Genet. 2017;13(3):e1006681.
44. Song S-K, Lee MM, Clark SE. POL and PLL1 phosphatases are CLAVATA1 signaling intermediates required for Arabidopsis shoot and floral stem cells. Development. 2006;133(23):4691–8.
45. Long JA, Moan EI, Medford JI, Barton MK. A member of the KNOTTED class of homeodomain proteins encoded by the STM gene of Arabidopsis. Nature. 1996;379:66–9.
46. Gallois J-L, Woodward C, Reddy GV, Sablowski R. Combined SHOOT MERISTEMLESS and WUSCHEL trigger ectopic organogenesis in Arabidopsis. Development. 2002;129:3207–17.
47. Lenhard M, Jürgens G, Laux T. The WUSCHEL and SHOOTMERISTEMLESS genes fulfil complementary roles in Arabidopsis shoot meristem regulation. Development. 2002;129:3195–206.
48. Stuurman J, Jäggi F, Kuhlemeier C. Shoot meristem maintenance is controlled by a GRAS-gene mediated signal from differentiating cells. Genes Dev. 2002;16:2213–8.
49. Engstrom EM, Andersen CM, Gumulak-Smith J, Hu J, Orlova E, Sozzani R, et al. Arabidopsis homologs of the Petunia HAIRY MERISTEM gene are required for maintenance of shoot and root indeterminacy. Plant Physiol. 2011;155(2):735–50.
50. Zhou Y, Liu X, Engstrom EM, Nimchuk ZL, Pruneda-Paz JL, Tarr PT, et al. Control of plant stem cell function by conserved interacting transcriptional regulators. Nature. 2015;517(7534):377–80.
51. Busch W, Miotk A, Ariel FD, Zhao Z, Forner J, Daum G, et al. Transcriptional control of a plant stem cell niche. Dev Cell. 2010;18(5):849–61.
52. Yadav RK, Perales M, Gruel J, Ohno C, Heisler M, Girke T, et al. Plant stem cell maintenance involves direct transcriptional repression of differentiation program. Mol Syst Biol. 2013;9:654.
53. Stahl Y, Wink RH, Ingram GC, Simon R. A signaling module controlling the stem cell niche in Arabidopsis root meristems. Curr Biol. 2009;19(11):909–14.
54. Shimizu N, Ishida T, Yamada M, Shigenobu S, Tabata R, Kinoshita A, et al. BAM 1 and RECEPTOR-LIKE PROTEIN KINASE 2 constitute a signaling pathway and modulate CLE peptide-triggered growth inhibition in Arabidopsis root. New Phytol. 2015;208(4):1104–13.
55. Zhang Y, Jiao Y, Liu Z, Zhu YX. ROW1 maintains quiescent centre identity by confining WOX5 expression to specific cells. Nat Commun. 2015;6:6003.
56. Han P, Li Q, Zhu Y-X. Mutation of Arabidopsis BARD1 causes meristem defects by failing to confine WUSCHEL expression to the organizing center. Plant Cell. 2008;20(6):1482–93.
57. Mähönen AP, Ten Tusscher K, Siligato R, Smetana O, Díaz-Triviño S, Salojärvi J, et al. PLETHORA gradient formation mechanism separates auxin responses. Nature. 2014;515(7525):125–9.

58. Ding Z, Friml J. Auxin regulates distal stem cell differentiation in Arabidopsis roots. Proc Natl Acad Sci. 2010;107(26):12046–51.
59. Nakajima K, Sena G, Nawy T, Benfey PN. Intercellular movement of the putative transcription factor SHR in root patterning. Nature. 2001;413(6853):307–11.
60. Cui H, Levesque MP, Vernoux T, Jung JW, Paquette AJ, Gallagher KL, et al. An evolutionarily conserved mechanism delimiting SHR movement defines a single layer of endodermis in plants. Science. 2007;316(5823):421–5.
61. Sozzani R, Cui H, Moreno-Risueno MA, Busch W, Van Norman JM, Vernoux T, et al. Spatiotemporal regulation of cell-cycle genes by SHORTROOT links patterning and growth. Nature. 2010;466(7302):128–32.
62. Welch D, Hassan H, Blilou I, Immink R, Heidstra R, Scheres B. Arabidopsis JACKDAW and MAGPIE zinc finger proteins delimit asymmetric cell division and stabilize tissue boundaries by restricting SHORT-ROOT action. Genes Dev. 2007;21(17):2196–204.
63. Gordon SP, Chickarmane VS, Ohno C, Meyerowitz EM. Multiple feedback loops through cytokinin signaling control stem cell number within the Arabidopsis shoot meristem. Proc Natl Acad Sci. 2009;106(38):16529–34.
64. Zhao Z, Andersen SU, Ljung K, Dolezal K, Miotk A, Schultheiss SJ, et al. Hormonal control of the shoot stem-cell niche. Nature. 2010;465(7301):1089–92.
65. Chickarmane VS, Gordon SP, Tarr PT, Heisler MG, Meyerowitz EM. Cytokinin signaling as a positional cue for patterning the apical-basal axis of the growing Arabidopsis shoot meristem. Proc Natl Acad Sci. 2012;109(10):4002–7.
66. Reinhardt D, Mandel T, Kuhlemeier C. Auxin regulates the initiation and radial position of plant lateral organs. Plant Cell. 2000;12(4):507–18.
67. Luo L, Zeng J, Wu H, Tian Z, Zhao Z. A molecular framework for auxin-controlled homeostasis of shoot stem cells in Arabidopsis. Mol Plant. 2018;11(7):899–913.
68. Kirch T, Simon R, Grunewald M, Werr W. Dornröschen/enhancer of shoot regeneration. Plant Cell. 2003;15(March):694–705.
69. Dello Ioio R, Nakamura K, Moubayidin L, Perilli S, Taniguchi M, Morita MT, et al. A genetic framework for the control of cell division and differentiation in the root meristem. Science. 2008;322(5906):1380–4.
70. Grieneisen VA, Xu J, Marée AFM, Hogeweg P, Scheres B. Auxin transport is sufficient to generate a maximum and gradient guiding root growth. Nature. 2007;449(7165):1008–13.
71. Di Mambro R, De Ruvo M, Pacifici E, Salvi E, Sozzani R, Benfey PN, et al. Auxin minimum triggers the developmental switch from cell division to cell differentiation in the *Arabidopsis* root. Proc Natl Acad Sci. 2017;114(36):E7641–9.
72. Dello Ioio R, Linhares FS, Scacchi E, Casamitjana-Martinez E, Heidstra R, Costantino P, et al. Cytokinins determine Arabidopsis root-meristem size by controlling cell differentiation. Curr Biol. 2007;17(8):678–82.
73. Zhang W, Swarup R, Bennett M, Schaller GE, Kieber JJ. Cytokinin induces cell division in the quiescent center of the arabidopsis root apical meristem. Curr Biol. 2013;23(20):1979–89.
74. Moubayidin L, DiMambro R, Sozzani R, Pacifici E, Salvi E, Terpstra I, et al. Spatial coordination between stem cell activity and cell differentiation in the root meristem. Dev Cell. 2013;26(4):405–15.
75. Ubeda-Tomás S, Federici F, Casimiro I, Beemster GTS, Bhalerao R, Swarup R, et al. Gibberellin signaling in the endodermis controls Arabidopsis root meristem size. Curr Biol. 2009;19(14):1194–9.
76. Gonzalez-Garcia M-P, Vilarrasa-Blasi J, Zhiponova M, Divol F, Mora-Garcia S, Russinova E, et al. Brassinosteroids control meristem size by promoting cell cycle progression in Arabidopsis roots. Development. 2011;138(5):849–59.
77. Hobe M, Müller R, Grünewald M, Brand U, Simon R. Loss of CLE40, a protein functionally equivalent to the stem cell restricting signal CLV3, enhances root waving in Arabidopsis. Dev Genes Evol. 2003;213(8):371–81.
78. Hirakawa Y, Shinohara H, Kondo Y, Inoue A, Nakanomyo I, Ogawa M, et al. Non-cell-autonomous control of vascular stem cell fate by a CLE peptide/receptor system. Proc Natl Acad Sci. 2008;105(39):15208–13.
79. Hirakawa Y, Kondo Y, Fukuda H. TDIF peptide signaling regulates vascular stem cell proliferation via the WOX4 Homeobox gene in Arabidopsis. Plant Cell. 2010;22(8):2618–29.

Axial Stem Cells and the Formation of the Vertebrate Body

André Dias and Rita Aires

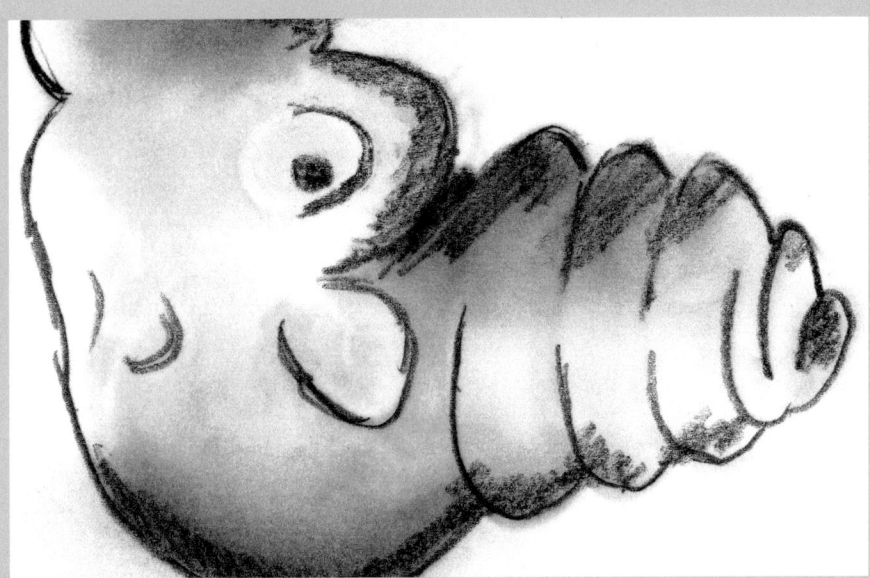

Contents

8.1 From a Fertilized Egg to the Gastrulating Embryo – 133
8.1.1 Early Postimplantation Development – 133
8.1.2 Gastrulation – 134

8.2 Axial Extension and Formation of the Vertebrate Body – 137
8.2.1 The Axial Stem Cell Niche – 138

André Dias and Rita Aires contributed equally to this chapter.

© Springer Nature Switzerland AG 2020
G. Rodrigues, B. A. J. Roelen (eds.), *Concepts and Applications of Stem Cell Biology*,
Learning Materials in Biosciences, https://doi.org/10.1007/978-3-030-43939-2_8

8.2.2	Primary and Secondary Body Formation – 138	
8.2.3	Axial Progenitor Cells – 139	

8.3 Molecular Mechanisms Controlling Axial Progenitors – 140
8.3.1 Maintenance of the Axial Stem Cell Niche – 140
8.3.2 Fate Determination in Axial Progenitors – 143
8.3.3 Regional Specification of Axial Progenitors – 145
8.3.4 Extinction of NMPs and the Cessation of Axial Extension – 146

8.4 Mechanisms of Axial Extension and the Evolution of Body Shape Diversity – 146

8.5 In Vitro Generation of NMPs – 148
8.5.1 ESCs-Derived NMPs – 148
8.5.2 EpiSCs-Derived NMPs – 149
8.5.3 Fate Determination of in Vitro–Generated NMPs – 150
8.5.4 Derivation of Human NMPs – 150

8.6 Self-Organization of Stem Cells into Embryo-Like Structures – 151

8.7 Online Resources/Protocols – 152

References – 153

8

Axial Stem Cells and the Formation of the Vertebrate Body

Summary
Development of a whole multicellular complex organism from a single cell is not only an evolutionary triumph, but also the most daunting and formidable of tasks. The organism's entire body plan is laid down in a series of intricate and interconnected events that comprise various levels of organization, from intracellular processes to vast morphogenetic tissue movements. This means that the embryo's early symmetries must be gradually broken and that most of the initial cell potency needs to be progressively lost so that the body can increase in complexity and, ultimately, achieve its final form. In this chapter, using the mouse embryo as the chief model organism, we will address the formation of the vertebrate embryo from the perspective of the axial progenitor cells that are responsible for generating and patterning the tissues that will compose the postoccipital body structures.

What Will You Learn in This Chapter?
This chapter was designed to provide an overview of axial stem cells and their role in the formation of the vertebrate body. We begin by presenting a small overview of early embryonic development, with a particular emphasis on gastrulation and formation of the primitive streak. Next, we will focus on the axial progenitor cells that have stem cell-like properties and are responsible for making the different tissues that compose the postoccipital regions of the vertebrate body. We will then revisit old concepts, related to the formation of the vertebrate body axis, and combine them with the new ideas and advances in this field of developmental biology. We will also explore the regions where axial stem cells reside and dissect the molecular mechanisms that control their self-renewal, fate, and extinction. Finally, we will explore how these cells can be derived in vitro from several sources, as well as their potential future use.

8.1 From a Fertilized Egg to the Gastrulating Embryo

8.1.1 Early Postimplantation Development

In mammals, after fertilization, the egg undergoes successive rounds of cell division without growth, reaching the eight-cell stage. Then, after a process of compaction, which will result in a solid mass called morula, the embryo undergoes additional cleavages and cell fate decisions until it reaches the blastocyst stage [1–3] – for more information about these early events see ▶ Chap. 2. After that, in the mouse embryo, around embryonic day (E) 4.0 (4 days post-coitum), implantation occurs as the embryo attaches to the uterine wall and establishes the first maternal-fetal connection [4]. At this stage, it is possible to distinguish two different cell compartments, the trophectoderm and the inner cell mass (ICM). Cells from the trophectoderm will be necessary for implantation and they will give rise to important extraembryonic tissues [1–3]. The ICM contains pluripotent cells that undergo a second lineage separation driven by Fgf signaling, to produce the visceral endoderm (positive for *Gata4/6*) and the epiblast (expressing *Nanog* and *Oct3/4*), which will give rise to the embryo

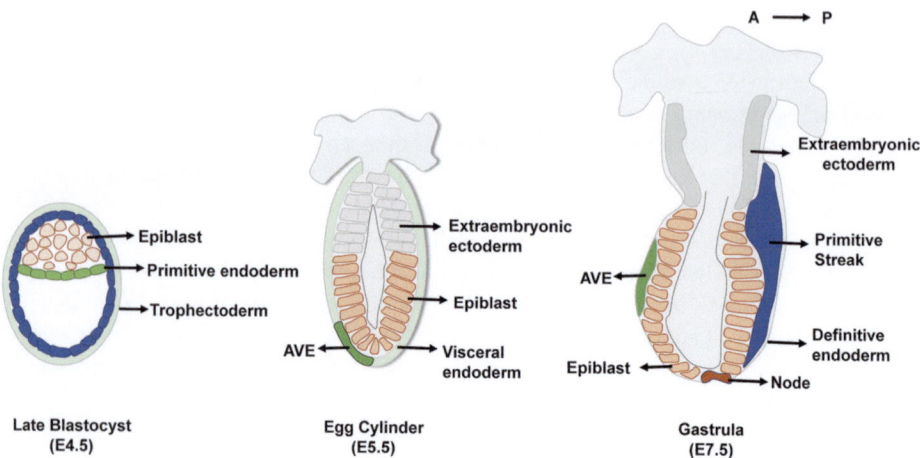

Fig. 8.1 Early postimplantation embryonic development. Schematic representation of postimplantation mouse embryonic development, from the late blastocyst until the gastrula stage, highlighting key structures and tissues. A anterior, AVE anterior visceral endoderm, E embryonic day, P posterior

proper [1–3]. After implantation, the mouse conceptus architecture undergoes dramatic transformations. Particularly, the epiblast changes morphologically to produce the egg cylinder (whereas most mammals develop as a flat structure) at around E5.5 [1, 5]. At this stage, the first signs of an anterior-posterior axis are evident with the formation of the anterior visceral endoderm (AVE), which expresses Nodal and Wnt inhibitors. The AVE derives from the distal visceral endoderm that migrates to the prospective anterior side of the embryo, shortly after being induced at the distal end of the embryo [1, 5]. At this stage, the molecular mechanisms controlling the anterior-posterior patterning of the embryo mostly rely on interactions between β-catenin and *Cripto* [1, 3, 5, 6]. Genetic experiments in the mouse and grafting experiments using other vertebrate model organisms showed that the AVE is involved in two main processes: the production of head structures (especially the forebrain) and the induction and control of primitive streak (PS) formation in the opposite side of the egg cylinder, which will break the radial symmetry in the embryo and will mark the onset of gastrulation [1, 3, 4] (Fig. 8.1).

8.1.2 Gastrulation

Beginning at about E6.0, the mouse embryo starts undergoing gastrulation. This is a process whereby concerted cell proliferation, migration, differentiation, and changes in cell shape and adhesion properties, among other morphogenetic events, will convert the two-layered embryo into a more complex structure composed by three embryonic germ layers: ectoderm, mesoderm, and definitive endoderm [1, 7]. Each of these layers will give rise to specific types of tissues. The ectoderm or the "outer layer" will generate the animal's epidermis and nervous system, whereas the mesoderm (or the "middle layer") will provide the skeletomuscular system, connective tissues, and contribute to the formation of internal organs such as the heart, the kidneys, or the muscular layers of the intestine. Finally, the innermost tissue, the endoderm, will produce the epithelial lining of the gut and respiratory system, besides playing a major part in the formation of digestive organs such as the liver and the pancreas [8]. During gastrulation, the PS is formed in the posterior epiblast region and will remain there until around E9.0 [1]. Formation of the PS involves a variety of cellular and molecular interactions. *Nodal* (see ▶ Box 8.1) is responsible for sorting epiblast cells towards the posterior part of the embryo to generate the PS [9, 10]. Canonical Wnt signaling through Wnt3 is also required for PS induction, maintenance, and for the transcriptional activation of *Brachyury* (T) in the newly formed mesoderm [11–13]. PS induction also requires the expression of Wnt and Nodal inhibitors from the AVE to concentrate Nodal and Wnt/β-catenin signaling in the posterior epiblast. Accordingly, loss of these inhibitors (e.g., Cerl1 and Lefty1) resulted in the production of ectopic/enlarged PS [1]. After PS formation, epiblast cells will then undergo an epithelial to mesenchymal transition (EMT) as they ingress through this new structure and will give rise to the mesoderm and definitive endoderm [14]. This process requires Fgf signaling (e.g., *Fgf8*), as its inactivation resulted in an accumulation of cells in the epiblast [15]. Similarly, absence of *Crumbs2* also resulted in trapped Sox2$^+$ cells in the PS [16]. This newly formed structure in the epiblast is thus characterized by a gene expression profile composed of genes belonging to or responding to these signaling pathways, including *Nodal, Wnt3, Wnt3a, Axin2, Lefty2, Fgf8, T,* and various others like *Snail, Crumbs2,* and *Sp5* [4, 17–20]. Epiblast cells located anterior to the newly formed PS are not affected by its activity and therefore remain within the epiblast layer, eventually giving rise to the ectoderm [1].

Box 8.1   Important Genes and Signalling Pathways Discussed in This Chapter

Axin2

Axin2 belongs to the AXIN family of proteins and is known to play important roles in the regulation of β-catenin stability involved in the canonical Wnt signalling pathway.

Cdx genes

Cdx homeobox genes belong to the *ParaHox* gene cluster and encode several important transcription factors that control axial patterning.

Cerl1

This gene encodes a cytokine and functions as an antagonist of the TGF-βfamily. It is involved in vertebrate head and heart induction and in the formation and patterning of the PS.

Cyp26a1

Cyp26a1 is one of the cytochrome P450 enzymes known to degrade retinoic acid, thus limiting its activity in a tissue-, time-, and dose-specific manner.

Cripto

This gene encodes an epidermal growth factor related protein that is essential to promote Nodal signalling pathway. *Cripto* is required for germ-layer formation and the correct positioning of the anterior-posterior axis.

Fgf signalling

Fibroblast growth factors (Fgfs) are extracellular signalling peptides that are known to play key roles in a variety of biological processes, including embryonic development. *Fgf4* and *Fgf8* are specifically involved in the regulation of cell proliferation and differentiation (e.g., limb development).

Gata genes

These genes are members of a specific family of zinc-finger transcription factors characterized by their ability to bind to the DNA sequence "GATA." *Gata 4* and *Gata 6* are important during embryogenesis, as they play key roles in regulating cellular differentiation.

Gdf11

Gdf11 is a secreted factor member of the transforming growth factor-β superfamily of signalling molecules. It binds to the TGF-β Receptor I (TGFβRI) (also known as *Alk5*) and activin type II receptors ActRIIA or ActRIIB. *Alk5* then activates *Smad2* and/or *Smad3* that, in turn, control gene expression upon interacting with *Smad4* and entering the nucleus.

Hox genes

Hox genes constitute a large family of genes encoding homeodomain-containing transcription factors. Mammalian genomes typically include 39 *Hox* genes, which are organized in four chromosomal clusters (identified from A to D). Each gene in one cluster has equivalent genes, or paralogues, in one or more of the other clusters, occupying relative similar positions within them. *Hox* genes are classified in 13 such groups, designated as paralogue groups (PGs). *Hox* gene inactivation typically causes homeotic transformations, whereby one body segment is converted into the identity of another.

Lefty genes

Lefty genes encode antagonists of *Nodal* activity, which prevents the interaction between *Nodal* and its receptors. *Lefty-1* is crucial during gastrulation, confining Nodal activity to the future PS region. It is also involved in the establishment of left-right asymmetry.

Lin28 genes

Lin28a and *Lin28b* encode RNA-binding proteins with important roles in the maturation of several microRNAs. They have an essential role in organismal growth and metabolism, tissue development, somatic reprogramming, and cancer.

Mesogenin1

Mesogenin1 is a key regulator of paraxial (but not dorsal) [MOU6] presomitic mesoderm formation and differentiation, possibly through the regulation of T-box transcription factors expression.

Nanog

Nanog is a DNA binding homeobox transcription factor that plays key roles in the regulation of cell proliferation, renewal, and pluripotency.

Nodal

Nodal is a signalling molecule of the TGF-β superfamily. Binding of Nodal to specific surface receptors triggers a signalling pathway involving recruitment and activation of the SMAD transcription factors family. It plays key roles in PS formation and in the establishment of left-right symmetry.

Oct4 (Pou5f1)

Embryonic stem cells are governed by a core of key transcription factors including *Oct4*. This gene encodes a protein essential for pluripotency and self-renewal properties during embryonic development. Additionally, it has recently been shown to play a key role controlling vertebrate trunk length.

Raldh2

This protein belongs to the aldehyde dehydrogenase family of proteins and it is one of the key enzymes that catalyzes the synthesis of retinoic acid from retinaldehyde.

Sall4

Sall4 is a member of the *Sall* family of zinc-finger transcription factors and is a key regulator of the pluripotency transcriptional network. This protein also has important roles in cell cycle progression and in the lineage commitment of blastocyst cells.

Snail

Snail is a zinc finger transcriptional repressor, known to be involved in the induction of EMT and formation and maintenance of embryonic mesoderm.

Sox2

Sox2 is a member of the SRY-related HMG-box (SOX) family of transcription factors that plays critical roles in a variety of developmental processes. It is involved in maintaining the self-renewal of neural progenitor cells, thus playing a key role in neural tube formation.

Brachyury (T)

The T-box transcription factor Brachyury (T) plays several key roles during embryonic development, particularly in mesoderm formation and differentiation by transcriptional regulation of important mesoderm-associated genes. It is also crucial for notochord development.

Tbx6

This gene is a member of the conserved family of the T-box transcription factors. It is involved in the regulation of mesoderm formation and differentiation and in left/right axis determination.

Wnt signaling

Wnt proteins are involved in key intercellular signaling with multiple crucial roles during embryonic development. Wnt3 regulates AP patterning in the early embryo and PS formation. Wnt3a is necessary for mesoderm production, as it controls progenitor cell fate during axial elongation. Wnt5a can induce both the canonical (β-catenin dependent) and noncanonical Wnt pathways, and it plays key roles in the regulation of cell fate and patterning during embryogenesis.

Generally, gastrulation is a continuous process that occurs simultaneously with PS formation. It encompasses the inactivation of epiblast genes, such as *Oct4* and *Sox2*, and the acquisition of mesoderm and endoderm-specific factors [18]. This gradual and sequential loss of cells from the epiblast sheet as cells ingress through the PS is compensated by the high proliferation rates observed during these stages. In the end, the overall net result is the generation of a force that passively pulls lateral epiblast cells towards the PS, while its overall epithelial integrity is maintained [21].

8.2 Axial Extension and Formation of the Vertebrate Body

Aside from extensive mesoderm and endoderm formation, the first stages of gastrulation also include substantial growth. However, the overall shape of the embryo is generally maintained. The first major large-scale tissue reorganization starts around E7.75 with the thickening, flattening, and folding of the anterior half of the embryo to build the head folds [22]. In the meantime, new tissue starts to be continuously and progressively added at the embryo's posterior end and will generate all the remainder postoccipital structures from head to the tail, i.e., the anterior-posterior (AP) axis [23]. This process of axial extension appears to be conserved in most vertebrates and depends on the proliferation of specialized progenitor cells with stem cell-like properties that reside initially near the PS and later in the tailbud [23]. In the mouse embryo, axial elongation lasts until E13.5, ceasing shortly before the last somites are formed [23–25].

8.2.1 The Axial Stem Cell Niche

A stem cell niche can be roughly defined as a tissue providing a specific microenvironment that promotes stem cell maintenance and differentiation, thus controlling their self-renewable state [23, 26]. The search for the vertebrate axial stem cell niche started long ago. In 1924, Spemann and Mangold showed that the dorsal blastopore lip of the early gastrula of the newt had the ability to induce the formation of a secondary body axis when transplanted onto the opposite side of a similarly staged embryo [27]. This particular region was termed as the "organizer", since it contains a group of cells with the ability to induce a new fate in neighboring cells and to pattern the induced tissues [23, 26]. The equivalent region to the blastopore lip in avian embryos, the Hensen's node, was also shown to be capable of organizing a secondary body axis when transplanted to the lateral side in a host chick embryo [26]. Later, these types of experiments were expanded to mammals. When pieces from the leading edge of the PS of rabbit embryos were transplanted into early chicken embryos, a duplication of the anterior-posterior axis of the host was also observed [28, 29]. In 1994, it was finally demonstrated that the mouse node organizes patterning during gastrulation, as node grafts made in a posterior-lateral region of a host mouse embryo of the same developmental stage resulted in the induction of a second neural tube and in the formation of ectopic somites [30].

8.2.2 Primary and Secondary Body Formation

Axial extension is a complex and dynamic process that starts at the end of gastrulation and will result in the formation of the postoccipital region of the head, neck, trunk, and tail structures [23, 31]. In 1925, Holmdahl proposed that the formation of the vertebrate body axis would entail two separate and distinct processes [32]. Primary body formation was defined to encompass head, neck, and trunk development, as these structures rely on the PS to build their mesodermal components. On the other hand, tail formation was defined as secondary body because, instead of the PS, it relies on a mass of cells (a "blastema") located in the tailbud [31, 32] (◘ Fig. 8.2). Later studies in the chick embryo showed that grafts of a region containing the remnants of the node, which in 1937 Pasteels named "the chordo-neural hinge (CNH)," contributed to the neural tube floor plate and to the caudal part of the notochord [33]. More recent work with mouse embryos has shown that indeed there are substantial differences in the way that the embryo makes the primary and secondary bodies [31]. At the beginning of axial elongation

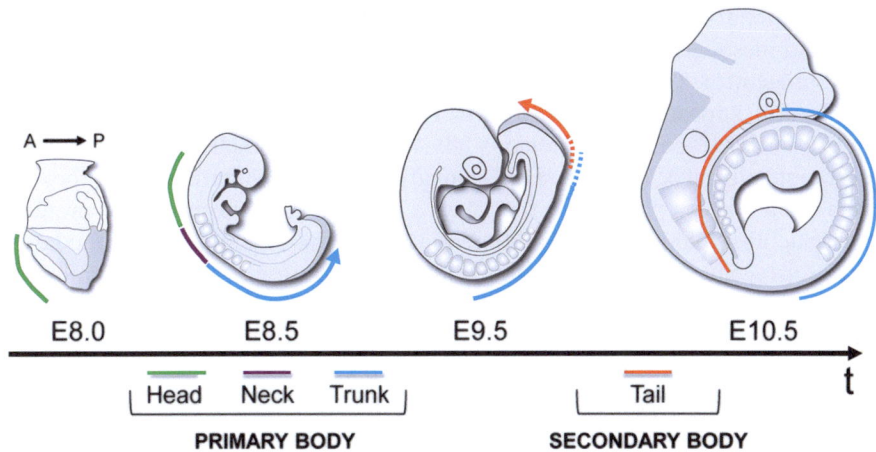

Fig. 8.2 Primary and secondary body formation. Schematic representation of the mouse embryonic development during axial elongation. Primary and secondary body structures are highlighted in different colors. A anterior, E embryonic day, P posterior

(E8.0), axial stem cells reside in the node-streak border (NSB – a region that comprises the caudal end of the node and the more rostral part of the PS) and in the caudal lateral epiblast (CLE) [25, 34]. Homotopic grafts of these two regions showed that cells in the NSB will give rise to ventral neural tube, somites, notochord and will also populate the CNH [25, 34]. Cells from the CLE will colonize mostly the dorsal part of the neural tube, somites, and will also contribute to the CNH (however to a less extent than cells from the NSB) [25, 34]. Cells in other regions of the PS (excluding the NSB) will only give rise to mesoderm whereas the ones lateral to the NSB do not seem to have stem cell-like properties as they are *en route* for differentiation in the axis [25, 34].

Generally, during primary body formation, axial progenitors that reside in the caudal epiblast epithelium will generate neural tissues by primary neurulation, i.e., they first form an open neural plate that bends on itself and then closes to form the neural tube. In contrast, mesodermal tissues are produced through the process of gastrulation/ingression, involving an EMT as they go through the PS [23]. However, as the PS regresses and the epiblast closes, long-term axial progenitors (see below 2.4) must be re-localized to the CNH so that the embryo can engage in secondary body formation [31, 35]. During tail formation, the neural tube is thought to be formed by secondary neurulation, whereupon it is extended as a rod that then cavitates to form a tube [36]. Thus, formation of mesodermal tissues during the last stages of extension no longer depends on the PS [23].

8.2.3 Axial Progenitor Cells

Axial progenitors are cells with stem cell-like properties. Recent efforts trying to characterize these cells showed the presence of transcriptionally dynamic heterogeneous subpopulations located in the progenitor region [31, 37]. These subpopulations can be distinguished by their potency and fate. Clonal analysis suggests the presence of axial progenitors that behave as dual-fated tissue stem cells, giving rise to neural and mesodermal tissues [38]. Therefore, these cells were termed neuro-mesodermal progenitors (NMPs) since they seem to be in a bipotent state [25, 38, 39]. Of these, several appear to be positive for T, Sox2, and Nkx1-2, as these genes are expressed in NMP-containing regions [40–44]. Although their emergence from the pluripotent epiblast cells during axial elongation is still a matter of debate, they seem to be a bona fide cell population that plays a key role in the formation of more posterior structures [25, 31, 45]. The differentiation of this subpopulation of axial progenitor cells seems to occur depending on their localization in the caudal epiblast, as these cells have extensive plasticity to respond to signals they receive from their environment [40]. NMPs that are initially present in the caudal epiblast and also persist in the CNH were designated long-termed axial progenitors, as they are present during long periods throughout axial elongation. Indeed, CNH grafts can undergo multiple passages through host embryos, by participating in long distances of axial elongation and successfully repopulating their CNH [25, 34, 46]. In contrast, axial progenitor cells that reside in more posterior regions of the CLE were shown to only have potential to generate mesodermal tissues [40]. Nevertheless, these cells seem to have some degree of plasticity within the mesodermal lineage, as they can generate lateral and paraxial mesoderm – therefore they were termed as lateral and paraxial mesoderm progenitors (LPMPs) [40] (◘ Fig. 8.3).

8.3 Molecular Mechanisms Controlling Axial Progenitors

During axial extension, different types of progenitors proliferate in a controlled manner, while also having to be instructed in the types of structures they should generate. This balance between self-renewal and differentiation during development of the embryonic axis is under tight regulation by several factors and signaling pathways [47] (◘ Fig. 8.4).

8.3.1 Maintenance of the Axial Stem Cell Niche

Maintenance of a proper stem cell niche is crucial not only for the proliferation of axial progenitors but also for keeping their stem-like properties during the entire duration of axial extension. The continuity of gene expression observed in the caudal epiblast and in the CNH hints at the existence of certain pathways and factors that are important for keeping progenitors from terminal differentiation [45].

8.3.1.1 The Pluripotency-associated Gene Network

An active pluripotency-associated network seems to be essential for proper axial progenitor function during all stages of axial extension in the mouse embryo. The first pluripotency-associated gene to have been described to have an impact in axial elongation was *Oct4* [48]. This factor is at the center of the pluripotency network; and not only is this gene one of the four original factors used to reprogram cells into pluripotent states, but its activity is also sufficient to reactivate the pluripotency network in

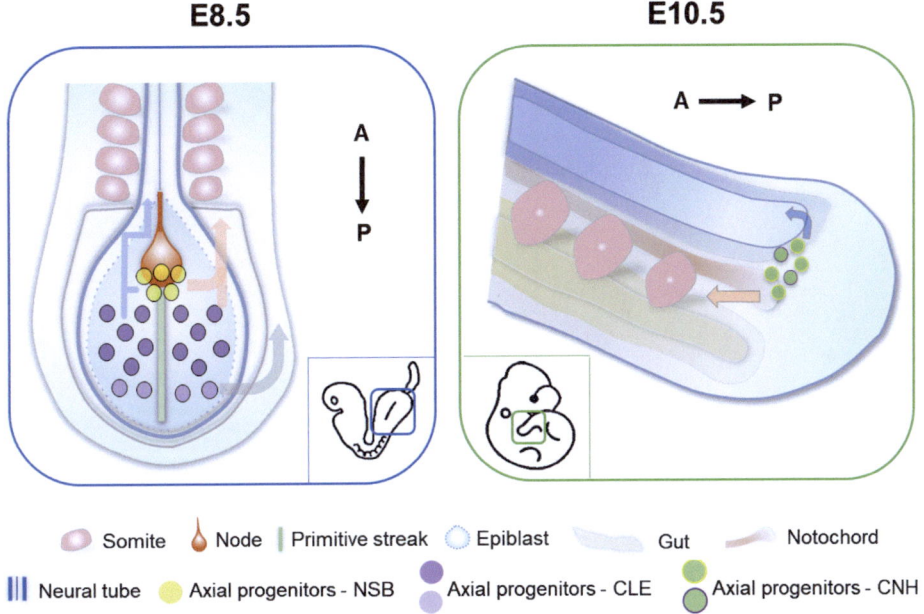

Fig. 8.3 Axial stem zone. Schematic representation of the caudal epiblast (E8.5) and tailbud (E10.5), highlighting the localization of the different axial progenitor cells and the different types of tissues that they generate. A anterior, CLE caudal lateral epiblast, CNH chordo-neural hinge, E embryonic day, NSB node-streak border, P posterior

the late epiblast [49, 50]. Oct4 seems to be a key regulator of extension through the trunk region, as conditional mutants for this gene generate embryos with dramatic shortenings of the trunk – yet developing perfectly specified tails [48]. Conversely, gain-of-function analyses showed that sustained Oct4 activity in axial progenitors was sufficient to dramatically increase the number of thoracic segments in the trunk while delaying transition into tail formation [51]. However, downregulation of *Oct4* is critical for the progression of axial extension after trunk formation, as its ectopic expression in the tail has deleterious effects in caudal embryonic structures [51, 52]. *Lin28* genes are another set of factors associated with stem-cell fate that were shown to have a role in axial extension [53–55]. However, these appear to act in a very different compartment than *Oct4*, as overexpression of *Lin28a*, *Lin28b*, or both was enough to stimulate tail growth up to seven extra segments; conversely, its conditional inactivation was enough to disrupt tail development [54, 55]. Similarly to *Lin28* genes, *Sall4* seems necessary for NMP maintenance at the tail level, since its conditional knock-out generates truncations at the level of the trunk to tail transition that correlate with sharp decreases in NMP numbers [56].

8.3.1.2 Wnt Signaling

Several members of the Wnt signaling pathway are expressed in axial progenitor-containing regions in chick and mouse [57, 58]. Among these, *Wnt3a* and *Wnt5a* seem to be the most relevant for axial extension in the mouse embryo. Absence of

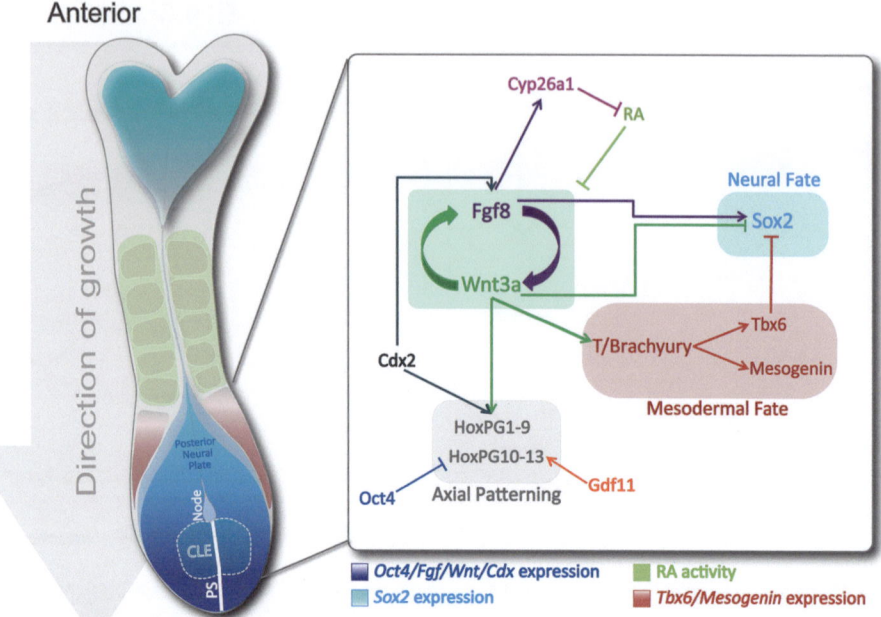

Fig. 8.4 Key molecular players involved in axial progenitor regulation. Left: Schematic representation of a day E8.5 mouse embryo depicting the expression patterns of several factors and the location of the node, CLE, and PS in the embryonic caudal region. Right: Genetic circuitry of relevant factors acting in axial progenitor maintenance, fate determination, and axial patterning

Wnt3a or conditional mutations of its downstream effector, β-catenin, induce severe axial truncations caudal to the forelimbs [59–61]. In contrast, overexpression of Wnt3a or β-catenin in progenitor-containing regions resulted in the absence of a recognizable neural tube or mesoderm, and in an accumulation of undifferentiated cells in the embryonic epiblast [62, 63].

Mutations in *Wnt5a*, on the other hand, affect specifically tail extension and other outgrowing structures of the embryo, like the limbs [64]. Interestingly, *Wnt5a* expression is absent from *Wnt3a* mutant embryos, which shows the importance of *Wnt3a* in the activation and/or maintenance of expression of other members of the Wnt signaling family [59]. Additionally, *Wnt3a* and *Wnt8a* activate expression of *Fgf8* in the mouse embryo [65, 66], which consolidates the expression of both pathways in the axial stem cell niche (see below).

8.3.1.3 Fgf Signaling

Similarly to the Wnt pathway, many components of the Fgf signaling pathway are expressed in axial progenitor-containing regions and are known to play important roles during axis extension. Absence of *Fgfr1* or *Fgf8* results in gastrulation defects involving lack of proper mesoderm formation, thus completely arresting axis extension [15, 67–69]. When conditionally knocked-out later in development, double mutations in *Fgf8* and *Fgf4* generate mouse embryos that are dramatically truncated after the first few somites, while displaying no increase in cell death or defects in cell migration [70, 71]. These findings indicate an essential role of Fgf in the maintenance of the progenitor state, as well as in mesoderm formation.

8.3.1.4 The Cdx family of genes

Cdx genes are likewise essential for the maintenance of the axial progenitor niche. In mammals, the *Cdx* family of genes is composed of three genes, *Cdx1*, *Cdx2*, and *Cdx4*. These genes show a certain degree of redundancy among them, and all three *Cdx* genes are expressed in axial progenitor-containing regions [72]. However, changes in the allelic composition of *Cdx2*, particularly in the absence of any other *Cdx* family members, originate axial truncations [73–75]. This factor was shown to be involved in the transcriptional activation of several components of the Wnt and Fgf signaling pathway during postcranial axial extension, specifically *Fgf8* and *Wnt5a* [76]. Importantly, stimulation of both pathways was sufficient to partly rescue the truncations observed in *Cdx2* mutants [72, 74].

8.3.1.5 Retinoic Acid

Exposure of the developing embryo to exogenous retinoic acid (RA) or to RA inhibitors is known to cause severe homeotic transformations [77, 78]. RA activity, particularly in the axial stem zone, seems to be of key importance in higher vertebrates, as it is thought to be crucial for the expansion of the NMP population before extension through the trunk and tail [79, 80]. RA is needed for the very early stages of axial extension in mouse, since embryos mutant for Raldh2 exhibit several axial defects, truncate at the forelimb level, and cannot undergo axial rotation [81]. Interestingly, these embryos can be rescued with RA administration until E8.25 (sixth somite stage), which indicates that RA is only essential for a very limited time during these very first stages of axial extension [82, 83]. Conversely, absence of the RA catabolizing enzyme Cyp26a1 generates severe axial truncations in mouse and chick embryos, particularly at the tail level [84], which can be rescued by genetic ablation of the RA receptor gamma [84, 85]; however, in the zebrafish, inhibiting RA synthesis does not generate an axial extension phenotype [79]. Yet, protection of the progenitor pool from RA through Cyp26-dependent elimination appears to be somewhat of an evolutionarily conserved trait in chordates, as both RA administration and inhibition of Cyp26 variants arrest axial extension in amphioxus [86]. One possible important function of RA might be to limit the expression domains of *Fgf8*, *Wnt3a*, and *Wnt8a*, and therefore, the control of the axial progenitor pool size and location [63, 87, 88].

8.3.2 Fate Determination in Axial Progenitors

In higher vertebrates, lineage commitment is a gradual process that begins shortly after fertilization and lasts until the end of embryonic development. NMPs are a particularly interesting case study among axial progenitors, as they are thought to constitute a population of bipotent cells that contribute with both neural and mesodermal derivatives during axial extension, thus expressing markers for both lineages. Besides their crucial role in progenitor survival and maintenance of the axial stem cell niche, Fgf and Wnt signaling are also necessary for the activation of important genes for fate determination in this specific type of progenitors [80].

Wnt signaling is involved in the regulation of the balance between paraxial mesoderm and neuroectoderm production from NMPs. Prolonged exposure to this pathway seems to drive progenitors towards mesodermal fates [41, 60, 63]. In fact, Wnt promotes the expression and maintenance of *T/Brachyury*, which has a crucial role in the mesoderm formation [39, 89–92]. This gene, together with active Wnt signaling, will then regulate other important genes involved in mesoderm specification, particularly *Tbx6* and *Mesogenin* [39, 47, 93, 94]. Loss of any of these genes generates severe axial truncations, lack of posterior mesoderm formation, or even the presence of ectopic neural tubes in prospective mesodermal compartments [93, 95–97]. Conversely, mutants for *Wnt3a* or components of the canonical Wnt signaling pathway show expansion of neural tissue at the expense of paraxial mesoderm [40, 59, 61, 63, 89]. Lineage tracing experiments in *Wnt3a* mutants have shown that the ectopic neural tubes rise from cells initially fated to become mesoderm [63]. This idea is further reinforced by lineage tracing in β-catenin conditional mutants, in which their characteristically expanded neural tubes derive from cells that previously expressed the mesoderm marker *T/Brachyury* [63]. Transplantation studies in the mouse have also shown that inactivation of β-catenin induces cells to undergo neural tube incorporation, even when initially fated to become mesoderm [40]. Altogether, these experiments demonstrate that the absence of canonical Wnt signaling in NMPs induces these cells to produce neural derivatives.

Neural lineage commitment in NMPs is achieved by the activation of *Sox2* through the caudal epiblast-specific N1 enhancer: cells fated to become neural tube sustain N1 enhancer activity, thus expressing *Sox2* and becoming part of the posterior neural plate. Both Fgf and Wnt signaling act synergistically to activate the N1 enhancer in the proper developmental time and place through the binding of several of its downstream effectors to specific sequences within this element [98, 99]. *Wnt3a*, in particular, acts as a negative regulator of *Sox2* expression, as its overexpression specifically impairs formation of neural derivatives [41, 43, 62]. *Tbx6 and Msgn1* somehow complement the effect of *Wnt3a* by ensuring the complete suppression of the neural transcription program in mesoderm-fated cells [99]. Interestingly, in the mouse tail bud, this activity of Tbx6 on the restriction of the neural lineage seems to occur directly in NMPs rather than in their mesodermal derivatives [97].

Members of the pluripotency network were also shown to have a role in lineage fate determination, but whether that role is direct or indirect remains mostly unknown. Overexpression of *Oct4* seems to bias NMPs into the neural fate and, consequently, to an excess neural tube formation [51, 54]. In contrast, *Sall4* appears to be involved in nascent mesoderm formation by controlling Wnt and Fgf signaling cascades, since *Sall4* conditional mutants downregulate *Wnt3a* and *Fgf8* while promoting *Sox2* expression [56].

Gdf11 signaling also seems to be involved in fate decision, as mutations or inhibition of the pathway will lead to an increase in neural tissue at the expense of mesoderm in the embryonic tail. *Gdf11* mutants have up to 20% more $Sox2^+/T^+$ positive cells (the consensus molecular signature of NMPs) and show a decrease in the expression of mesodermal genes, while neural genes display an overall upregulation. However, most of the excess neural tissue appears to stem from its accelerated proliferation and differentiation [54].

8.3.3 Regional Specification of Axial Progenitors

Throughout axial extension, progenitors must be instructed on the types of structure that each AP level requires. That way, axial progenitors and their derivatives must somehow acquire a specific positional identity. This is provided mainly by *Hox* genes, genes from the *Cdx* family, and the Gdf11 signaling pathway.

8.3.3.1 *Cdx* Gene Family

In addition to their role in axial progenitor maintenance, *Cdx* genes also provide important AP patterning cues [100]. In the mouse, inactivation of these genes generates homeotic transformations, albeit somewhat less extensive than those observed in mutants for *Hox* genes [73, 101–103] (see below). These homeotic transformations likely result from alterations in *Hox* gene activity, since Cdx proteins have been described to control their expression through a feed-forward loop in the caudal embryonic growth zone [73, 76, 104].

8.3.3.2 Hox Genes

In vertebrates, expression of anterior *Hox* genes is triggered by both canonical Wnt signaling and *Cdx2* activity in the early embryo [76, 104]. Posterior-most *Hox* genes such as those of HoxPG10-13 seem to be under transcriptional repression by Oct4 and require additional Gdf11 signaling to be activated [51, 105–108].

A hallmark of *Hox* genes is their characteristic temporal and spatial activation in the three germ layers, in a sequence that follows their order within the clusters – a phenomenon known as collinearity [109, 110]. Members of the Hox paralogous group (HoxPG)1 to HoxPG4 are the first to be expressed [111, 112], followed by the remaining HoxPGs in a progressive temporal activation at the embryonic posterior end. This temporal collinearity is translated into a spatial collinearity by the consecutive activation of successively more posterior *Hox* genes in the continuously emerging tissues during extension of the axis [113]. Ultimately, each axial level will be characterized by a unique combination of active *Hox* genes – the "*Hox* code." Thus, specific sets of Hox genes encode for each particular axial level, thereby providing spatial patterning cues to axial progenitors and their derivatives for the making of appropriate axial structures [110, 113]. This way, HoxPG1 is associated with the anterior-most part of the embryo, whereas HoxPG6 is normally linked to thoracic segments and HoxPG10 and HoxPG11 specify lumbar and sacral segments, respectively. Interestingly, axial truncations induced by *Cdx2* inactivation can be rescued by *Hoxa5* and *Hoxb8*, which shows that certain *Hox* genes may still have the potential to promote axial progenitor pool homeostasis [74].

8.3.3.3 Gdf11 Signaling

Gdf11 is expressed in the axial progenitor niche throughout axial extension, but particularly in the embryonic tail bud where it is maintained until the end of axis elongation [114]. Signaling through Gdf11/Smad is known to activate directly *Hoxd11* expression and genetic experiments have also shown that mutations in the Gdf11 receptor ActRIIB display multiple patterning defects, including mild homeotic transformations [105, 115]. In the mouse, complete inactivation of *Gdf11* delays the trunk to tail transition, whereas expression of a constitutively active form of the Gdf11 receptor Alk5 leads to dramatic shortenings in the trunk region [107]. In both cases, these changes are concomitantly associated with displacement of the cloaca and hindlimbs, as well as posterior shifts in the activation of 5' *Hox* genes like those of *HoxPG10, HoxPG11,* and *HoxPG13* [54, 106, 107].

8.3.4 Extinction of NMPs and the Cessation of Axial Extension

Ultimately, the end of embryonic extension in vertebrates will depend on the timing of NMP extinction in the tail bud. Termination of this process is usually associated with the progressive loss of factors that protect axial progenitors from terminal differentiation in the axial progenitor territory, specifically *Wnt3a* and *Fgf8*. In the mouse, these factors start becoming downregulated in the tail bud 48h before extension arrest and disappear completely as the last somites are formed [34, 40]. Although the mechanistic details determining NMP extinction and axial extension arrest are still not fully understood, there are many factors known to specifically affect tail development in vertebrates. RA has been proposed to control this process; however, most of the experiments showing that effect were performed in chick embryos [42] [107]. In mice, while RA generates axial truncations when ectopically provided [68, 108], axial termination occurs normally in rescued embryos lacking *Raldh2* [109].

In recent years, various studies have been hinting at genes from the HoxPG13, specifically *Hoxb13* and *Hoxc13,* as important for axial cessation in the mouse embryo (◘ Fig. 8.5, bottom). Interestingly, these two genes are not only the last *Hox* to be activated during development in vertebrates but they are also the only *Hox* genes whose expression is restricted to the tailbud [116]. Mutations in *Hoxb13* generate animals with longer tails mostly due to a decrease in apoptosis and maintenance of a highly proliferative state in tail tissues [117]. Conversely, overexpression of *Hoxb13* and *Hoxc13* specifically truncate the tail mainly through decreased proliferation and increased cell death through apoptosis in the neural tube and axial progenitor area [54, 118]. This process of axial growth arrest seems to be coordinated by Gdf11 signaling which, besides its role in the trunk to tail transition and axial patterning, controls axial cessation through the concomitant activation of HoxPG13 genes and downregulation of the expression of tail NMP pluripotency network members, Lin28a and Lin28b [54]. Yet, whether these effects are direct or indirect remains to be shown.

8.4 Mechanisms of Axial Extension and the Evolution of Body Shape Diversity

Vertebrates are one of evolution's most successful organisms. Their incredible versatility greatly relies on the immense variation of vertebrate body shape and size. However, regardless of the species, all vertebrate embryos share most developmental processes and genetic networks, which include the general mechanisms of axial extension [31]. So how can one highly conserved process of elongation generate the diversity of body plans observed in this clade?

The different proportions of head, neck, trunk, and tail can be ultimately achieved through coordination between tissue growth at the caudal end, the rate of embryonic segment formation and the time spent by axial progenitors in a given configuration

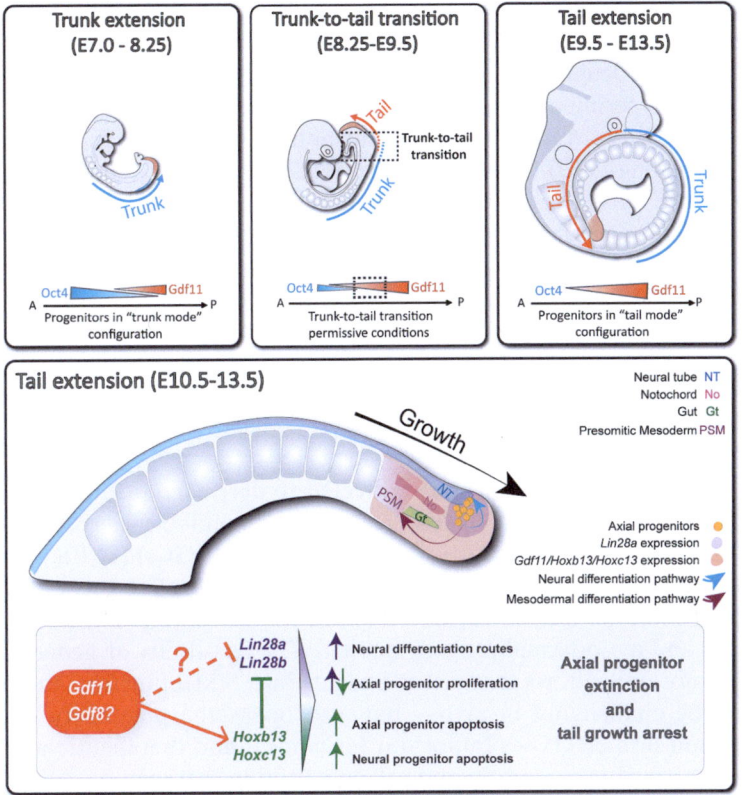

Fig. 8.5 Oct4 and Gdf11 define trunk length and the onset of the trunk to tail transition. (Top) During the early stages of axial extension (E8.25), Oct4 maintains axial progenitors in a "trunk mode" configuration, even in the presence of Gdf11 signaling. As Oct4 is progressively downregulated during axial elongation, Gdf11 signaling becomes prevalent and triggers the onset of the trunk to tail transition around E9.0. During tail development, progenitors express Gdf11 and are kept in a "tail mode" configuration until the end of axial extension (E13.5). (Bottom) Schematic view of the molecular players, their expression patterns, and the gene network regulating tail extension and its arrest during mouse development. E embryonic day

("trunk mode" vs. "tail mode") [51, 54, 80, 119]. In this context, the mechanisms controlling the developmental timing of the transitions between the different areas in the body will determine their relative contribution to the final body plan.

In the mouse, loss- and gain-of-function experiments have demonstrated that the posterior boundary of the trunk region – and thus, the transition from trunk to tail – is mainly controlled by *Oct4* and *Gdf11* (◘ Fig. 8.5, Top). Their genetic interplay provides a relatively simple mechanism explaining the origin of body plan diversity in other vertebrates. In fact, *Oct4* was shown to still be expressed in the posterior-most trunk regions in corn snake embryos but not in their tailbuds, even well after the transition into tail development [51]. Additionally, the axial level of *Gdf11* expression shows a direct correlation with the position of the trunk to tail transition in several vertebrate species [51, 120].

Variability in tail sizes, on the other hand, is most likely related with the onset of mechanisms that extinguish NMPs. As discussed above, RA is an intriguing case in which there is an apparent contradiction in RA deployment during axis cessation in mouse and chicken [42, 81]. This could, perhaps, reflect different ways to generate tails with varying lengths in the two model organisms: RA could function as a method to swiftly terminate axial extension in short-tail species like chicken; whereas RA absence would lead to a gradual exhaustion of tailbud NMPs and progressive tail tapering in long-tailed vertebrates. Likewise, the spatial and temporal association of *Hoxb13* and *Hoxc13* gene expression with the final stages of tail extension in mouse and chicken embryos suggests a conserved role of these genes in axis termination [42, 54, 116, 117] (◘ Fig. 8.5, Bottom). Ultimately, more studies are needed to show if genes that can affect the formation of the mouse tail (e.g., *Lin28* genes) also play a role during tail extension and cessation in other vertebrate embryos.

8.5 In Vitro Generation of NMPs

In recent years, much effort has been directed to derive axial stem cells, particularly NMPs, from embryonic stem cells (ESCs), epiblast-derived stem cells (EpiSCs) or even from induced pluripotent stem cells (iPSCs) (◘ Fig. 8.6). These studies aimed to recapitulate the sequence of events leading to the appearance of NMPs in the embryo and to explore this cell population's potential to generate both neural and mesodermal progeny in vitro. The possibility of generating bipotent NMPs not only allows for further insights into NMP function and biology – which can be challenging in vivo – but also opens the possibility of directed differentiation into cell types important for clinical and therapeutic applications such as spinal cord regeneration and cell substitution therapies.

8.5.1 ESCs-Derived NMPs

Wnt and Fgf signaling seem to have a determinant role in the in vitro differentiation of NMPs from ESCs. NMP-like cells can be derived by subjecting mouse ESCs to bFGF (basic fibroblast growth factor) and Wnt3a, or Wnt signaling agonists such as the Glycogen synthase kinase-3 (GSK-3) inhibitor CHIR99021 [121, 122]. Wnt signaling, in particular, was shown to be fundamental to this process, as even a brief exposure to Wnt3a or CHIR in the presence of bFGF was sufficient to activate mesodermal markers such as *Cdx2*, *T/Brachyury*, and *Tbx6* [121]. Under these conditions, and similarly to their in vivo counterparts, ESCs-derived NMPs display a transcriptomic program that is a combination of neural and mesodermal-associated factors, yield both neural and mesodermal derivatives under appropriate culture conditions and are able to incorporate efficiently in embryos, giving rise to both neural tube and paraxial mesoderm [121]. Importantly, this combination of factors was sufficient to generate neural cells, which were molecularly similar to the ones found in the spinal cord, particularly regarding their AP identity [121, 122]. In fact, these cells expressed genes associated with thoracic and lumbar spinal cord identities such as posterior *Hox* genes (*HoxPG6-10*) and low levels of anterior neural markers such as *Otx2*. Interestingly, there seems to be a sensitive period for the induction of NMPs from ESCs, which can correspond to the acquisition of an intermediate EpiSC identity by ESCs [121, 122] (see below). This appears to corroborate the view that ES cell differ-

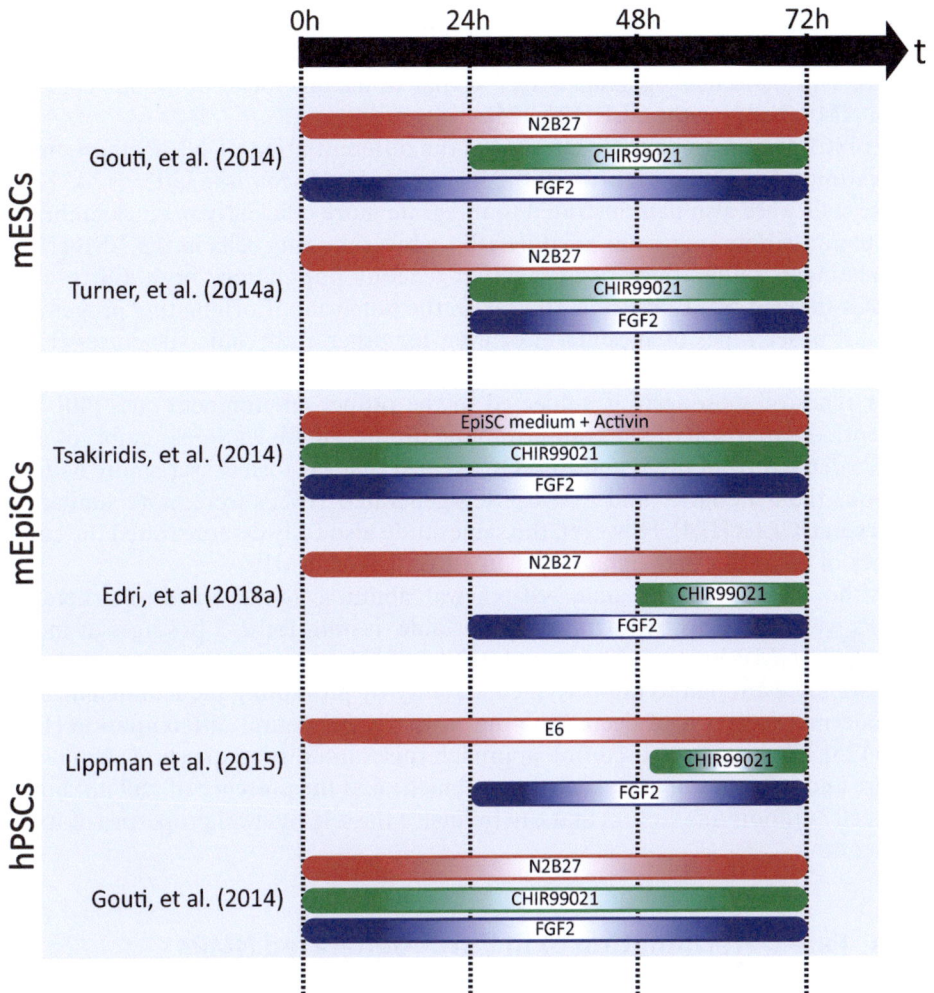

Fig. 8.6 Summary of the protocols used for the in vitro generation of NMPs from mESCs, mEpiSCs, and hPSCs. Red bars: media used, green bars: timing of Wnt inhibitor CHIR99021 administration, blue bars: Fgf regime. For further details, refer to the original publications (on the left side)

entiation into NMPs recapitulates the corresponding developmental trajectory, namely the gradual loss of stemness and passage through intermediate stages of potency seen during development.

8.5.2 EpiSCs-Derived NMPs

EpiSCs are derived from postimplantation epiblast cells, which will give rise to all embryonic structures with the exception of supporting tissues such as the placenta and yolk sac [123]. While still pluripotent, these cells are thought to represent a later stage in the characteristic gradual loss of pluripotency during development. Similarly to ESCs, subjecting EpiSCs to bFGF and CHIR yielded cells co-expressing *Sox2* and *T*/

Brachyury that also expressed genes observed in embryonic axial progenitor-containing regions (such as *Fgf8*, *Cyp26a1*, and *Oct4*) and that exhibited posterior *Hox* gene activation. This molecular signature is very similar to the one found in the late epiblast, when NMPs reside in the CLE [124, 125].

EpiSCs appear to be more efficient in the differentiation of NMPs than direct derivation from ESCs, even when the former resulted from the latter [124, 125]. These cells were also demonstrated to integrate more efficiently in chick embryos and to contribute to greater axial lengths, while retaining cells in the NSB [125]. Interestingly, rather than being a homogeneous population, both ESCs- and EpiSCs-derived NMP-like cells still retain the potential of originating progenitor cells for other types of mesoderm or even for other embryonic structures [125]. This mimics the dynamics in the embryo, where cells in the CLE can contribute to other types of mesoderm if subjected to the proper environment cues [40]. An in-depth comparison of the transcriptomic profiles of E8.25 mouse embryos and NMPs derived from ESCs and EpiSCs revealed that the former were more heterogeneous than the latter and that EpiSCs-generated NMPs were more similar to embryonic CLEs [124]. However, the same study also fully demonstrated the complexity in the molecular signature of in vitro derived NMPs.

Although containing some self-renewal abilities, so far in vitro generated NMPs were found to be intrinsically unstable, lasting for 2–3 passages at most, and stable NMP lines could not be established. However, adding node-like cells appears to be enough to improve their stability by promoting the maintenance of mesodermal factors and decreasing the tendency for neural differentiation [121, 124, 125]. This in vitro co-culture approach raises interesting questions about the nature and influence of the axial stem-cell niche and the presence of still unknown non-cell autonomous factors that can influence the self-renewal properties of axial progenitors.

8.5.3 Fate Determination of in Vitro–Generated NMPs

Fate determination of in vitro–derived NMPs is mainly regulated by levels of Wnt signaling [43]. Wnt signaling attenuation or ablation of β-catenin resulted in the downregulation or complete absence of *T*/*Brachyury* and strong expression of *Sox2*. In contrast, elevated Wnt signaling promoted the exit of pluripotency and formation of NMPs and mesendodermal (ME) progenitors, as demonstrated by the presence of cells co-expressing *Sox2* and *T*/*Brachyury* or *Foxa2* and *T*/*Brachyury*, respectively [43]. When grafted in mouse embryos, both types of progenitors were able to integrate into their respective tissue, which further demonstrates both the functional competence of EpiSCs-derived NMPs and ME precursors [43].

8.5.4 Derivation of Human NMPs

The role of Fgf and Wnt signaling seems to be evolutionarily conserved, as these factors are also sufficient to generate NMPs from human ESCs (hESCs), human pluripotent stem cells (hPSCs), and human-induced pluripotent stem cells (iPSCs) [121, 126]. Similarly to the NMP-derivation from mouse ESCs or EpiSCs, culturing hESCs or hPSCs with bFGF/CHIR yielded cells that had a NMP-like molecular signature. These cells expressed both neural and mesodermal markers and could be differentiated into posterior spinal cord and mesodermal derivatives [121]. However, in these conditions, the NMP state was only maintained for 3 days. The homogeneity and stability of NMP cultures could be improved by pretreating hPSCs with FGF8b, followed by incubation with FGF8b/CHIR supplemented media [126]. In contrast to mouse NMP culture, most hPSCs-derived NMPs co-expressed SOX2 and T/BRACHYURY and were stable for up to 7 days. Again, under these conditions, addition of the Wnt agonist CHIR was enough to induce NMPs with spinal cord identity, as seen by downregulation of hindbrain-associated genes and activation of anterior HOX gene expression. However, when combined with Fgf8b, HOX collinear activation was accelerated and more caudal genes such as HOXC6, HOXC8, and HOXC9 were sequentially expressed. Finally, lumbar and sacral identities (HOXD10, HOXD11) could only be obtained by supplementing Fgf8b/CHIR-cultured hPSCs with GDF11, which parallels the in vivo activation of caudal *HOX* genes during the vertebrate trunk to tail transition [126]. Interestingly, the full collinear expression of *HOX* genes observed in hPSCs-derived NMPs depended on the time cells spent under FGF8/CHIR conditions [126]. Using RA to induce neuronal differentiation and, thus, arrest NMP progression at various time points, a uniform progression of expression from anterior to posterior *HOX* genes could be recapitulated. This way, differentiation of hPSCs into neuroectoderm through a stable NMP intermediate allowed for a deterministic *HOX* patterning, thus instructing different regional identities that followed the regionalization observed within the embryonic spinal cord.

8.6 Self-Organization of Stem Cells into Embryo-Like Structures

Self-organization is a process by which an ensemble of independent and isolated elements spontaneously develops into a large-scale ordered structure [5, 127]. Recent advances in stem cell biology enabled the self-organization of stem cells into embryo-like structures or into 3D structures – the so-called organoids – that recapitulate key characteristics of real organs. For example, mouse ESCs were shown to cooperate with extraembryonic trophoblast stem cells (TSCs) and form structures that morphologically and transcriptionally resemble early blastocysts during implantation [128, 129]. AP patterning and recapitulation of key morphogenetic events of gastrulation (e.g., EMT and mesoderm specification) in these embryo-like structures was achieved when mouse ESCs, together with TSCs, were also combined with extra-embryonic-endoderm (XEN) cells [130].

Embryoid bodies are 3D aggregates of pluripotent stem cells and therefore illicit the self-organizing capability and potential of these cells when cultured in vitro [5, 127]. Interestingly, when a pulse of CHIR is added between 48 and 72h to a clump of around 250 mouse ESCs, in the presence of N2B27 medium, a localized T expression emerges reproducibly at one end of these structures – termed gastruloids – which undergo gastrulation-like events and elongation similar to the posterior region of the mouse embryo [131, 132]. In the absence of extra-embryonic tissues, gastruloids exhibit a pattern that resembles the spatiotemporal gene expression of postoccipital neural, mesodermal, and endodermal derivatives of the mouse embryo during axial extension [132]. Since these 3D structures display the three major body axes and most of the hallmarks of axial gene regulatory systems, including the collinear *Hox* gene expression along the AP axis, they can thus be used as a complementary system to study mouse axial elongation [132].

8.7 Online Resources/Protocols

- *Compilation of defined conditions for propagation and manipulation of mouse ESCs* [133] contains several protocols for derivation, propagation, genetic modification, and primary differentiation of embryonic stem cells. ▶ https://dev.biologists.org/content/develop/146/6/dev173146.full.pdf

Recently, the development of new techniques to explore, at the single cell level, the mouse transcriptome opened new avenues to further our knowledge of embryonic development. The resulting datasets (see below) are of public access.
- *3D gene expression database of the gastrulating mouse embryo* [134] contains the spatially resolved transcriptome of single-cell populations at defined positions in the germ layers during development from pre- to late-gastrulation stages (~E5.5 – E7.5). ▶ http://egastrulation.sibcb.ac.cn
- *Single-cell molecular map of mouse gastrulation and early organogenesis* [135] contains the transcriptome profile of 116,312 single cells, from mouse embryos collected at nine sequential time points ranging from 6.5 to 8.5 days postfertilization, and a created molecular map of cellular differentiation from pluripotency towards all major embryonic lineages. ▶ https://marionilab.cruk.cam.ac.uk/MouseGastrulation2018/
- *Mouse organogenesis single-cell atlas* [136] contains the transcriptome profile of around two million single-cells, derived from 61 embryos staged between 9.5 and 13.5 days of gestation. ▶ http://atlas.gs.washington.edu/mouse-rna

---

Take-Home Messages

The most important points discussed in this chapter are summarized here:
- After early embryonic development, vertebrate embryos elongate their AP axis by the gradual and successive addition of tissue to its caudal end, in a process designated as axial extension.
- Axial extension is powered by the proliferation of a dedicated population of cells with stem-cell like properties – the axial progenitors.
- There are several types of axial progenitors and their localization changes throughout axial extension. In the caudal epiblast, they will generate primary body structures (head, neck, and trunk), whereas in the tailbud, they will form secondary body tissues (tail).
- Axial progenitors are tightly regulated during development by several factors and signaling pathways, namely factors that maintain the axial stem cell niche, that are involved in their fate decision and that provide AP patterning cues.
- Recently, there have been efforts to derive and culture axial progenitors in vitro from both mouse and human, and from various cellular sources such as ESCs, EpiSCs, and iPSCs.

Acknowledgments The authors would like to thank Moisés Mallo for the helpful insights, comments, and suggestions to this chapter; and to the remaining members of the Mallo lab for all the unparalled support and constant companionship.

References

1. Arnold SJ, Robertson EJ. Making a commitment: cell lineage allocation and axis patterning in the early mouse embryo. Nat Rev Mol Cell Biol. 2009;10(2):91–103.
2. Rossant J, Tam PPL. Blastocyst lineage formation, early embryonic asymmetries and axis patterning in the mouse. Development. 2009;136(5):701–13.
3. Takaoka K, Hamada H. Cell fate decisions and axis determination in the early mouse embryo. Development. 2012;139(1):3–14.
4. Tam PPL, Loebel DAF. Gene function in mouse embryogenesis: get set for gastrulation. Nat Rev Genet. 2007;8(May):368–81.
5. Shahbazi MN, Zernicka-Goetz M. Deconstructing and reconstructing the mouse and human early embryo. Nat Cell Biol. 2018;20(8):878–87.
6. Morkel M, Huelsken J, Wakamiya M, Ding J, van de Wetering M, Clevers H, et al. Beta-catenin regulates Cripto- and Wnt3-dependent gene expression programs in mouse axis and mesoderm formation. Development. 2003;130(25):6283–94.
7. Tam PP, Loebel DA, Tanaka SS. Building the mouse gastrula: signals, asymmetry and lineages. Curr Opin Genet Dev. 2006;16(4):419–25.
8. Wolpert L, RSP B, Brockes J, Jessell TM, Lawrence P, Meyerowitz E. Principles of development. 1st ed. London: Oxford University Press; 1998.
9. Varlet I, Collignon J, Robertson EJ. Nodal expression in the primitive endoderm is required for specification of the anterior axis during mouse gastrulation. Development. 1997;124(5):1033–44.
10. Conlon FL, Lyons KM, Takaesu N, Barth KS, Kispert A, Herrmann B, et al. A primary requirement for nodal in the formation and maintenance of the primitive streak in the mouse. Development. 1994;120(7):1919–28.
11. Barrow JR, Howell WD, Rule M, Hayashi S, Thomas KR, Capecchi MR, et al. Wnt3 signaling in the epiblast is required for proper orientation of the anteroposterior axis. Dev Biol. 2007;312(1):312–20.
12. Liu P, Wakamiya M, Shea MJ, Albrecht U, Behringer RR, Bradley A. Requirement for Wnt3 in vertebrate axis formation. Nat Genet. 1999;22(4):361–5.
13. Huelsken J, Vogel R, Brinkmann V, Erdmann B, Birchmeier C, Birchmeier W. Requirement for beta-catenin in anterior-posterior axis formation in mice. J Cell Biol. 2000;148(3):567–78.
14. Tam PP, Behringer RR. Mouse gastrulation: the formation of a mammalian body plan. Mech Dev. 1997;68(1–2):3–25.
15. Sun X, Meyers EN, Lewandoski M, Martin GR. Targeted disruption of Fgf8 causes failure of cell migration in the gastrulating mouse embryo. Genes Dev. 1999;13(14):1834–46.
16. Ramkumar N, Omelchenko T, Silva-Gagliardi NF, McGlade CJ, Wijnholds J, Anderson KV. Crumbs2 promotes cell ingression during the epithelial-to-mesenchymal transition at gastrulation. Nat Cell Biol. 2016;18(12):1281–91.
17. Ben-Haim N, Lu C, Guzman-Ayala M, Pescatore L, Mesnard D, Bischofberger M, et al. The nodal precursor acting via activin receptors induces mesoderm by maintaining a source of its convertases and BMP4. Dev Cell. 2006;11(3):313–23.
18. Pfister S, Steiner KA, Tam PPL. Gene expression pattern and progression of embryogenesis in the immediate post-implantation period of mouse development. Gene Expr Patterns. 2007;7(5):558–73.
19. Robb L, Tam PPL. Gastrula organiser and embryonic patterning in the mouse. Semin Cell Dev Biol. 2004;15(5):543–54.
20. Tortelote GG, Rivera-Pérez J. a. Wnt3 function in the epiblast is required for the maintenance but not the initiation of gastrulation in mice. Dev Biol. 2013;130(29):9492–9.
21. Williams M, Burdsal C, Periasamy A, Lewandoski M, Sutherland A. Mouse primitive streak forms in situ by initiation of epithelial to mesenchymal transition without migration of a cell population. Dev Dyn. 2012;241(2):270–83.
22. Sutherland AE. Tissue morphodynamics shaping the early mouse embryo. Semin Cell Dev Biol. 2016;55:89–98.

23. Stern CD, Charité J, Deschamps J, Duboule D, Durston AJ, Kmita M, et al. Head-tail patterning of the vertebrate embryo: one, two or many unresolved problems? Int J Dev Biol. 2006;50(1):3–15.
24. Tam PPL, Tan SS. The somitogenetic potential of cells in the primitive streak and the tail bud of the organogenesis-stage mouse embryo. Development. 1992;115(3):703–15.
25. Wilson V, Olivera-Martinez I, Storey KG. Stem cells, signals and vertebrate body axis extension. Development. 2009;136(12):2133.
26. Martinez Arias A, Steventon B. On the nature and function of organizers. Development. 2018;145(5)
27. Spemann H, Mangold H. Über Induktion von Embryonalanlagen durch Implantation artfremder Organisatoren. Arch für Mikroskopische Anat und Entwicklungsmechanik. 1924;100(3-4):599–638.
28. Waddington C. Experiments on the development of chick and duck embryos, cultivated in vitro. Phil Trans R Soc Lond B. 1932;221:179–230.
29. Waddington CH. Principles of embryology. London: Georg Allen Unwin; 1954.
30. Beddington RSP. Induction of a second neural axis by the mouse node. Development. 1994;120(3):613–20.
31. Aires R, Dias A, Mallo M. Deconstructing the molecular mechanisms shaping the vertebrate body plan. Curr Opin Cell Biol. 2018;55:81–6.
32. Holmdahl DE. Experimentelle Untersuchungen uber die Lage der Grenze primarer und sekundarer Korperentwicklung beim Huhn. Anat Anz. 1925;59:393–6.
33. Catala M, Teillet MA, Le Douarin NM. Organization and development of the tail bud analyzed with the quail-chick chimaera system. Mech Dev. 1995;51(1):51–65.
34. Cambray N, Wilson V. Two distinct sources for a population of maturing axial progenitors. Development. 2007;134(15):2829–40.
35. Griffith CM, Wiley MJ, Sanders EJ. Anatomy and embryology review article the vertebrate tail bud: three germ layers from one tissue. Anat Embryol (Berl). 1992;185:101–13.
36. Schoenwolf GC. Tail (end) bud contributions to the posterior region of the chick embryo. J Exp Zool. 1977;201(2):227–45.
37. Wymeersch FJ, Skylaki S, Huang Y, Watson JA, Economou C, Marek-Johnston C, et al. Transcriptionally dynamic progenitor populations organised around a stable niche drive axial patterning. Development. 2019;146(1):1–16.
38. Tzouanacou E, Amélie W, Filip JW, Valerie W, Jean-François N. Redefining the progression of lineage segregations during mammalian embryogenesis by clonal analysis. Dev Cell. 2009;17:365–76.
39. Koch F, Scholze M, Wittler L, Schifferl D, Sudheer S, Grote P, et al. Antagonistic activities of Sox2 and brachyury control the fate choice of neuro-mesodermal progenitors. Dev Cell. 2017;42(5):514–526.e7.
40. Wymeersch FJ, Huang Y, Blin G, Cambray N, Wilkie R, Wong FCK, et al. Position-dependent plasticity of distinct progenitor types in the primitive streak. elife. 2016;5(JANUARY2016):1–28.
41. Martin BL, Kimelman D. Canonical Wnt signaling dynamically controls multiple stem cell fate decisions during vertebrate body formation. Dev Cell. 2012;22(1):223–32.
42. Olivera-Martinez I, Harada H, Halley PA, Storey KG. Loss of FGF-dependent mesoderm identity and rise of endogenous retinoid signalling determine cessation of body axis elongation. PLoS Biol. 2012;10(10):e1001415.
43. Tsakiridis A, Huang Y, Blin G, Skylaki S, Wymeersch F, Osorno R, et al. Distinct Wnt-driven primitive streak-like populations reflect in vivo lineage precursors. Development. 2014;141(6):1209–21.
44. Rodrigo Albors A, Halley PA, Storey KG. Lineage tracing of axial progenitors using Nkx1-2CreER T2 mice defines their trunk and tail contributions. Development. 2018;145(19)
45. Henrique D, Abranches E, Verrier L, Storey KG. Neuromesodermal progenitors and the making of the spinal cord. Development. 2015;142(17):2864–75.
46. Cambray N, Wilson V. Axial progenitors with extensive potency are localised to the mouse chordoneural hinge. Development. 2002;129(20):4855–66.
47. Gouti M, Delile J, Stamataki D, Kleinjung J, Wilson V, Briscoe J, et al. A gene regulatory network balances neural and mesoderm specification during vertebrate trunk development. Dev Cell. 2017;41(3):243–261.e7.
48. DeVeale B, Brokhman I, Mohseni P, Babak T, Yoon C, Lin A, et al. Oct4 is required ~E7.5 for proliferation in the primitive streak. PLoS Genet. 2013;9(11)

49. Osorno R, Tsakiridis A, Wong F, Cambray N, Economou C, Wilkie R, et al. The developmental dismantling of pluripotency is reversed by ectopic Oct4 expression. Development. 2012;139(13):2288–98.
50. Takahashi K, Yamanaka S. Induction of pluripotent stem cells from mouse embryonic and adult fibroblast cultures by defined factors. Cell [Internet]. 2006 [cited 2014 May 23];126(4):663–76.
51. Aires R, Jurberg AD, Leal F, Nóvoa A, Cohn MJ, Mallo M. Oct4 is a key regulator of vertebrate trunk length diversity. Dev Cell. 2016;38(3):262–74.
52. Economou C, Tsakiridis A, Wymeersch FJ, Gordon-Keylock S, Dewhurst RE, Fisher D, et al. Intrinsic factors and the embryonic environment influence the formation of extragonadal teratomas during gestation early development. BMC Dev Biol. 2015;15(1):1–15.
53. Shyh-Chang N, Daley GQ. Lin28: primal regulator of growth and metabolism in stem cells. Cell Stem Cell. 2013;12(4):395–406.
54. Aires R, de Lemos L, Nóvoa A, Jurberg AD, Mascrez B, Duboule D, et al. Tail bud progenitor activity relies on a network comprising Gdf11, Lin28, and Hox13 genes. Dev Cell. 2019;48(3):383–395.e8.
55. Robinton DA, Chal J, Lummertz da Rocha E, Han A, Yermalovich AV, Oginuma M, et al. The Lin28/let-7 pathway regulates the mammalian caudal body axis elongation program. Dev Cell. 2019;48(3):396–405.e3.
56. Tahara N, Kawakami H, Chen KQ, Anderson A, Yamashita Peterson M, Gong W, et al. Sall4 regulates neuromesodermal progenitors and their descendants during body elongation in mouse embryos. Development. 2019;146(14):dev177659.
57. Parr BA, Shea MJ, Vassileva G, McMahon AP. Mouse Wnt genes exhibit discrete domains of expression in the early embryonic CNS and limb buds. Development. 1993;119(1):247–61.
58. Olivera-Martinez I, Storey KG. Wnt signals provide a timing mechanism for the FGF-retinoid differentiation switch during vertebrate body axis extension. Development. 2007;134(11):2125–35.
59. Takada S, Stark KL, Shea MJ, Vassileva G, McMahon JA, McMahon AP. Wnt-3a regulates somites and tailbud formation in the mouse embryo. Genes Dev. 1994;8:174–89.
60. Aulehla A, Wiegraebe W, Baubet V, Wahl MB, Deng C, Taketo M, et al. A beta-catenin gradient links the clock and wavefront systems in mouse embryo segmentation. Nat Cell Biol. 2008;10(2):186–93.
61. Yoshikawa Y, Fujimori T, McMahon AP, Takada S. Evidence that absence of Wnt-3a signaling promotes neuralization instead of paraxial mesoderm development in the mouse. Dev Biol. 1997;183(2):234–42.
62. Jurberg AD, Aires R, Nóvoa A, Rowland JE, Mallo M. Compartment-dependent activities of Wnt3a/β-catenin signaling during vertebrate axial extension. Dev Biol. 2014;394(2):253–63.
63. Garriock RJ, Chalamalasetty RB, Kennedy MW, Canizales LC, Lewandoski M, Yamaguchi TP. Lineage tracing of neuromesodermal progenitors reveals novel Wnt-dependent roles in trunk progenitor cell maintenance and differentiation. Development. 2015;142(9):1628–38.
64. Yamaguchi TP, Bradley A, McMahon AP, Jones S. A Wnt5a pathway underlies outgrowth of multiple structures in the vertebrate embryo. Development. 1999;126(6):1211–23.
65. Aulehla A, Wehrle C, Brand-Saberi B, Kemler R, Gossler A, Kanzler B, et al. Wnt3a plays a major role in the segmentation clock controlling somitogenesis. Dev Cell. 2003;4(3):395–406.
66. Cunningham TJ, Kumar S, Yamaguchi TP, Duester G. Wnt8a and Wnt3a cooperate in the axial stem cell niche to promote mammalian body axis extension. Dev Dyn. 2015;244(6):797–807.
67. Ciruna B, Rossant J. FGF signaling regulates mesoderm cell fate specification and morphogenetic movement at the primitive streak. Dev Cell. 2001;1(1):37–49.
68. Guo Q, Li JYH. Distinct functions of the major Fgf8 spliceform, before and during mouse gastrulation. Development. 2007;134(12):2251–60.
69. Wahl MB, Deng C, Lewandowski M, Pourquié O. FGF signaling acts upstream of the NOTCH and WNT signaling pathways to control segmentation clock oscillations in mouse somitogenesis. Development. 2007;134(22):4033–41.
70. Naiche LA, Holder N, Lewandoski M. FGF4 and FGF8 comprise the wavefront activity that controls somitogenesis. Proc Natl Acad Sci. 2011;108(10):4018–23.
71. Boulet AM, Capecchi MR. Signaling by FGF4 and FGF8 is required for axial elongation of the mouse embryo. Dev Biol. 2012;371(2):235–45.
72. van Rooijen C, Simmini S, Bialecka M, Neijts R, van de Ven C, Beck F, et al. Evolutionarily conserved requirement of Cdx for post-occipital tissue emergence. Development. 2012;139(14):2576–83.

73. van den Akker E, Forlani S, Chawengsaksophak K, de Graaff W, Beck F, Meyer BI, et al. Cdx1 and Cdx2 have overlapping functions in anteroposterior patterning and posterior axis elongation. Development. 2002;129(9):2181–93.
74. Young T, Rowland JE, van de Ven C, Bialecka M, Novoa A, Carapuco M, et al. Cdx and Hox genes differentially regulate posterior axial growth in mammalian embryos. Dev Cell. 2009;17(4):516–26.
75. van Nes J, de Graaff W, Lebrin F, Gerhard M, Beck F, Deschamps J. The Cdx4 mutation affects axial development and reveals an essential role of Cdx genes in the ontogenesis of the placental labyrinth in mice. Development. 2006;133(3):419–28.
76. Amin S, Neijts R, Simmini S, van Rooijen C, Tan SC, Kester L, et al. Cdx and T Brachyury co-activate growth signaling in the embryonic axial progenitor niche. Cell Rep. 2016;17(12):3165–77.
77. Kessel M, Gruss P. Homeotic transformations of murine vertebrae and concomitant alteration of Hox codes induced by retinoic acid. Cell. 1991;67(1894)
78. Kessel M. Respecification of vertebral identities by retinoic acid. Development. 1992;115(2):487–501.
79. Berenguer M, Lancman JJ, Cunningham TJ, Dong PDS, Duester G. Mouse but not zebrafish requires retinoic acid for control of neuromesodermal progenitors and body axis extension. Dev Biol. 2018;441(1):127–31.
80. Steventon B, Martinez AA. Evo-engineering and the cellular and molecular origins of the vertebrate spinal cord. Dev Biol. 2017;432(1):3–13.
81. Niederreither K, Subbarayan V, Chambon P, Dollé P. Embryonic retinoic acid synthesis is essential for early mouse post- implantation development letter. Nat Genet. 1999;21:444–8.
82. Mic FA, Haselbeck RJ, Cuenca AE, Duester G. Novel retinoic acid generating activities in the neural tube and heart identified by conditional rescue of Raldh2 null mutant mice. Development. 2002;129(9):2271–82.
83. Niederreither K, Vermot J, Schuhbaur B, Chambon P, Dollé P. Embryonic retinoic acid synthesis is required for forelimb growth and anteroposterior patterning in the mouse. Development. 2002;129(15):3563–74.
84. Abu-Abed S, Dolle P, Metzger D, Wood C, MacLean G, Chambon P, et al. Developing with lethal RA levels: genetic ablation of Rarg can restore the viability of mice lacking Cyp26a1. Development. 2003;130(7):1449–59.
85. Abu-Abed S. The retinoic acid-metabolizing enzyme, CYP26A1, is essential for normal hindbrain patterning, vertebral identity, and development of posterior structures. Genes Dev. 2001;15(2):226–40.
86. Carvalho JE, Theodosiou M, Chen J, Chevret P, Alvarez S, De Lera AR, et al. Lineage-specific duplication of amphioxus retinoic acid degrading enzymes (CYP26) resulted in sub-functionalization of patterning and homeostatic roles. BMC Evol Biol. 2017;17(1):1–23.
87. Cunningham TJ, Brade T, Sandell LL, Lewandoski M, Trainor PA, Colas A, et al. Retinoic acid activity in undifferentiated neural progenitors is sufficient to fulfill its role in restricting Fgf8 expression for somitogenesis. PLoS One. 2015;10(9):1–15.
88. Zhao X, Duester G. Effect of retinoic acid signaling on Wnt/β-catenin and FGF signaling during body axis extension. Gene Expr Patterns. 2009;9(6):430–5.
89. Yamaguchi TP, Takada S, Yoshikawa Y, Wu N, McMahon AP. T (Brachyury) is a direct target of Wnt3a during paraxial mesoderm specification. Genes Dev. 1999;13(24):3185–90.
90. Wilson V, Beddington R. Expression of T protein in the primitive streak is necessary and sufficient for posterior mesoderm movement and somite differentiation. Dev Biol. 1997;192(1):45–58.
91. Stott D, Kispert A, Herrmann BG. Rescue of the tail defect of Brachyury mice. Genes Dev. 1993;7(2):197–203.
92. Galceran J, Hsu SC, Grosschedl R. Rescue of a Wnt mutation by an activated form of LEF-1: regulation of maintenance but not initiation of Brachyury expression. Proc Natl Acad Sci U S A. 2001;98(15):8668–73.
93. Nowotschin S, Ferrer-Vaquer A, Concepcion D, Papaioannou VE, Hadjantonakis AK. Interaction of Wnt3a, Msgn1 and Tbx6 in neural versus paraxial mesoderm lineage commitment and paraxial mesoderm differentiation in the mouse embryo. Dev Biol. 2012;367(1):1–14.
94. Chalamalasetty RB, Dunty WC, Biris KK, Ajima R, Iacovino M, Beisaw A, et al. The Wnt3a/β-catenin target gene Mesogenin1 controls the segmentation clock by activating a Notch signalling program. Nat Commun. 2011;2(1):312–90.

95. Chapman DL, Papaioannou VE. Three neural tubes in mouse embryos with mutations in the T-box gene Tbx6. Nature. 1998;391(1991):695–7.
96. Chapman DL, Agulnik I, Hancock S, Silver LM, Papaioannou VE. Tbx6, a mouse T-box gene implicated in paraxial mesoderm formation at gastrulation. Dev Biol. 1996;180(2):534–42.
97. Javali A, Misra A, Leonavicius K, Acharyya D, Vyas B, Sambasivan R. Co-expression of Tbx6 and Sox2 identifies a novel transient neuromesoderm progenitor cell state. Development. 2017;144(24):4522–9.
98. Takemoto T, Uchikawa M, Kamachi Y, Kondoh H. Convergence of Wnt and FGF signals in the genesis of posterior neural plate through activation of the Sox2 enhancer N-1. Development. 2006;133(2):297–306.
99. Takemoto T, Uchikawa M, Yoshida M, Bell DM, Lovell-Badge R, Papaioannou VE, et al. Tbx6-dependent Sox2 regulation determines neural or mesodermal fate in axial stem cells. Nature. 2011;470(7334):394–8.
100. Deschamps J, van Nes J. Developmental regulation of the Hox genes during axial morphogenesis in the mouse. Development. 2005;132(13):2931–42.
101. Chawengsaksophak K, James R, Hammond VE, Köntgen F, Beck F. Homeosis and intestinal tumours in Cdx2 mutant mice. Nature. 1997;386(6620):84–7.
102. Chawengsaksophak K, de Graaff W, Rossant J, Deschamps J, Beck F. Cdx2 is essential for axial elongation in mouse development. Proc Natl Acad Sci U S A. 2004;101(20):7641–5.
103. Subramanian V, Meyer BI, Gruss P. Disruption of the murine homeobox gene Cdx1 affects axial skeletal identities by altering the mesodermal expression domains of Hox genes. Cell. 1995;83(4):641–53.
104. Neijts R, Amin S, van Rooijen C, Deschamps J. Cdx is crucial for the timing mechanism driving colinear Hox activation and defines a trunk segment in the Hox cluster topology. Dev Biol. 2017;422(2):146–54.
105. Gaunt SJ, George M, Paul YL. Direct activation of a mouse Hoxd11 axial expression enhancer by Gdf11/Smad signalling. Dev Biol. 2013;383(1):52–60.
106. Mcpherron AC, Lawler AM, Lee S. Regulation of anterior / posterior patterning of the axial skeleton by growth/differentiation factor 11. Nature. 1999;22(july):1–5.
107. Jurberg AD, Aires R, Varela-Lasheras I, Nóvoa A, Mallo M. Switching axial progenitors from producing trunk to tail tissues in vertebrate embryos. Dev Cell. 2013;25(5):451–62.
108. McPherron AC, Huynh TV, Lee SJ. Redundancy of myostatin and growth/differentiation factor 11 function. BMC Dev Biol. 2009;9(1):1–9.
109. Duboule D, Dollé P. The structural and functional organization of the murine HOX gene family resembles that of Drosophila homeotic genes. EMBO J. 1989;8(5):1497–505.
110. Kmita M, Duboule D. Organizing axes in time and space; 25 years of colinear tinkering. Science (80-). 2003;301(5631):331–3.
111. Forlani S. Acquisition of Hox codes during gastrulation and axial elongation in the mouse embryo. Development. 2003;130(16):3807–19.
112. Iimura T, Pourquié O. Collinear activation of Hoxb genes during gastrulation is linked to mesoderm cell ingression. Nature. 2006;442(7102):568–71.
113. Deschamps J, Duboule D. Embryonic timing, axial stem cells, chromatin dynamics, and the Hox clock. Genes Dev. 2017;31(14):1406–16.
114. Nakashima M, Toyono T, Akamine A, Joyner A. Expression of growth/differentiation factor 11, a new member of the BMP/TGFbeta superfamily during mouse embryogenesis. Mech Dev. 1999;80(2):185–9.
115. Oh SP, Li E. The signaling pathway mediated by the type IIB activin receptor controls axial patterning and lateral asymmetry in the mouse. Genes Dev. 1997;11(14):1812–26.
116. Denans N, Iimura T, Pourquié O. Hox genes control vertebrate body elongation by collinear Wnt repression. elife. 2015;2015(4):1–33.
117. Economides KD, Zeltser L, Capecchi MR. Hoxb13 mutations cause overgrowth of caudal spinal cordand tail vertebrae. Dev Biol. 2003;256(2):317–30.
118. van de Ven C, Bialecka M, Neijts R, Young T, Rowland JE, Stringer EJ, et al. Concerted involvement of Cdx/Hox genes and Wnt signaling in morphogenesis of the caudal neural tube and cloacal derivatives from the posterior growth zone. Development. 2011;138(16):3451–62.
119. Gomez C, Özbudak EM, Wunderlich J, Baumann D, Lewis J, Pourquié O. Control of segment number in vertebrate embryos. Nature. 2008;454(7202):335–9.

120. Matsubara Y, Hirasawa T, Egawa S, Hattori A, Suganuma T, Kohara Y, et al. Anatomical integration of the sacral-hindlimb unit coordinated by GDF11 underlies variation in hindlimb positioning in tetrapods. Nat Ecol Evol. 2017;1(9):1392–9.
121. Gouti M, Tsakiridis A, Wymeersch FJ, Huang Y, Kleinjung J, Wilson V, et al. In vitro generation of neuromesodermal progenitors reveals distinct roles for wnt signalling in the specification of spinal cord and paraxial mesoderm identity. PLoS Biol. 2014;12(8)
122. Turner DA, Hayward PC, Baillie-Johnson P, Rue P, Broome R, Faunes F, et al. Wnt/ -catenin and FGF signalling direct the specification and maintenance of a neuromesodermal axial progenitor in ensembles of mouse embryonic stem cells. Development. 2014;141(22):4243–53.
123. Tesar PJ, Chenoweth JG, Brook FA, Davies TJ, Evans EP, Mack DL, et al. New cell lines from mouse epiblast share defining features with human embryonic stem cells. Nature. 2007;448(7150):196–9.
124. Edri S, Hayward P, Jawaid W, Martinez AA. Neuro-mesodermal progenitors (NMPs): a comparative study between pluripotent stem cells and embryo-derived populations. Development. 2019;146(12)
125. Edri S, Hayward P, Baillie-Johnson P, Steventon BJ, Martinez AA. An epiblast stem cell-derived multipotent progenitor population for axial extension. Development. 2019;146(10)
126. Lippmann ES, Williams CE, Ruhl DA, Estevez-silva MC, Chapman ER, Coon JJ, et al. Deterministic HOX Patterning in Human Pluripotent Stem Cell-Derived Neuroectoderm. Stem Cell Reports. 2015;4(4):632–44.
127. Simunovic M, Brivanlou AH. Embryoids, organoids and gastruloids: new approaches to understanding embryogenesis. Dev. 2017;144(6):976–85.
128. Harrison SE, Sozen B, Christodoulou N, Kyprianou C, Zernicka-Goetz M. Assembly of embryonic and extraembryonic stem cells to mimic embryogenesis in vitro. Science (80-). 2017;356(6334)
129. Rivron NC, Frias-Aldeguer J, Vrij EJ, Boisset JC, Korving J, Vivié J, et al. Blastocyst-like structures generated solely from stem cells. Nature. 2018;557(7703):106–11.
130. Sozen B, Amadei G, Cox A, Wang R, Na E, Czukiewska S, et al. Self-assembly of embryonic and two extra-embryonic stem cell types into gastrulating embryo-like structures. Nat Cell Biol. 2018;20(8):979–89.
131. Turner DA, Girgin M, Alonso-Crisostomo L, Trivedi V, Baillie-Johnson P, Glodowski CR, et al. Anteroposterior polarity and elongation in the absence of extraembryonic tissues and of spatially localised signalling in gastruloids: mammalian embryonic organoids. Dev. 2017;144(21):3894–906.
132. Beccari L, Moris N, Girgin M, Turner DA, Baillie-Johnson P, Cossy AC, et al. Multi-axial self-organization properties of mouse embryonic stem cells into gastruloids. Nature. 2018;562(7726):272–6.
133. Mulas C, Kalkan T, von Meyenn F, Leitch HG, Nichols J, Smith A. Correction: defined conditions for propagation and manipulation of mouse embryonic stem cells. Development. 2019;146(7) https://doi.org/10.1242/dev.173146.
134. Peng G, Suo S, Cui G, Yu F, Wang R, Chen J, et al. Molecular architecture of lineage allocation and tissue organization in early mouse embryo. Nature. 2019;572(7770):528–32.
135. Pijuan-Sala B, Griffiths JA, Guibentif C, Hiscock TW, Jawaid W, Calero-Nieto FJ, et al. A single-cell molecular map of mouse gastrulation and early organogenesis. Nature. 2019;566(7745):490–5.
136. Cao J, Spielmann M, Qiu X, Huang X, Ibrahim DM, Hill AJ, et al. The single-cell transcriptional landscape of mammalian organogenesis. Nature. 2019;566(7745):496–502.

Skeletal Muscle Development: From Stem Cells to Body Movement

Marianne Deries, André B. Gonçalves, and Sólveig Thorsteinsdóttir

Contents

9.1 Anatomy of Skeletal Muscle – 161

9.2 Skeletal Muscle Development at a Glance – 163

9.3 Muscle Stem Cells: Where Do They Come From? – 164

9.4 Onset of Skeletal Muscle Development – 166
9.4.1 Triggers of Myogenic Differentiation – 166
9.4.2 Keeping the Balance Between Differentiation and Self-renewal of MuSCs in Space and Time – 169

Marianne Deries and André B. Gonçalves contributed equally to this chapter.

© Springer Nature Switzerland AG 2020
G. Rodrigues, B. A. J. Roelen (eds.), *Concepts and Applications of Stem Cell Biology*,
Learning Materials in Biosciences, https://doi.org/10.1007/978-3-030-43939-2_9

9.5		Primary Myogenesis: Construction of the Skeletal Muscle Pattern – 170
9.5.1		Primary Myogenesis – 170
9.5.2		Innervation – 172
9.5.3		Myogenesis and the Development of Muscle Connective Tissue Including Tendons – 172
9.6		Secondary Myogenesis: The Growth of the Embryonic Muscle Pattern – 174
9.6.1		Secondary Myogenesis – 174
9.6.2		Change in MuSC Identity – 175
9.7		Muscle Development and Regeneration After Birth – 176
9.7.1		Perinatal Muscle Growth – 176
9.7.2		The Setting Aside of Satellite Cells – 176
9.7.3		Skeletal Muscle Regeneration – 178
9.8		How Does the Study of Muscle Stem Cells in Development Contribute to the Study of Skeletal Muscle Diseases? – 179
		References – 181

9

Skeletal Muscle Development: From Stem Cells to Body Movement

What Will You Learn in This Chapter?
- In this chapter, the example of skeletal muscle development will be used to learn the following:
- How and where muscle stem cells (MuSCs) arise.
- How myogenesis starts in the embryo.
- How MuSCs interact with each other and with their neighbours, be it differentiated muscles cells or other cell types, and how they respond to the cues around them.
- How some MuSCs can stay undifferentiated and proliferative while others differentiate and construct the muscle tissue.
- How skeletal muscle is constructed progressively and in separate phases.
- How MuSC characteristics change over developmental time in harmony with their changing environment.
- How some MuSCs stay undifferentiated, even in the formed tissue after birth, and are set aside as quiescent MuSCs in the adult.
- How skeletal muscle becomes a highly regenerative tissue that responds to exercise or damage by activating its quiescent MuSCs which repair the muscle.
- How certain muscle diseases, such as muscular dystrophies, seriously affect muscle construction and function.
- How many outstanding questions regarding the development and plasticity of skeletal muscle remain and why they need to be addressed to fully understand this amazing tissue and to develop therapeutic strategies for muscle-related diseases.

9.1 Anatomy of Skeletal Muscle

In vertebrates, there are three types of muscles: (1) smooth muscles which are found in internal organs and around blood vessels, (2) cardiac muscle of the heart, responsible for the pumping of our blood, and (3) skeletal muscles which are attached to our bones and enable us to breathe, move, maintain our posture and produce heat. While smooth and cardiac muscles are tuned by the autonomous nervous system, the contraction of skeletal muscles is controlled by our will, through the action of motor neurons. Thus, skeletal muscles enable us to walk, run, jump as well as dance and smile: in essence, express who we are.

Skeletal muscle cells are large multinucleated cells surrounded by a specialised extracellular matrix called basement membrane (◘ Fig. 9.1a). They are called muscle fibres because they have a long cylindrical shape, lie parallel to each other within a muscle and generally stretch from tendon to tendon (◘ Fig. 9.1a). Connective tissue organises the superstructure of the muscle: it provides routes for blood vessels and nerve fibres and organises muscle fibres in different subunits. Considering the connective tissue layers from internal to external: individual muscle fibres are wrapped by the endomysium, the perimysium organises them into bundles (or fascicles) and the epimysium surrounds the whole muscle (◘ Fig. 9.1a). These three sheaths are continuous with each other and extend to attach directly to the bone or to join the tendon which anchors to the bone or cartilage. They can, therefore, transmit forces generated by the contraction of the fibres.

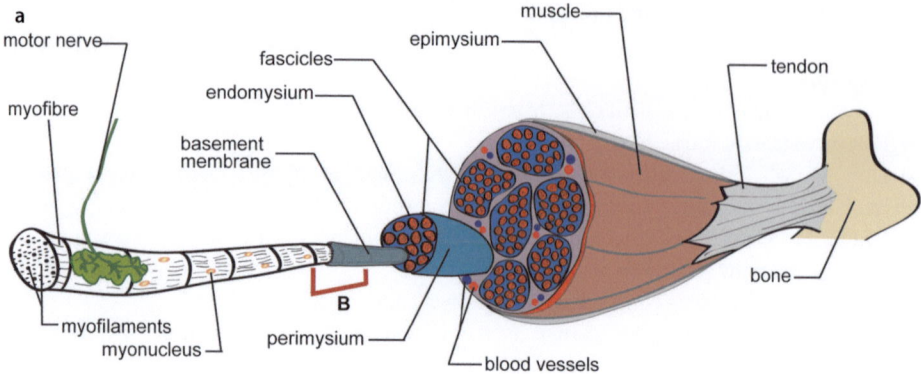

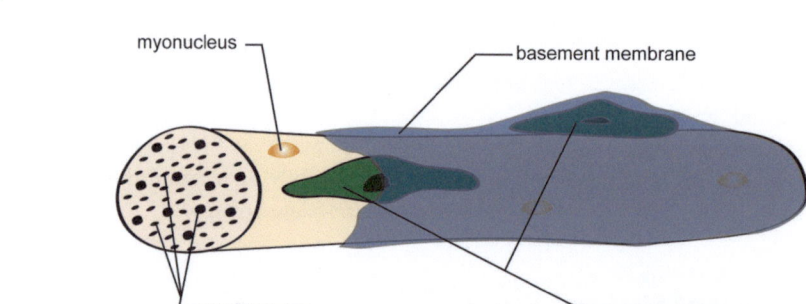

Fig. 9.1 Muscle anatomy. **a** Adult skeletal muscles are part of the musculo-skeletal system which is composed of bone, tendon attaching the bone to the muscle, nerves which enable inputs to order contraction, the skeletal muscle and the blood vessels. The whole muscle is highly organised by its connective tissues (endomysium, perimysium and epimysium) wrapping muscle structures. The muscle cell itself, called myofibre or muscle fibre, is a multinucleated cell containing myofilaments which enable its contraction. It is wrapped by its basement membrane. **b** Muscle stem cells (MuSCs; green) lie underneath the basement membrane of the myofibre, scattered along the fibre as satellites (hence their designation as satellite cells). Myonucleus refers to the nuclei of the muscle fibre

Muscle fibres are highly specialised cells with their many nuclei situated in the periphery of the fibres. The sarcoplasm (cytoplasm of the muscle) is packed with contractile units called the sarcomeres. Within the sarcomeres, thick and thin filaments (containing, respectively, the myosin and the actin filaments) alternate and make the striation of the muscle visible by microscopy. This particular organisation of the filaments is essential for the contraction of the cell. Skeletal muscle contraction is triggered and controlled by motor nerves (Fig. 9.1a). When a signal is sent by some specific motor nerves, the muscle fibres within a certain muscle contract in synchrony, enabled by some specific superstructure of the muscle, and produce the movement.

Close examination of a muscle fibre reveals the presence of small elongated cells lying between the muscle fibre and its basement membrane (Fig. 9.1b). These are adult muscle stem cells (MuSCs), also called satellite cells [1] because of their position near each muscle fibre. In healthy adult muscles, these MuSCs remain quiescent. However, if the muscle is injured, they are activated and proliferate. Subsequently, some of them differentiate and repair the muscle, while others return to a quiescent state under the basement membrane.

9.2 Skeletal Muscle Development at a Glance

Skeletal muscles occupy a huge volume in our bodies and a great number of cells are needed to construct them. Differentiation of the first skeletal muscle cells starts very early in embryonic development (blue cells in ◘ Fig. 9.2a) and this process proceeds at a steady pace thereafter [2, 3]. This progressive build-up of skeletal muscle is achieved through a tightly regulated balance between proliferation and differentiation of MuSCs. In essence, at each stage of development, some MuSCs differentiate, while others keep proliferating, generating enough cells to maintain skeletal muscle differentiation and growth until adulthood. Towards the end of embryonic development and before foetal development starts, all muscle groups have formed (blue cells in ◘ Fig. 9.2b), have been innervated and connected to the skeleton via tendons.

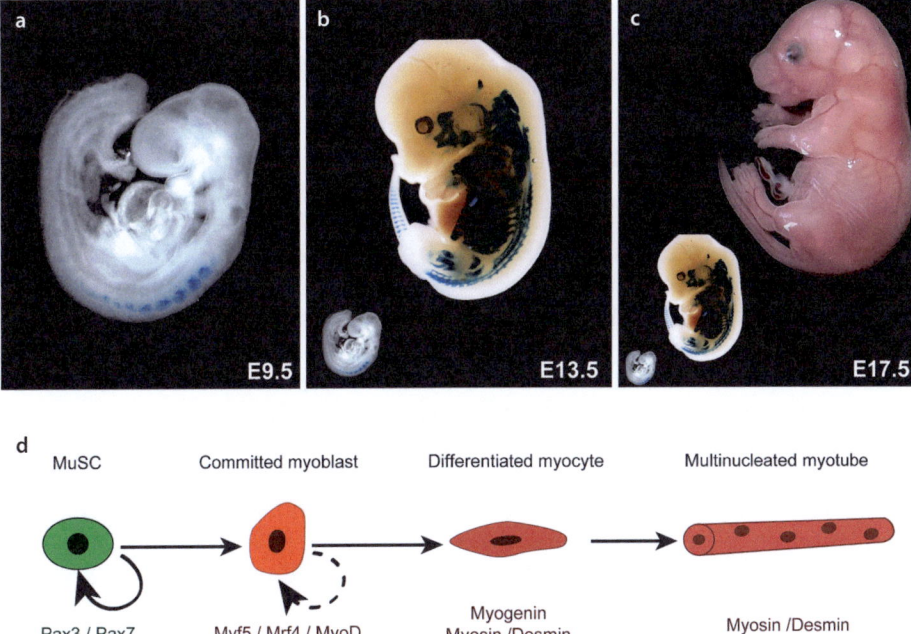

◘ **Fig. 9.2** Myogenesis goes through distinct steps. **a** Skeletal muscle development starts early in development (E8.5-E9.0 in the mouse). Myogenesis is first triggered in the dorso-medial lip of the dermomyotome and cells that activate *Myf5* expression (blue; seen through X-gal staining of heterozygous $Myf5^{nLacZ}$ mice; courtesy of S. Tajbakhsh) enter the myotome as committed myoblasts and differentiate into myotomal myocytes. **b** Towards the end of primary myogenesis (here shown for E13.5 in the mouse), the basic muscle pattern has formed. By E14.5, these 'miniature muscles' have been innervated and connected to tendons. Muscles (blue) are visualised with X-gal staining in mice with the myosin light chain type 3F (*Mlc3f*) transgene; (courtesy of M. Buckingham). Inset shows the relative size of the E13.5 compared to the E9.5 embryos. **c** Mouse foetus at E17.5, showing the increase in size between E9.5 (first inset), E13.5 (second inset) and E17.5. **d** Myogenesis goes through distinct steps: a Pax3- and/or Pax7-expressing MuSC turns on Myf5, Mrf4 and/or MyoD, designated myogenic regulatory factors (MRFs), which turns them into committed myoblasts. These myoblasts can divide a few times, but after they upregulate the MRF Myogenin, a differentiation factor, they exit the cell cycle and start expressing muscle structural proteins such as myosins and desmin. These differentiated myocytes then fuse with each other forming multinucleated myotubes, which mature into muscle fibres. During foetal and perinatal stages, myocytes can also fuse with existing myofibres, increasing their size

During foetal development, each muscle grows tremendously through the addition of differentiated cells to the muscle pattern established during embryonic development, and in tune with the growth of the foetus (Fig. 9.2c). Growth then continues after birth, until adulthood.

MuSCs are multipotent stem cells, which in the trunk and limbs are characterised by the expression of Pax3 and/or Pax7 (Fig. 9.2d; [3, 4]). MuSCs enter the myogenic differentiation programme when they start expressing a specific set of transcription factors called the myogenic regulatory factors (MRFs). MRFs are members of the MyoD family of myogenic basic helix–loop–helix (bHLH) transcription factors and their expression leads to myogenic determination and differentiation [5, 6]. In vertebrates, there are four MRFs named Myf5, Mrf4 (also known as Myf6), MyoD and Myogenin [7]. When a MuSC turns on the expression of Myf5, MyoD or Mrf4, it becomes committed to myogenesis and is termed a myoblast. Committed myoblasts can divide a few times (Fig. 9.2d). However, after entering the committed state, myoblasts upregulate the fourth MRF, Myogenin, which leads to their exit from the cell cycle, the synthesis of muscle structural proteins such as desmin and myosins and their terminal differentiation (Fig. 9.2d). Finally, terminally differentiated myocytes fuse with each other or to existing muscle fibres (Fig. 9.2d), thus contributing to muscle growth.

9.3 Muscle Stem Cells: Where Do They Come From?

MuSCs of the trunk and limbs are derived from transient embryonic structures called somites. Somites are formed progressively and in pairs early in development, one somite on each side of the neural tube and the notochord [8]. They form from the unsegmented presomitic mesoderm, budding off at regular intervals as metameric spheres of epithelial cells encompassing a central cavity, named somitocoel (Fig. 9.3a; [9]).

Soon after epithelial somites form, they give rise to different compartments. The ventral portion of each somite undergoes an epithelium to mesenchyme transition to form the sclerotome, a compartment containing the precursors of the vertebrae and ribs (Fig. 9.3b; [10]). Later, the dorsal-most part of the sclerotome, which is called the syndetome [11], is specified into tendon precursors which will form the tendons attaching the axial muscles to the vertebrae and ribs.

Cells in the dorsal somite remain epithelial and form a compartment designated the dermomyotome (Fig. 9.3b). All MuSCs (except those of the head) are derived from the dermomyotomes present along the rostro-caudal axis from neck to tail. Dermomyotomal cells are multipotent and, although their major derivative is MuSCs, certain regions or cells of each dermomyotome give rise to other cell types, such as precursors of the dorsal dermis, smooth muscle, endothelia and brown fat [12–15].

The dermomyotome is an epithelial sheet whose four extremities curl into four contiguous lips - defined as dorso-medial, ventro-lateral, rostral and caudal lips - surrounding what is termed the central dermomyotome. Dermomyotomal cells express the transcription factors Pax3 and/or Pax7, which mark their myogenic potential [16–19]. The dermomyotomal epithelium is lined dorsally by a basement membrane which prevents precocious myogenic differentiation in the dermomyotome

Skeletal Muscle Development: From Stem Cells to Body Movement

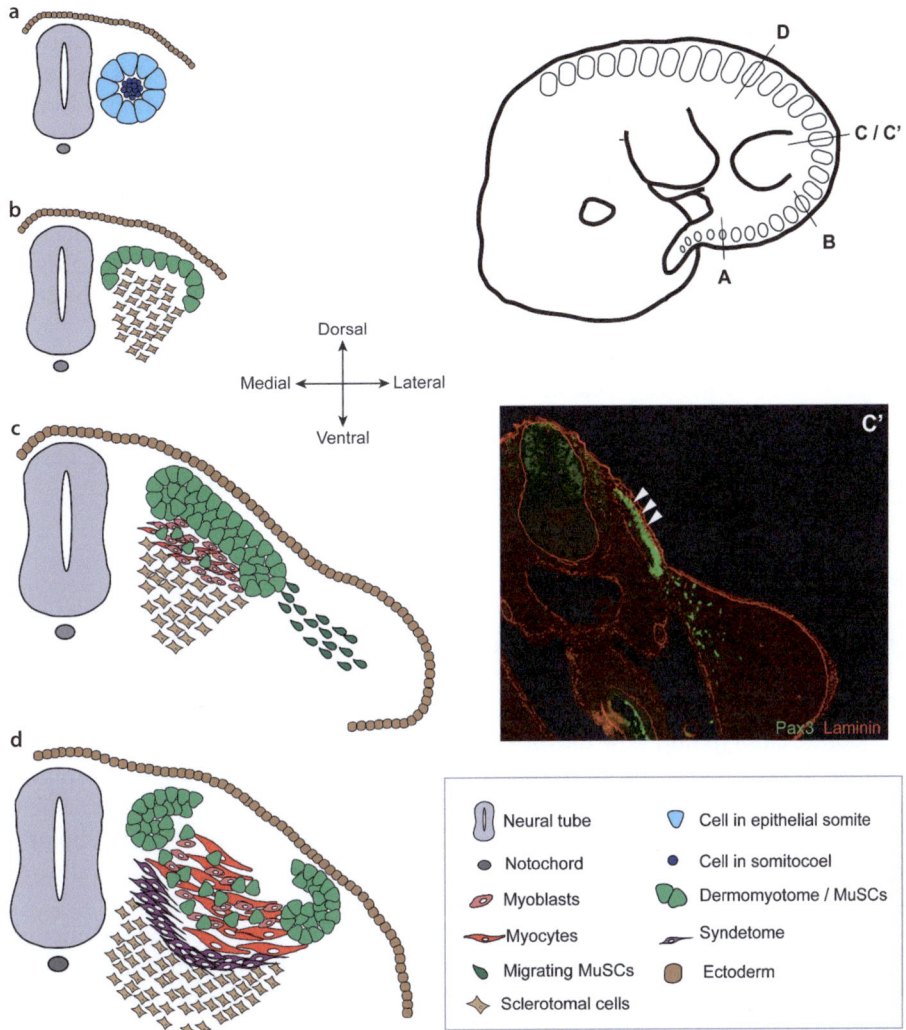

Fig. 9.3 Embryonic origin of skeletal muscle. Schematic illustration of somite development along the caudal to the rostral axis at four different levels of an E11.5 mouse embryo (see drawing, top right). Since embryos at this stage are more developed in their rostral part than their caudal region, this analysis is also a temporal one. **a** Epithelial somites are rosettes of epithelioid cells (blue) with their apical side turned towards a central somitocoel which contains a few mesenchymal cells (dark blue). **b** The ventral portion of the somite de-epithelializes to form the sclerotome (brown), while the dorsal portion remains epithelial and is designated the dermomyotome (green). **c** The dermomyotome originates MuSCs from its edges, and these MuSCs enter the area ventral to the dermomyotome, where they differentiate, first into myoblasts (pink cells) and later into myotomal myocytes (red cells). At limb level, the ventro-lateral lip of the dermomyotome originates Pax3-positive MuSCs (green cells) that migrate to the limb bud. **c'** Transverse section of a mouse embryo stained by immunofluorescence showing Pax3-positive (green) cells in the neural tube and in the dermomyotome which includes MuSCs and Pax3-positive MuSCs migrating from the dermomyotome into the limb bud. Laminin (red) lines the dermomyotomal epithelium (arrowheads), contributing to maintaining its non-differentiated state. Laminin also stains all other basement membranes present at this stage (e.g. ectoderm, neural tube). **d** The central dermomyotome dissociates and many MuSCs (green) colonise the myotome. Some remain proliferative to compose the resident pool of stem cells, while others differentiate giving rise to skeletal muscles. Communication between the myotome and the dorsal sclerotome induces the syndetome (purple)

(🔵 Fig. 9.3c'; [20]). As the dermomyotome grows, MuSCs delaminate in synchronous waves from the four dermomyotomal lips and colonise the area underneath to form the myotome, where myogenic differentiation starts (🔵 Fig. 9.3c). These are the dermomyotome lip-derived MuSCs (🔵 Table 9.1). At limb levels, MuSCs from the ventro-lateral dermomyotomal lip develop in a different way because they delaminate and migrate towards the limb bud (🔵 Fig. 9.3c') and only differentiate upon arrival to their target sites [21]. Dermomyotome-derived MuSCs also migrate to form the diaphragm and tongue [22, 23]. The dermomyotome eventually de-epithelializes, releasing proliferative MuSCs into the myotome (🔵 Fig. 9.3d). The MuSCs that migrate to the limbs, diaphragm and tongue and the MuSCs derived from the central dermomyotome are designated embryonic MuSCs (🔵 Table 9.1).

Once in their muscle mass (myotome, limb muscle or other) MuSCs have two possible fates: either they activate the myogenic differentiation programme and differentiate, or they remain proliferative to maintain the MuSC pool within the muscle mass. Some of these proliferative MuSCs come to differentiate later during development while others are put aside as quiescent adult MuSCs [16–19].

9.4 Onset of Skeletal Muscle Development

9.4.1 Triggers of Myogenic Differentiation

The segmentally organised myotomes are the first skeletal muscles to form in the vertebrate embryo and how they form has been the subject of intensive study. Early studies showed that extrinsic signals coming from the neighbouring tissues play key roles in activating MRF expression and consequently triggering myogenic differentiation (🔵 Fig. 9.4a; [14, 24, 25]). Here, we will focus on the dorso-medial lip of the dermomyotome, which is where myogenesis starts in amniote embryos, and we will see how multiple signals converge to trigger its onset.

Recent studies in the chick embryo have shown that neural crest cells migrating from the dorsal neural tube and passing the dorso-medial lip of the dermomyotome contribute to trigger myogenesis. Migrating neural crest cells were shown to express Delta-like ligand 1 (Dll1) which binds to Notch receptors on cells in the dorso-medial lip, leading to a transient activation of Notch which culminates in Myf5 activation in these cells (🔵 Fig. 9.4b; [26, 27]). Migrating neural crest cells also express a transmembrane heparan sulphate proteoglycan, named glypican 4 (GLP4), on their cell surface, which carries Wnts from the dorsal neural tube to the dorso-medial dermomyotomal lip (🔵 Fig. 9.4b; [28]). Studies in both chick and mouse have shown that Wnt signalling in the dorso-medial lip synergizes with Sonic hedgehog (Shh) signalling to activate Myf5 expression (🔵 Fig. 9.4c; [29]). Shh is secreted from the notochord and neural tube floor plate, travels through the sclerotome and activates Gli2 and Gli3 in the cells of the dorso-medial lip (🔵 Fig. 9.4c; [30, 31]). Shh is, however, not able to trigger the myogenic differentiation programme by itself. Rather, it seems that Wnts (Wnt1, 3a, 4) from the dorsal neural tube and ectoderm act through β-catenin which synergises with Shh signalling to activate Myf5 in the dorso-medial lip (🔵 Fig. 9.4c; [29]). β-catenin also induces the expression of Noggin, a bone mor-

Table 9.1 MuSC terminology

MuSC type (/alternative name)	Pax gene expression	Stage (in mouse)	Location	Arises from	Differentiates into
MuSCs from dermomyotome lips / founder MuSCs[a]	First none[a], then Pax3	E8.5/9.0–E11.0	Trunk	Dorso-medial, rostral, caudal and ventro-lateral dermomyotome lips (in trunk)	Myotome
Embryonic MuSCs (trunk) / founder MuSCs[b]	Pax3/Pax7	E11.0–E14.5	Trunk	Central dermomyotome	Primary (embryonic) myofibres in trunk
Embryonic MuSCs (limbs) / founder MuSCs[3]	Pax3	E10.5–E14.5	Limb-levels	Ventro-lateral dermomyotome lip (limb levels)	Primary (embryonic) myofibres in limbs
Foetal MuSCs	Pax7	E14.5–birth	Trunk and limbs	Embryonic MuSCs[b]	Secondary (foetal) myofibres. Also contribute to the growth of all myofibres.
Perinatal MuSCs / juvenile satellite cells	Pax7	Birth–P21	Trunk and limbs	Foetal MuSCs[b]	Contribute to the growth of all myofibres. Generate some new myofibres.
Adult MuSCs / Satellite cells	Pax7	P21 onwards	Trunk and limbs	Perinatal MuSCs[b]	Enter quiescence. Activated upon growth, exercise or injury and contribute to muscle repair. After repair, some of them reenter quiescence.

Simplified terminology of the different types of dermomyotome-derived MuSCs (head MuSCs are not included). Six major MuSCs types have been defined, but it is important to appreciate that each MuSC type is heterogeneous and that the significance of this heterogeneity is presently unclear (based on [2, 3, 90, 93])

[a]The first cells to differentiate from the dermomyotome lips do not express Pax3. They may not be true muscle stem cells

[b]Although the current view is that one MuSC type develops into the next type, the possibility that different subpopulations within the dermomyotome originate the different MuSC types cannot be excluded

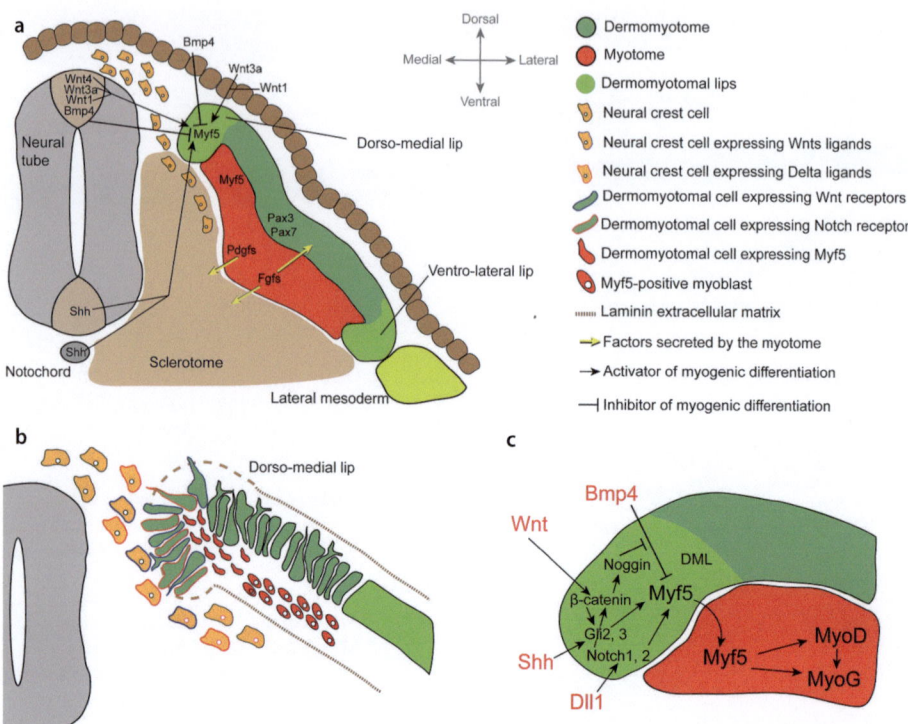

Fig. 9.4 Induction of myotome formation by neighbouring tissues. **a** Global overview of the extrinsic signals that trigger myogenesis in the dorso-medial lip of the dermomyotome. A combination of factors is thought to lead to Myf5 activation in this region. Neural crest cells migrate from the dorsal neural tube and bring the Notch ligand Dll1 and Wnts on their cell surface to the cells of the dorso-medial lip as they migrate past this region (see B for more details). In parallel, Shh secreted from the notochord and the floor plate of the neural tube acts together with Wnt and Notch signalling to upregulate Myf5 in this region. In contrast, Bmp4 secreted by the neural tube and ectoderm acts as a repressor of myogenesis. As myogenesis progresses, cells in the myotome (red) produce Pdgfs that regulate sclerotome differentiation and Fgfs, which induce syndetome formation and the dissociation of the central dermomyotome. **b** Migratory neural crest cells expressing Delta and carrying Wnt ligands from the dorsal neural tube briefly bind to filopodia extended by cells in the dorso-medial lip which have Notch and Frizzled receptors. These two signals, as well as Shh coming from the notochord and floor plate (see A), induce Myf5 expression and dermomyotomal cells delaminate as committed myoblasts (red cells) into the myotomal area. The brown dashed lines represent the dermomyotomal and myotomal basement membranes, which are discontinuous. Notably, the holes in the basement membrane appear to enable cell–cell interactions between dermomyotomal cells and the neural crest cells. **c** Schematic summary of the interplay between the different signalling pathways involved in the induction of myogenesis in the dorso-medial lip (bright green). Dll1 induces Notch activity which brings β-catenin from the membrane to the cytoplasm in the cells of the dorso-medial lip, Wnt signalling acts through β-catenin to induce Noggin and promotes Gli2 and Gli3 expression and Shh signalling in turn activates Gli2 and Gli3. The result is that attenuated Bmp4 signalling lifts the block on Myf5 transcription, while Notch, Wnt and Shh signalling appear to cooperate to activate Myf5. The Myf5-expressing myoblasts then enter the myotome (red) where they express additional MRFs and differentiate into myotomal myocytes

phogenetic protein (Bmp) antagonist [32]. Bmp4 secreted from the dorsal ectoderm and neural tube normally represses myogenesis [33]. However, through the activation of Noggin in the dorso-medial lip, Bmp signalling is specifically blocked in this region and allows for Myf5 activation (◘ Fig. 9.4c). In summary, Notch, Wnt and Shh signalling converge in cells of the dorso-medial dermomyotomal lip where they appear to collaborate to counteract myogenic repressors and activate myogenesis (◘ Fig. 9.4c).

Myogenesis continues in the dorso-medial lip and meanwhile also starts in the ventro-lateral, rostral and caudal dermomyotomal lips [34–36]. As new myoblasts enter the myotome and differentiate, the myotome grows in both the dorsal–ventral and medial–lateral direction. The myotome itself also starts influencing its neighbours, for example by producing platelet-derived growth factors (Pdgfs) and fibroblast growth factors (Fgfs) [37–39]. Pdgfs influence differentiation in the sclerotome (◘ Fig. 9.4a; [40]), while Fgfs induce syndetome specification [39] and act on the central dermomyotome, promoting its subsequent de-epithelialization (◘ Fig. 9.4a; [41]). This de-epithelialization brings the proliferative Pax3- and/or Pax7-positive embryonic MuSCs into the myotome [16–19].

The extrinsic signals that trigger myogenesis in other sites of the embryo are not exactly the same [14, 42]. Nevertheless, even if the details differ, the Wnt, Shh, Fgf, Bmp and Notch signalling pathways are consistently involved in both trunk and limb myogenesis. An interesting difference is that during limb muscle development differentiated myocytes fuse into myotubes faster than those of the myotome [43], most likely because they are evolutionary more recent and are not restrained by the developmental programme specific to the myotome [14, 44].

Head muscle development is very different. Head mesoderm, from which head MuSCs derive, does not express Pax3 [45] and Pax7 is only upregulated after developing muscle masses have formed [18, 19, 46]. Nevertheless, as in trunk and limbs, myogenic differentiation goes through the MRFs, but their specification involves the transcription factors, Tbx1 and/or Pitx2, suggesting that myogenesis in the head took a different evolutionary route from trunk (and limb) myogenesis [23].

9.4.2 Keeping the Balance Between Differentiation and Self-renewal of MuSCs in Space and Time

Skeletal muscle development from here on proceeds through different steps which take place at different time points and do not have the same role [47]. First, embryonic MuSCs undergo primary myogenesis which sets the basic muscle pattern (between the E11.0 and E14.5 in the mouse). During secondary myogenesis, this basic pattern is used as a scaffold for tremendous muscle growth (from E14.5 to birth in the mouse) [48, 49]. Finally, muscle growth after birth is driven by perinatal MuSCs (also called juvenile satellite cells; ◘ Table 9.1) which are not yet quiescent. We will look at these different steps in more detail in the next section.

However, one obvious question that arises when studying any step of myogenesis is how some MuSCs are kept in an undifferentiated state within the developing muscles while others are induced to enter myogenesis. In other words, how do some MuSCs avoid differentiation in an environment that promotes entry into the myogenic programme? This balance between proliferation and differentiation is regulated by the Notch signalling pathway [50, 51]. During early stages of myogenesis in the trunk and limbs, differentiating myoblasts express Notch ligands which can bind to Notch receptors expressed on MuSCs and maintain their undifferentiated and proliferative state [52–54]. At later stages of myogenesis, not only myoblasts, but also the forming myofibres express Notch ligands [55]. Indeed, when the Notch intracellular domain (NICD), the part of the Notch receptor that enters the nucleus to activate target genes is overexpressed in MuSCs, they remain undifferentiated and proliferative until late foetal development and no skeletal muscles form [56]. Importantly, this Notch signalling is not transient (which leads to Myf5 activation in the dorso-medial lip in the chick; [26]), but sustained. Sustained Notch signalling is thus the single most important pathway in maintaining the pool of MuSCs throughout myogenic development.

9.5 Primary Myogenesis: Construction of the Skeletal Muscle Pattern

9.5.1 Primary Myogenesis

Primary myogenesis is essential to organise the basic muscle pattern and set the connection between muscles and their tendons and nerves. In the trunk, primary myogenesis starts after myotome development, when MuSCs from the dissociating dermomyotome enter the myotome (Fig. 9.5a; [57, 58]; Table 9.1). In the limbs, and other regions receiving migrating dermomyotome-derived MuSCs, it starts after the arrival of the MuSCs to their target sites (Fig. 9.5a; [59]; Table 9.1).

It is interesting to note that the environment of primary myogenesis is different from that of the later myogenesis steps. Obviously, as myogenesis advances, the tissues surrounding the developing muscles also advance in their development and, therefore, change their repertoire of secreted factors. Thus, MuSC identity changes over time and these environmental cues appear to play a major role in this change (Fig. 9.5b; [50, 60]; Table 9.1).

In the limb, primary myoblasts (also called embryonic myoblasts) differentiate from MuSCs expressing Pax3 ([61]; Table 9.1). In the trunk, the same presumably holds true, although since Pax7 is expressed earlier in the trunk (i.e. in the central dermomyotome), primary myoblasts are probably also derived from cells co-expressing Pax3 and Pax7 ([19], Table 9.1). Curiously, embryonic MuSCs and myoblasts express Hox genes, which later stage myogenic cells do not [62]. It is interesting to speculate that these Hox genes are important to construct the muscle pattern during primary myogenesis. Multiple fusion events between differentiating myoblasts and/or the existing myocytes quickly generate the first primary myotubes (Fig. 9.5a). These primary myotubes extend from tendon to tendon, are fully differentiated and can contract. However, they are few in number and have a small cross-sectional area (Fig. 9.5c; [63]). Although the primary myotubes are multinucleated, their nuclei are centred in the cell.

One other striking difference between primary myogenesis and later stages of myogenesis is that there is a total absence of laminin matrix around the primary

Skeletal Muscle Development: From Stem Cells to Body Movement

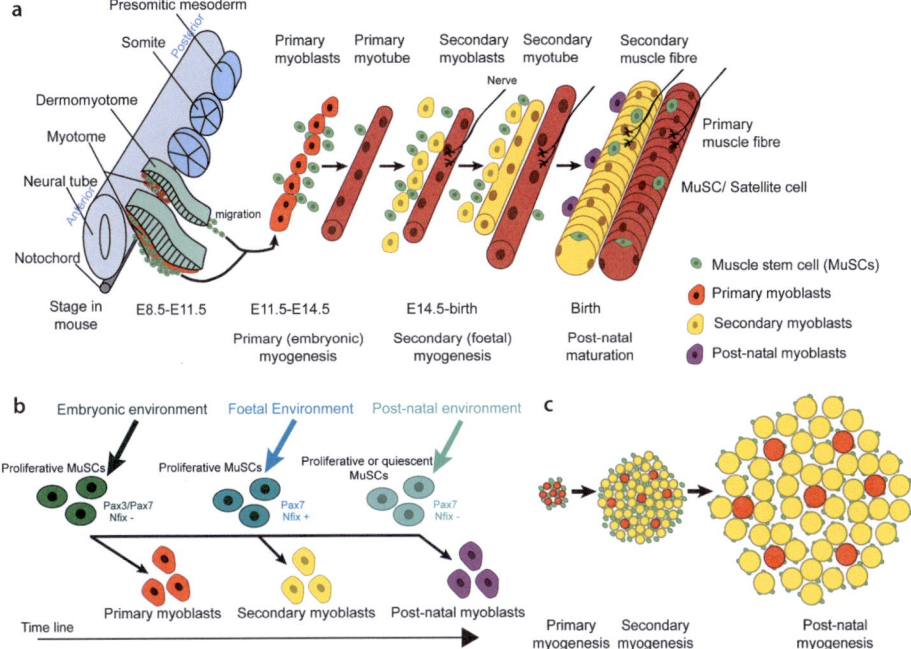

Fig. 9.5 Summary of skeletal muscle development. **a** All the muscles of the vertebrate body (except those of the head) are derived from MuSCs originating from the dermomyotomes of the somites. MuSCs (green) either undergo long-range migration and differentiate after reaching their target site or translocate underneath the dermomyotome and differentiate there to form the segmented myotomes which will later transform into axial muscles. MuSCs either stay proliferative or differentiate into primary (red) then secondary (yellow) myoblasts and finally into post-natal (perinatal or adult) myoblasts, depending on when during development they enter the differentiation programme. Primary myoblasts fuse with each other forming multinucleated primary myotubes (red), then secondary myoblasts first fuse with each other forming secondary myotubes (yellow) and then with all the existing myotubes, increasing their size. During post-natal stages, myotubes mature into myofibres and MuSCs present at those stages come to enter quiescence and occupy a position as satellite cells along the fibres. **b** MuSCs develop different stage-specific identities. Tissue complexity increases over time, which leads to a constantly changing environment. For example, in developing skeletal muscle, other cell types such as in-growing blood vessels and nerves send different cues to their muscle neighbours at different developmental stages. MuSCs are sensitive to their surrounding environment and their identity changes as development proceeds. **c** Transverse view of skeletal muscle during the three steps of muscle growth during myogenesis: During primary myogenesis, primary myofibres (red) develop to make the muscle pattern of the musculo-skeletal system. During secondary myogenesis, the number of fibres increases drastically as secondary myoblasts form secondary myotubes (yellow) around the primary myotubes. This phase defines the number of fibres of each muscle of the adult. In late foetal perinatal stages, growth is normally only by cell-mediated hypertrophy where each fibre grows by addition (fusion) of myoblasts but the number of fibres stays the same

muscle fibres [64]. Indeed, primary myotubes not only lack a basement membrane but they do not express any laminin receptors [65]. The developmental significance of this fact is presently not known, but it is possible that since primary myotubes are contacted by nerves and surrounded by mesenchymal cells, it is important that no extracellular matrix barrier prevents this communication.

9.5.2 Innervation

As soon as primary myogenesis is underway, nerves invade the presumptive muscle masses [66]. Sensory neurons derive from the neural crest cells which migrate to form the dorsal root ganglia, located in pairs on each side of the neural tube. The body of the neurons is in the dorsal root ganglia and they extend their axons and dendrites from there. Motor neurons exit the ventral part of the neural tube and mix with the sensory neurons to form the spinal nerves and they migrate together towards their targets [67]. The axons of motor neurons migrate to their target through precise paths. Their growth cones sense the different cell surface-bound molecules along the path and are attracted or repulsed by them, allowing these 'seeker heads' to find the way [68]. As nerves grow, neural crest cell precursors of Schwann cells migrate along just after them [69]. Once at their target sites, they differentiate into Schwann cells. The whole peripheral nervous system is thus in place to innervate the muscles.

When motor neurons contact muscle fibres, an intricate communication between the two cells takes place to form the specialised area called the synapse (Fig. 9.6a–c; [70–72]). In the synapse region, the nerve terminal and the muscle endplate become specialised areas within both cells, so that neural signals come to command the contraction of the muscle cell (Fig. 9.6c).

9.5.3 Myogenesis and the Development of Muscle Connective Tissue Including Tendons

Muscle anchors to tendons at each end of the muscle fibre and muscle membrane receptors such as integrin α7β1 [73–75], or dystroglycan [76] play a role in organising these connections. Muscle connective tissue including tendons and muscle cells derive from different lineages, but both tissues are essential for each other's development [77–79]. Indeed, the communication between muscle and tendon progenitors has been reported early in development. For example, fibroblast growth factor 4 (Fgf4) secreted by the myotome is necessary for the induction of *Scleraxis* expression (a marker of tendinocyte specification) in the axial system [80]. In the limb, Fgf4 and Fgf8 secreted by the tips of muscle fibres allow the specification of tendon cells [81, 82]. In reverse, if tendinocytes are absent, or if they do not express the transcription factor Tcf4, muscle development is impaired [77, 79]. Therefore, muscle connective tissue, tendon and muscle tissue grow and organise themselves together [11, 77]. Communication between tendon cells and muscle cells ensures their synchronous development, and at the end of primary myogenesis, both are well organised (Fig. 9.7).

In conclusion, through primary myogenesis, primary muscle fibres form, are innervated and come to stretch from tendon to tendon, where they attach to the developing cartilage. They can be considered 'miniature muscles' which serve as a template for the construction of the definitive muscle. It is on top of this template that secondary myogenesis will take place (Fig. 9.5a, c).

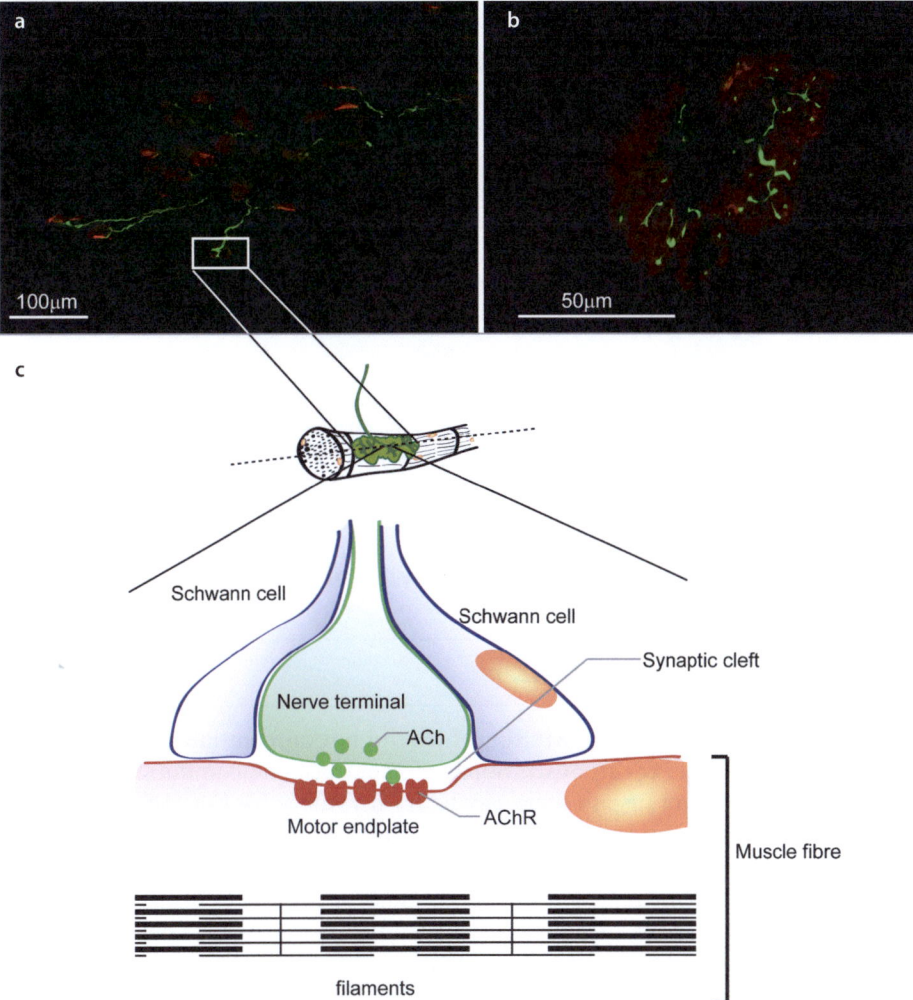

Fig. 9.6 The synapse. **a** Whole mount preparation of the diaphragm of an adult mouse. Motor neurons – stained by immunohistochemistry with antibodies against neurofilament and synaptophysin (both green) – contact the muscle cells at very specific areas called the endplates – stained with α-bungarotoxin (red) which binds to acetylcholine receptors. **b** Section of the extensor digitorum longus muscle of an adult mouse stained by immunohistochemistry (as in a.) to show a motor endplate and its nerve terminal. **c** The synapse is the specialised area where the motor nerve axon contacts the muscle fibre to send signals and control its contraction. The nerve terminal releases neurotransmitters (acetylcholine (ACh)) in the synaptic cleft, which binds to ACh receptors (AChR), located in the motor endplate of the muscle fibre

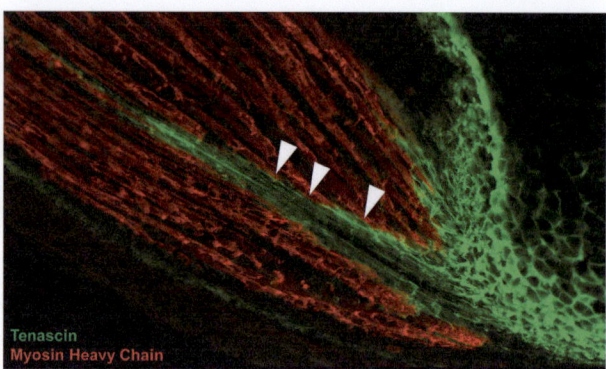

Fig. 9.7 Muscle fibres and tendons. At the end of primary myogenesis (E14.5 in the mouse embryo), muscles have connected to the bone via tendons. Here, a longitudinal section of a forelimb muscle and its tendon are stained by immunohistochemistry with antibodies against tenascin (green) an extracellular matrix component present in tendons and against myosin heavy chain (red) which stains the muscle fibres. Note how the tendon matrix is continuous with the muscle connective tissue (arrowheads)

9.6 Secondary Myogenesis: The Growth of the Embryonic Muscle Pattern

9.6.1 Secondary Myogenesis

During secondary myogenesis, MuSCs which did not differentiate during primary myogenesis proliferate enormously. Some of them become committed myoblasts which differentiate and fuse to form secondary myotubes (Fig. 9.5a) while others stay proliferative. Interestingly, differentiation and fusion of secondary myoblasts start at the innervation point of the primary myotubes, which is located near their centre [63]. Secondary myotubes are initially smaller than primary myotubes and form all around them, using them as a scaffold (Fig. 9.5c). They then extend along the primary myotubes in both directions, to finally run the whole length of the muscle and insert into the tendons [63, 83, 84]. Secondary myotubes are numerous and make most of the adult muscle fibres (Fig. 9.5a, c). Towards the end of foetal development, the formation of new secondary myofibres slows down, as differentiated myoblasts start to preferentially fuse with all existing myofibres and increase their size. This pattern of growth continues after birth.

During secondary myogenesis, the myofibre basement membrane is progressively assembled. At first, this matrix is discontinuous [64] but at the end of foetal development it forms a thin sheet wrapped around the fibre (Fig. 9.8a). This basement membrane interacts with the muscle fibre through an adhesion complex which bridges the actin cytoskeleton of the myofibre with laminin, the major constituent of the basement membrane (Fig. 9.8b). Moreover, MuSCs become wrapped within this basement membrane which, together with the myofibre, constitutes their niche. Mutations in the proteins composing this adhesion complex linking muscle cells to the basement membrane lead to diseases called muscular dystrophies.

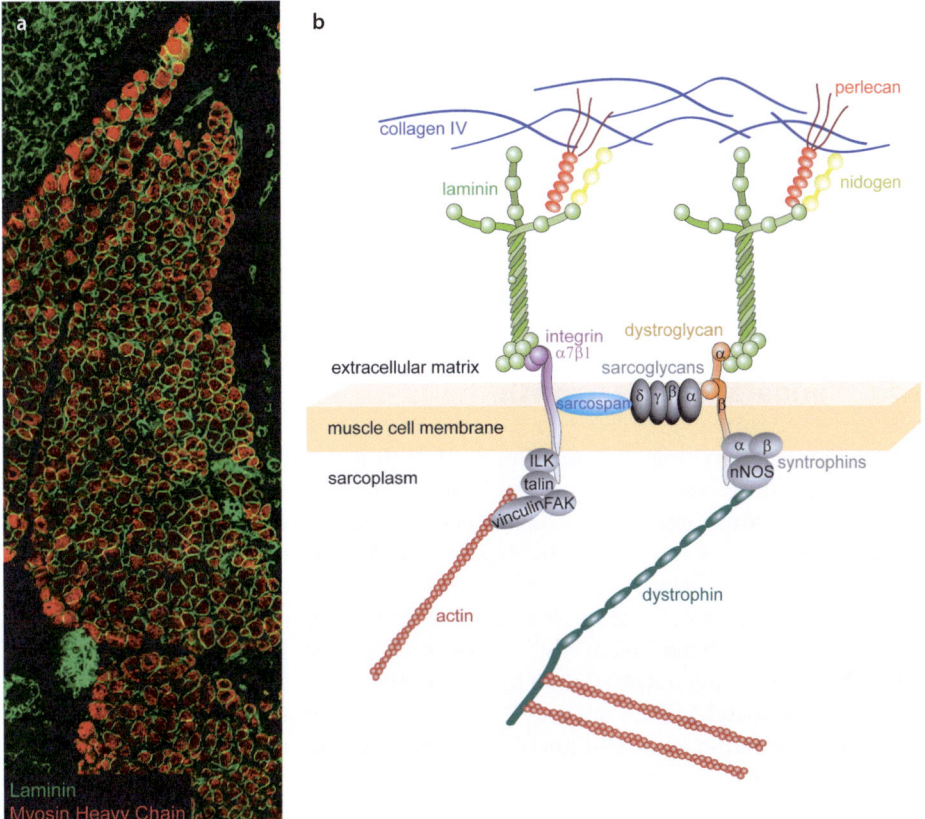

◘ **Fig. 9.8** Connection between the muscle fibre cytoskeleton and its basement membrane. **a** Transverse section of an E17.5 mouse foetus showing the deep back muscles stained with an antibody against myosin heavy chain (red) marking the muscle fibres and with an antibody recognising all muscle laminins (green). At this stage, a continuous laminin matrix surrounds each myofibre. **b** Schematic drawing showing the major proteins involved in the connection between the muscle cell cytoskeleton and its surrounding basement membrane [105, 109]. In adult muscle, the predominant laminin isoform is laminin 211

9.6.2 Change in MuSC Identity

As mentioned earlier, the profile of MuSCs producing secondary (or foetal) myoblasts is different from that of the profile of MuSCs producing primary (or embryonic) myoblasts (◘ Fig. 9.5b). In fact, foetal MuSCs express Pax7 and no longer express Pax3 (◘ Table 9.1) and they also express the transcription factor Nuclear factor one X (Nfix) (◘ Fig. 9.5b), which acts as a switch between embryonic and foetal MuSC identity [60]. In the limb, Pax7-positive, Pax3-negative cells arise later than the earlier Pax3-positive MuSCs [43] and first become detectable near the point where nerves enter the muscle masses [66]. Moreover, denervation affects secondary myogenesis more than primary myogenesis [85]. These two facts taken together raise the interesting possibility that nerves may be important to convert embryonic MuSCs into foetal MuSCs, which will later originate secondary myofibres.

If embryonic and foetal MuSCs show some differences in their profile, embryonic and foetal myoblasts show drastic differences. They respond differently to hormones and growth factors [86, 87] and their gene expression profile is very different [62]. For example, integrin-α7 and Pax7 are more expressed in foetal myoblasts than in embryonic myoblasts, whereas Pax3 and Paraxis are more expressed in embryonic myoblasts.

9.7 Muscle Development and Regeneration After Birth

9.7.1 Perinatal Muscle Growth

Skeletal muscle development continues after birth with an intense growth of muscle mass for the first 3 weeks in the mouse. Most secondary myofibres form before birth. Thus, muscle growth late in foetal development and early post-natal development primarily involves the proliferation of MuSCs and the differentiation of some of them into myoblasts which fuse with the existing myofibres, increasing their size (designated cell-mediated hypertrophy; [88]). Again, the identity of MuSCs changes from the foetal to the perinatal stage. Foetal MuSCs have a higher resistance to differentiation and a higher self-renewing potential than perinatal MuSCs [2, 89]. This makes sense as one considers that foetal MuSCs not only generate the myoblasts that form all the secondary myofibres within the foetal muscles but at the same time also build up a MuSC population that will support enormous muscle growth perinatally. However, as the environment switches from foetal to perinatal, MuSCs become progressively more prone to differentiation [90]. Perinatal MuSCs are located under the myofibre basement membrane and tend to divide asymmetrically ([88]; see Fig. 9.9a, b). This type of division produces a MuSC and a committed myoblast which differentiates and then fuses with the myofibre, contributing to its growth [88, 91].

9.7.2 The Setting Aside of Satellite Cells

Perinatal MuSC number remains relatively constant from birth until P14, indicating that asymmetric MuSC divisions are the norm in this period [88]. However, between P14 and P21, the number of perinatal MuSCs is progressively reduced, which correlates with slower proliferation rates and, by P21, MuSCs have entered quiescence [88]. The perinatal period thus starts with the growth of myofibres by cell-mediated hypertrophy and ends at P21 after which quiescent adult MuSCs or satellite cells (Table 9.1) have been set aside. Importantly, myofibre growth after P21 does not involve the addition of new cells to the fibre; rather, adult myofibres are thought to grow exclusively by protein synthesis [88, 92]. However, as we will see in the next section, if the muscle is injured, the quiescent satellite cells are activated and, through a process called muscle regeneration, can restore the structure and function of the damaged area.

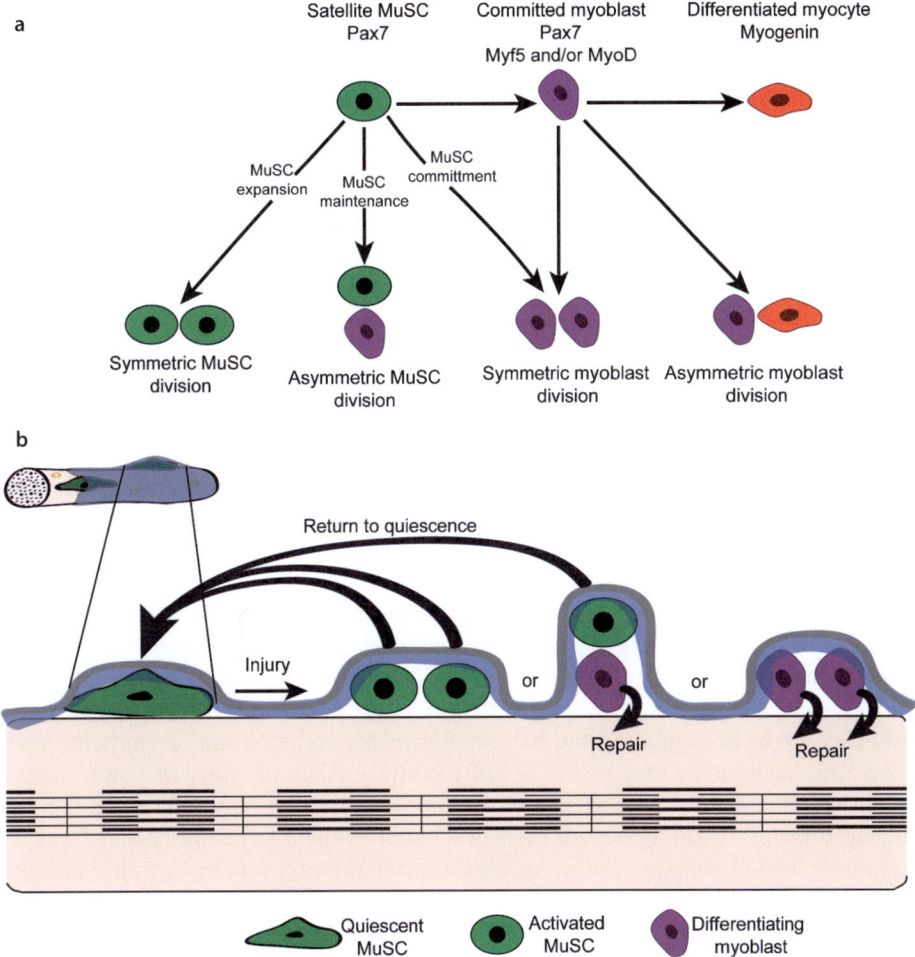

Fig. 9.9 Muscle regeneration. **a** Proposed modes of MuSC division and/or differentiation. MuSC can divide symmetrically into two MuSCs (MuSC expansion), they can divide asymmetrically into one MuSC and one myoblast (MuSC maintenance) or they can commit to myogenesis and divide into two myoblasts (MuSC commitment). MuSCs can also differentiate into myoblasts and give rise to a myocyte without dividing. Myoblasts can divide into either two myoblasts or one myoblast and one myocyte. Myoblasts can also differentiate without dividing. **b** Skeletal muscle is capable of regeneration. When the muscle experiences assault due to exercise, injury or disease, quiescent MuSCs – also called satellite cells – are activated and proliferate. These cells can either divide (1) symmetrically, giving rise to two MuSCs, which in turn can proliferate more, (2) asymmetrically, originating one MuSC and one committed myoblast, or (3) they initiate the differentiation programme and then divide symmetrically into two committed myoblasts. Committed myoblasts either differentiate without dividing or undergo a limited number of divisions before they undergo terminal differentiation and fuse with the existing myofibres. Importantly, MuSCs, which stay undifferentiated during the regeneration process, can return to a quiescent state, thus maintaining the pool of quiescent MuSCs ready for the next assault on the muscle

9.7.3 Skeletal Muscle Regeneration

Adult skeletal muscle has a remarkable ability to regenerate, even in mammals, and this property is due to the setting aside of quiescent adult MuSCs (the so-called satellite cells) during development. Quiescent satellite cells maintain a low metabolic state, are resistant to DNA damage and can retain their stem cell properties for a lifetime [93]. Upon injury, the damaged skeletal muscle releases factors, such as fibroblast and insulin-like growth factors, which induce satellite cell proliferation, while matrix metalloproteases degrade the extracellular matrix, releasing the satellite cells as well as extracellular matrix-bound mitotic factors [91, 93]. These activated adult MuSCs then rapidly migrate bidirectionally along the muscle basement membrane of the damaged fibres (designated 'ghost fibres'; [94]). There, they divide symmetrically (Fig. 9.9a, b) generating cells which become evenly distributed along the longitudinal axis of the ghost fibres [94]. Signals involving fibronectin - a component of the interstitial matrix which gets exposed upon injury - and Wnt7a promote these symmetric cell divisions that expand the MuSC pool [95, 96]. After an initial boost of self-renewing symmetric cell divisions, MuSCs start undergoing asymmetric divisions, where one cell stays in contact with the basement membrane and remains a MuSC, while the other upregulates MyoD and gets committed to myogenesis (Fig. 9.9a, b). These asymmetric divisions involve mechanisms that enhance the segregation of the two cell types. For example, the committed myoblast expresses Delta, which binds to Notch receptors on the MuSC, reinforcing its stem cell identity [97]. It is unclear what leads to this switch in types of divisions. Probably many factors play a role in regulating the balance between symmetric and asymmetric divisions. One such factor may be the size of the injury. A small injury may not 'need' a symmetric expansion of MuSCs, since asymmetric divisions may suffice to repair the damaged myofibre and replenish the MuSC pool, while a large injury may 'require' expansive MuSC divisions before myoblast differentiation sets in. Another alternative is that the MuSC pool that gets activated upon injury is heterogeneous and different subsets of MuSCs undergo symmetric versus asymmetric divisions. If this is the case, different signals would stimulate these two subtypes differently. It is known that adult MuSCs are heterogeneous [3, 90, 98], but it is presently not clear how this heterogeneity plays out in an injury setting.

Other modes of MuSC divisions have also been proposed [91, 98]. MuSC may start differentiating before dividing and thus give rise to two committed myoblasts (Fig. 9.9a, b), a situation which leads to a net loss of MuSCs. Furthermore, one of the committed myoblasts may differentiate faster than the other one, and thus not have time to divide, leading to fewer numbers of fusion-competent myoblasts available for muscle repair (Fig. 9.9a). These two latter modes of MuSC divisions appear to occur more often in aged skeletal muscles, possibly reflecting age-related changes in the MuSCs themselves and/or of the surrounding environment that stops being able to support the types of divisions that maintain the MuSC pool [91, 99].

After muscle regeneration has been completed in a healthy muscle, some MuSCs return to quiescence, thus replenishing the pool. Several mechanisms are involved in promoting this transition. For example, Delta on the muscle fibre binds to Notch receptors on the MuSCs and Notch signalling increases Pax7 expression, promoting the quiescent state [100–102]. These MuSCs entering quiescence also express Sprouty which counteracts Fgfs [103]. Interestingly, Notch signalling has recently been shown to also activate collagen V production by quiescent MuSCs which in turn binds to the calcitonin receptor on these cells, generating an autocrine loop that further reinforces the quiescent state [104].

9.8 How Does the Study of Muscle Stem Cells in Development Contribute to the Study of Skeletal Muscle Diseases?

Muscular dystrophies are inherited diseases that lead to muscle weakness and tissue degeneration [105, 106]. There are more than 30 different muscular dystrophies and they vary in severity. However, several muscular dystrophies are devastating diseases because of the tissue degeneration involved. Patients growing with these dystrophies gradually lose the ability to walk, will have difficulty breathing and eating, which may eventually lead to premature death [105, 106]. Many (but not all) of the muscular dystrophies are due to mutations in the proteins that connect the actin cytoskeleton of muscle fibres (and its associated MuSCs) to the extracellular matrix [107]. For example, to name only a few, mutations in the intracellular protein dystrophin (◘ Fig. 9.8b) lead to either Duchenne muscular dystrophy (DMD) or the milder Becker's muscular dystrophy (BMD). Mutations in the transmembrane sarcoglycans, which bind to dystroglycan (◘ Fig. 9.8b), or in enzymes which regulate the glycosylation of α-dystroglycan, lead to certain types of limb–girdle muscular dystrophies (LGMD), affecting primarily the pelvic and shoulder girdles [108]. Finally, mutations in the laminin α2 chain of laminin 211 (◘ Fig. 9.8b) lead to Merosin-deficient congenital muscular dystrophy 1A (MDC1A), which is characterized by severe muscle weakness from birth and usually leads to premature death [109].

There are presently no specific treatments for muscular dystrophies and therapy is limited to palliative care. Given the debilitating nature of these diseases, there is an urgent need to find ways to improve muscle function and halt tissue degeneration. A huge effort has gone into addressing how the proteins mutated in muscular dystrophies contribute to normal skeletal muscle development and function. One important conclusion from this work is that these so-called 'structural' proteins not only have structural functions but are also members of integrated signalling networks [105]. Another important conclusion is that several muscular dystrophies lead to changes in the pool of MuSCs. There is evidence for defects in the development of MuSCs in mouse models of DMD and MDC1A [64, 110], as well as perturbations in the regenerative response of these cells upon muscle injury [111, 112].

The study of normal skeletal muscle development and regeneration is important to be able to design therapies that address the underlying processes that go wrong in muscle diseases. The better we understand how the communication between cells occurs, what signalling pathways are used and how MuSCs adapt their response to the need of the tissue at each point of development and regeneration, the more likely we are to be able to design strategies to improve muscle function in a disease setting. This applies not only to the muscular dystrophies discussed

> **Take-Home Message**
>
> - MuSCs arise early in development, generate all the myonuclei of skeletal muscles while simultaneously maintaining a MuSC population which enters quiescence in adult muscles.
>   - MuSCs change over developmental time, acquiring characteristics which are appropriate for each stage of myogenesis, but how these changes are regulated is not well understood.
>   - Myogenesis is triggered by extrinsic factors, which act on MuSCs and activate transcription factors, the myogenic regulatory factors (the MRFs), leading to myoblast differentiation.
>   - Differentiated myocytes fuse to generate multinucleated myotubes which mature into muscle fibres.
> - The construction of skeletal muscle occurs in different phases over time, each phase playing a specific role in the construction of the tissue:
>   - Primary myogenesis generates the muscle pattern, i.e. 'miniature muscles' connected via muscle connective tissue including tendons to the skeleton and innervated by the peripheral nervous system.
>   - Secondary myogenesis builds on the pattern generated through primary myogenesis, leading to a several-fold increase in the number of muscle fibres within each muscle.
>   - During perinatal myogenesis, muscle fibres increase in size through the differentiation of myoblasts that fuse with the existing fibres.
>   - At the end of perinatal myogenesis, MuSCs enter quiescence and are called satellite cells.
> - Skeletal muscle has an amazingly high regenerative potential, which is due to the presence of adult, quiescent MuSCs, which activate, proliferate and repair the tissue upon injury.
> - Skeletal muscle diseases, such as mutations in the linkage between the muscle cell cytoskeleton and the surrounding extracellular matrix, lead to muscle weakness and can be extremely debilitating.
> - Increasing our knowledge on how MuSCs interact with their complex environment to construct a healthy tissue and successfully repair it will hopefully bring us closer to efficient treatments of muscle diseases.

briefly above, but also to other disease states that impact muscle function, such as cancer-induced cachexia and age-related sarcopenia.

Acknowledgements We thank the members of our group, particularly Gabriela Rodrigues, Luís Marques and Inês Antunes, for their contributions to this chapter, and multiple generations of students of the MSc in Evolutionary and Developmental Biology (▶ http://bed.campus.ciencias.ulisboa.pt/) for their interest in this topic. The MF20 and Pax3 antibodies were developed by DA Fischman and CP Ordahl, respectively, and were obtained from the Developmental Studies Hybridoma Bank, developed under the auspices of the NICHD and maintained by the University of Iowa, Department of Biology, Iowa City, IA52242, USA. The original data shown in figures in this chapter were obtained within projects financed by Fundação para a Ciência e a Tecnologia (FCT), Portugal (PTDC/SAU-BID/120130/2010) and Association Française contre les Myopathies (AFM) – Téléthon, France (project n° 19959). MD and ABG were supported by fellowships SFRH/BPD/65370/2009 and SFRH/BD/90827/2012 from FCT. ABG is an alumnus of the MSc in Evolutionary and Developmental Biology.

References

1. Mauro A. Satellite cell of skeletal muscle fibers. J Biophys Biochem Cytol. 1961;9:493–5.
2. Biressi S, Molinaro M, Cossu G. Cellular heterogeneity during vertebrate skeletal muscle development. Dev Biol. 2007;308(2):281–93.
3. Tajbakhsh S. Skeletal muscle stem cells in developmental versus regenerative myogenesis. J Intern Med. 2009;266(4):372–89.
4. Relaix F, Marcelle C. Muscle stem cells. Curr Opin Cell Biol. 2009;21(6):748–53.
5. Hollway G, Currie P. Vertebrate myotome development. Birth Defects Res C Embryo Today. 2005;75(3):172–9.
6. Emerson CP Jr. Embryonic signals for skeletal myogenesis: arriving at the beginning. Curr Opin Cell Biol. 1993;5(6):1057–64.
7. Pownall ME, Emerson CP Jr. Sequential activation of three myogenic regulatory genes during somite morphogenesis in quail embryos. Dev Biol. 1992;151(1):67–79.
8. Ordahl CP, Le Douarin NM. Two myogenic lineages within the developing somite. Development. 1992;114:339–53.
9. Christ B, Ordahl CP. Early stages of chick somite development. Anat Embryol (Berl). 1995;191(5): 381–96.
10. Monsoro-Burq AH. Sclerotome development and morphogenesis: when experimental embryology meets genetics. Int J Dev Biol. 2005;49(2–3):301–8.
11. Brent AE, Schweitzer R, Tabin CJ. A somitic compartment of tendon progenitors. Cell. 2003;113(2):235–48.
12. Buckingham M. Myogenic progenitor cells and skeletal myogenesis in vertebrates. Curr Opin Genet Dev. 2006;16(5):525–32.
13. Christ B, Huang R, Scaal M. Amniote somite derivatives. Dev Dyn. 2007;236(9):2382–96.
14. Deries M, Thorsteinsdóttir S. Axial and limb muscle development: dialogue with the neighbourhood. Cell Mol Life Sci. 2016;73(23):4415–31.
15. Thorsteinsdóttir S, Deries M, Cachaço AS, Bajanca F. The extracellular matrix dimension of skeletal muscle development. Dev Biol. 2011;354(2):191–207.
16. Ben-Yair R, Kalcheim C. Lineage analysis of the avian dermomyotome sheet reveals the existence of single cells with both dermal and muscle progenitor fates. Development. 2005;132(4):689–701.
17. Gros J, Manceau M, Thomé V, Marcelle C. A common somitic origin for embryonic muscle progenitors and satellite cells. Nature. 2005;435(7044):954–8.
18. Kassar-Duchossoy L, Giacone E, Gayraud-Morel B, Jory A, Gomes D, Tajbakhsh S. Pax3/Pax7 mark a novel population of primitive myogenic cells during development. Genes Dev. 2005;19(12):1426–31.
19. Relaix F, Rocancourt D, Mansouri A, Buckingham M. A Pax3/Pax7-dependent population of skeletal muscle progenitor cells. Nature. 2005;435(7044):948–53.
20. Bajanca F, Luz M, Raymond K, Martins GG, Sonnenberg A, Tajbakhsh S, et al. Integrin α6β1-laminin interactions regulate early myotome formation in the mouse embryo. Development. 2006;133(9):1635–44.
21. Buckingham M. How the community effect orchestrates muscle differentiation. BioEssays. 2003; 25(1):13–6.
22. Babiuk RP, Zhang W, Clugston R, Allan DW, Greer JJ. Embryological origins and development of the rat diaphragm. J Comp Neurol. 2003;455(4):477–87.
23. Sambasivan R, Kuratani S, Tajbakhsh S. An eye on the head: the development and evolution of craniofacial muscles. Development. 2011;138(12):2401–15.
24. Munsterberg AE, Lassar AB. Combinatorial signals from the neural tube, floor plate and notochord induce myogenic bHLH gene expression in the somite. Development. 1995;121(3):651–60.
25. Chang CN, Kioussi C. Location, location, location: signals in muscle specification. J Dev Biol. 2018;18:6(2).

26. Rios AC, Serralbo O, Salgado D, Marcelle C. Neural crest regulates myogenesis through the transient activation of NOTCH. Nature. 2011;473(7348):532–5.
27. Sieiro D, Rios AC, Hirst CE, Marcelle C. Cytoplasmic NOTCH and membrane-derived beta-catenin link cell fate choice to epithelial-mesenchymal transition during myogenesis. elife. 2016;24:5.
28. Serralbo O, Marcelle C. Migrating cells mediate long-range WNT signaling. Development. 2014;141(10):2057–63.
29. Borello U, Berarducci B, Murphy P, Bajard L, Buffa V, Piccolo S, et al. The Wnt/β-catenin pathway regulates Gli-mediated Myf5 expression during somitogenesis. Development. 2006;133(18):3723–32.
30. Borycki AG, Brunk B, Tajbakhsh S, Buckingham M, Chiang C, Emerson CP Jr. Sonic hedgehog controls epaxial muscle determination through Myf5 activation. Development. 1999;126(18):4053–63.
31. Gustafsson MK, Pan H, Pinney DF, Liu Y, Lewandowski A, Epstein DJ, et al. Myf5 is a direct target of long-range Shh signaling and Gli regulation for muscle specification. Genes Dev. 2002;16(1):114–26.
32. Marcelle C, Stark MR, Bronner-Fraser M. Coordinate actions of BMPs, Wnts, Shh and noggin mediate patterning of the dorsal somite. Development. 1997;124(20):3955–63.
33. Kablar B, Rudnicki MA. Skeletal muscle development in the mouse embryo. Histol Histopathol. 2000;15(2):649–56.
34. Venters SJ, Thorsteinsdóttir S, Duxson MJ. Early development of the myotome in the mouse. Dev Dyn. 1999;216(3):219–32.
35. Gros J, Scaal M, Marcelle C. A two-step mechanism for myotome formation in chick. Dev Cell. 2004;6(6):875–82.
36. Kalcheim C, Ben-Yair R. Cell rearrangements during development of the somite and its derivatives. Curr Opin Genet Dev. 2005;15(4):371–80.
37. deLapeyrière O, Ollendorff V, Planche J, Ott MO, Pizette S, Coulier F, et al. Expression of the Fgf6 Gene is restricted to developing skeletal muscle in the mouse embryo. Development. 1993;118(2):601–11.
38. Han JK, Martin GR. Embryonic expression of Fgf-6 is restricted to the skeletal muscle lineage. Dev Biol. 1993;158(2):549–54.
39. Brent AE, Braun T, Tabin CJ. Genetic analysis of interactions between the somitic muscle, cartilage and tendon cell lineages during mouse development. Development. 2005;132(3):515–28.
40. Vinagre T, Moncaut N, Carapuco M, Novoa A, Bom J, Mallo M. Evidence for a myotomal Hox/Myf cascade governing nonautonomous control of rib specification within global vertebral domains. Dev Cell. 2010;18(4):655–61.
41. Delfini MC, De La Celle M, Gros J, Serralbo O, Marics I, Seux M, et al. The timing of emergence of muscle progenitors is controlled by an FGF/ERK/SNAIL1 pathway. Dev Biol. 2009;333(2):229–37.
42. Francis-West PH, Antoni L, Anakwe K. Regulation of myogenic differentiation in the developing limb bud. J Anat. 2003;202(1):69–81.
43. Lee AS, Harris J, Bate M, Vijayraghavan K, Fisher L, Tajbakhsh S, et al. Initiation of primary myogenesis in amniote limb muscles. Dev Dyn. 2013;242(9):1043–55.
44. Deries M, Collins JJ, Duxson MJ. The mammalian myotome: a muscle with no innervation. Evol Dev. 2008;10(6):746–55.
45. Tajbakhsh S, Rocancourt D, Cossu G, Buckingham M. Redefining the genetic hierarchies controlling skeletal myogenesis: Pax-3 and Myf-5 act upstream of MyoD. Cell. 1997;89(1):127–38.
46. Nogueira JM, Hawrot K, Sharpe C, Noble A, Wood WM, Jorge EC, et al. The emergence of Pax7-expressing muscle stem cells during vertebrate head muscle development. Front Aging Neurosci. 2015;7:62.
47. Kelly AM, Zacks SI. The histogenesis of rat intercostal muscle. J Cell Biol. 1969;42:135–53.
48. Ross JJ, Duxson MJ, Harris AJ. Formation of primary and secondary myotubes in rat lumbrical muscles. Development. 1987;100(3):383–94.
49. Ontell M, Hughes D, Bourke D. Secondary myogenesis or normal muscle produces abnormal myotubes. Anat Rec. 1982;204:199–207.
50. Mourikis P, Sambasivan R, Castel D, Rocheteau P, Bizzarro V, Tajbakhsh S. A critical requirement for notch signaling in maintenance of the quiescent skeletal muscle stem cell state. Stem Cells. 2012;30(2):243–52.

51. Vasyutina E, Lenhard DC, Birchmeier C. Notch function in myogenesis. Cell Cycle. 2007;6(12):1451–4.
52. Hirsinger E, Malapert P, Dubrulle J, Delfini MC, Duprez D, Henrique D, et al. Notch signalling acts in postmitotic avian myogenic cells to control MyoD activation. Development. 2001;128(1):107–16.
53. Delfini MC, Hirsinger E, Pourquié O, Duprez D. Delta 1-activated notch inhibits muscle differentiation without affecting Myf5 and Pax3 expression in chick limb myogenesis. Development. 2000;127(23):5213–24.
54. Schuster-Gossler K, Cordes R, Gossler A. Premature myogenic differentiation and depletion of progenitor cells cause severe muscle hypotrophy in Delta1 mutants. Proc Natl Acad Sci U S A. 2007;104(2):537–42.
55. Beckers J, Clark A, Wunsch K, Hrabe De Angelis M, Gossler A. Expression of the mouse Delta1 gene during organogenesis and fetal development. Mech Dev. 1999;84(1–2):165–8.
56. Mourikis P, Gopalakrishnan S, Sambasivan R, Tajbakhsh S. Cell-autonomous Notch activity maintains the temporal specification potential of skeletal muscle stem cells. Development. 2012;139(24):4536–48.
57. Deries M, Gonçalves AB, Vaz R, Martins GG, Rodrigues G, Thorsteinsdóttir S. Extracellular matrix remodeling accompanies axial muscle development and morphogenesis in the mouse. Dev Dyn. 2012;241(2):350–64.
58. Deries M, Schweitzer R, Duxson MJ. Developmental fate of the mammalian myotome. Dev Dyn. 2010;239(11):2898–910.
59. Goulding M, Lumsden A, Paquette AJ. Regulation of Pax-3 Expression in the dermomyotome and its role in muscle development. Development. 1994;120(4):957–71.
60. Messina G, Biressi S, Monteverde S, Magli A, Cassano M, Perani L, et al. Nfix regulates fetal-specific transcription in developing skeletal muscle. Cell. 2010;140(4):554–66.
61. Hutcheson DA, Zhao J, Merrell A, Haldar M, Kardon G. Embryonic and fetal limb myogenic cells are derived from developmentally distinct progenitors and have different requirements for beta-catenin. Genes Dev. 2009;23(8):997–1013.
62. Biressi S, Tagliafico E, Lamorte G, Monteverde S, Tenedini E, Roncaglia E, et al. Intrinsic phenotypic diversity of embryonic and fetal myoblasts is revealed by genome-wide gene expression analysis on purified cells. Dev Biol. 2007;304(2):633–51.
63. Duxson MJ, Usson Y. Cellular insertion of primary and secondary myotubes in embryonic rat muscles. Development. 1989;107:243–51.
64. Nunes AM, Wuebbles RD, Sarathy A, Fontelonga TM, Deries M, Burkin DJ, et al. Impaired fetal muscle development and JAK-STAT activation mark disease onset and progression in a mouse model for merosin-deficient congenital muscular dystrophy. Hum Mol Genet. 2017;26(11):2018–33.
65. Cachaço AS, Pereira CS, Pardal RG, Bajanca F, Thorsteinsdóttir S. Integrin repertoire on myogenic cells changes during the course of primary myogenesis in the mouse. Dev Dyn. 2005;232(4):1069–78.
66. Hurren B, Collins JJ, Duxson MJ, Deries M. First neuromuscular contact correlates with onset of primary myogenesis in rat and mouse limb muscles. PLoS One. 2015;10(7):e0133811.
67. Marmigère F, Ernfors P. Specification and connectivity of neuronal subtypes in the sensory lineage. Nat Rev Neurosci. 2007;8(2):114–27.
68. Bonanomi D, Pfaff SL. Motor axon pathfinding. Cold Spring Harb Perspect Biol. 2010;2(3):a001735.
69. Jessen KR, Mirsky R, Lloyd AC. Schwann cells: development and role in nerve repair. Cold Spring Harb Perspect Biol. 2015;7(7):a020487.
70. Weatherbee SD, Anderson KV, Niswander LA. LDL-receptor-related protein 4 is crucial for formation of the neuromuscular junction. Development. 2006;133(24):4993–5000.
71. Kim N, Stiegler AL, Cameron TO, Hallock PT, Gomez AM, Huang JH, et al. Lrp4 is a receptor for Agrin and forms a complex with MuSK. Cell. 2008;135(2):334–42.
72. Burden SJ, Yumoto N, Zhang W. The role of MuSK in synapse formation and neuromuscular disease. Cold Spring Harb Perspect Biol. 2013;5(5):a009167.
73. Bao ZZ, Lakonishok M, Kaufman S, Horwitz AF. α7β1 integrin is a component of the myotendinous junction on skeletal muscle. J Cell Sci. 1993;106(Part 2):579–90.
74. Velling T, Collo G, Sorokin L, Durbeej M, Zhang H, Gullberg D. Distinct α7Aβ1 and α7Bβ1 integrin expression patterns during mouse development: α7A is restricted to skeletal muscle but

α7B is expressed in striated muscle, vasculature, and nervous system. Dev Dyn. 1996;207(4): 355–71.
75. van der Flier A, Gaspar AC, Thorsteinsdóttir S, Baudoin C, Groeneveld E, Mummery CL, et al. Spatial and temporal expression of the β1D integrin during mouse development. Dev Dyn. 1997;210(4):472–86.
76. Nawrotzki R, Willem M, Miosge N, Brinkmeier H, Mayer U. Defective integrin switch and matrix composition at alpha 7-deficient myotendinous junctions precede the onset of muscular dystrophy in mice. Hum Mol Genet. 2003;12(5):483–95.
77. Kardon G. Muscle and tendon morphogenesis in the avian hind limb. Development. 1998; 125(20):4019–32.
78. Chevallier A, Kieny M. On the role of the connective tissue in the patterning of the chick limb musculature. Wilhelm Roux Arch Dev Biol. 1982;191:277–80.
79. Mathew SJ, Hansen JM, Merrell AJ, Murphy MM, Lawson JA, Hutcheson DA, et al. Connective tissue fibroblasts and Tcf4 regulate myogenesis. Development. 2011;138(2):371–84.
80. Brent AE, Tabin CJ. FGF acts directly on the somitic tendon progenitors through the Ets transcription factors Pea3 and Erm to regulate scleraxis expression. Development. 2004;131(16): 3885–96.
81. Edom-Vovard F, Schuler B, Bonnin MA, Teillet MA, Duprez D. Fgf4 positively regulates scleraxis and tenascin expression in chick limb tendons. Dev Biol. 2002;247(2):351–66.
82. Eloy-Trinquet S, Wang H, Edom-Vovard F, Duprez D. Fgf signaling components are associated with muscles and tendons during limb development. Dev Dyn. 2009;238(5):1195–206.
83. Ontell M, Hughes D, Bourke D. Morphometric analysis of the developing mouse soleus muscle. Am J Anat. 1988;181:279–88.
84. Duxson MJ, Usson Y, Harris AJ. The origin of secondary myotubes in mammalian skeletal muscles: ultrastructural studies. Development. 1989;107:743–50.
85. Harris AJ. Embryonic growth and innervation of rat skeletal muscles. I. Neural regulation of muscle fibre numbers. Philos Trans R Soc Lond Ser B Biol Sci. 1981;293(1065):257–77.
86. Cossu G, Ranaldi G, Senni MI, Molinaro M, Vivarelli E. 'Early' mammalian myoblasts are resistant to phorbol ester-induced block of differentiation. Development. 1988;102(1):65–9.
87. Cusella-De Angelis MG, Molinari S, Le Donne A, Coletta M, Vivarelli E, Bouche M, et al. Differential response of embryonic and fetal myoblasts to TGFβ: a possible regulatory mechanism of skeletal muscle histogenesis. Development. 1994;120(4):925–33.
88. White RB, Bierinx AS, Gnocchi VF, Zammit PS. Dynamics of muscle fibre growth during postnatal mouse development. BMC Dev Biol. 2010;10:21.
89. Tierney MT, Gromova A, Sesillo FB, Sala D, Spenle C, Orend G, et al. Autonomous extracellular matrix remodeling controls a progressive adaptation in muscle stem cell regenerative capacity during development. Cell Rep. 2016;14(8):1940–52.
90. Tierney MT, Sacco A. Satellite cell heterogeneity in skeletal muscle homeostasis. Trends Cell Biol. 2016;26(6):434–44.
91. Dumont NA, Bentzinger CF, Sincennes MC, Rudnicki MA. Satellite cells and skeletal muscle regeneration. Compr Physiol. 2015;5(3):1027–59.
92. Ontell M, Kozeka K. Organogenesis of the mouse extensor digitorum logus muscle: a quantitative study. Am J Anat. 1984;171(2):149–61.
93. Wang YX, Dumont NA, Rudnicki MA. Muscle stem cells at a glance. J Cell Sci. 2014;127(21): 4543–8.
94. Webster MT, Manor U, Lippincott-Schwartz J, Fan CM. Intravital imaging reveals ghost fibers as architectural units guiding myogenic progenitors during regeneration. Cell Stem Cell. 2016;18(2):243–52.
95. Bentzinger CF, Wang YX, von Maltzahn J, Rudnicki MA. The emerging biology of muscle stem cells: implications for cell-based therapies. BioEssays. 2013;35(3):231–41.
96. Le Grand F, Jones AE, Seale V, Scime A, Rudnicki MA. Wnt7a activates the planar cell polarity pathway to drive the symmetric expansion of satellite stem cells. Cell Stem Cell. 2009;4(6):535–47.
97. Kuang S, Kuroda K, Le Grand F, Rudnicki MA. Asymmetric self-renewal and commitment of satellite stem cells in muscle. Cell. 2007;129(5):999–1010.
98. Kuang S, Gillespie MA, Rudnicki MA. Niche regulation of muscle satellite cell self-renewal and differentiation. Cell Stem Cell. 2008;2(1):22–31.
99. Blau HM, Cosgrove BD, Ho AT. The central role of muscle stem cells in regenerative failure with aging. Nat Med. 2015;21(8):854–62.

100. Fukada S, Yamaguchi M, Kokubo H, Ogawa R, Uezumi A, Yoneda T, et al. Hesr1 and Hesr3 are essential to generate undifferentiated quiescent satellite cells and to maintain satellite cell numbers. Development. 2011;138(21):4609–19.
101. Wen Y, Bi P, Liu W, Asakura A, Keller C, Kuang S. Constitutive Notch activation upregulates Pax7 and promotes the self-renewal of skeletal muscle satellite cells. Mol Cell Biol. 2012;32(12):2300–11.
102. Bjornson CR, Cheung TH, Liu L, Tripathi PV, Steeper KM, Rando TA. Notch signaling is necessary to maintain quiescence in adult muscle stem cells. Stem Cells. 2012;30(2):232–42.
103. Shea KL, Xiang W, LaPorta VS, Licht JD, Keller C, Basson MA, et al. Sprouty1 regulates reversible quiescence of a self-renewing adult muscle stem cell pool during regeneration. Cell Stem Cell. 2010;6(2):117–29.
104. Baghdadi MB, Castel D, Machado L, Fukada SI, Birk DE, Relaix F, et al. Reciprocal signalling by Notch-Collagen V-CALCR retains muscle stem cells in their niche. Nature. 2018;557(7707): 714–8.
105. Davies KE, Nowak KJ. Molecular mechanisms of muscular dystrophies: old and new players. Nat Rev Mol Cell Biol. 2006;7(10):762–73.
106. Mendell JR, Clark KR. Challenges for gene therapy for muscular dystrophy. Curr Neurol Neurosci Rep. 2006;6(1):47–56.
107. Van Ry PM, Fontelonga TM, Barraza-Flores P, Sarathy A, Nunes AM, Burkin DJ. ECM-related myopathies and muscular dystrophies: pros and cons of protein therapies. Compr Physiol. 2017;7(4):1519–36.
108. Laval SH, Bushby KM. Limb-girdle muscular dystrophies--from genetics to molecular pathology. Neuropathol Appl Neurobiol. 2004;30(2):91–105.
109. Gawlik KI, Durbeej M. Skeletal muscle laminin and MDC1A: pathogenesis and treatment strategies. Skelet Muscle. 2011;1(1):9.
110. Merrick D, Stadler LK, Larner D, Smith J. Muscular dystrophy begins early in embryonic development deriving from stem cell loss and disrupted skeletal muscle formation. Dis Model Mech. 2009;2(7–8):374–88.
111. Dumont NA, Wang YX, von Maltzahn J, Pasut A, Bentzinger CF, Brun CE, et al. Dystrophin expression in muscle stem cells regulates their polarity and asymmetric division. Nat Med. 2015;21(12):1455–63.
112. Van Ry PM, Minogue P, Hodges BL, Burkin DJ. Laminin-111 improves muscle repair in a mouse model of merosin-deficient congenital muscular dystrophy. Hum Mol Genet. 2014;23(2):383–96.

Cardiac Regeneration and Repair: From Mechanisms to Therapeutic Strategies

Vasco Sampaio-Pinto, Ana C. Silva, Perpétua Pinto-do-Ó, and Diana S. Nascimento

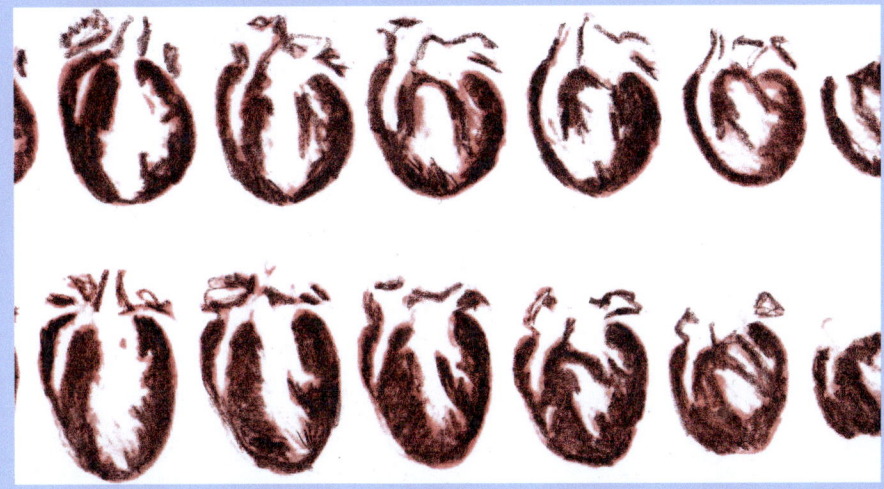

Contents

10.1 Epidemiology of Cardiovascular Disease – 189

10.2 Evolutionary Perspective on Cardiac Development, Physiology and Injury Response – 191

10.3 An Ontogenic Window for Cardiac Regeneration in Mammals – 193

10.4 Critical Synopsis on Neonatal Injury Models – 194

© Springer Nature Switzerland AG 2020
G. Rodrigues, B. A. J. Roelen (eds.), *Concepts and Applications of Stem Cell Biology*,
Learning Materials in Biosciences, https://doi.org/10.1007/978-3-030-43939-2_10

10.5		From Regeneration to Repair – Ontogenic Changes During Neonatal Life – 196
10.5.1		Cardiomyocytes – 196
10.5.2		Non-cardiomyocyte Heart Cell Subsets – 198
10.5.3		Extracellular Cardiac Microenvironment – 199
10.6		Past, Present and Future of Therapeutic Approaches – 201
		References – 204

Cardiac Regeneration and Repair: From Mechanisms to Therapeutic ...

> **Summary**
> Cardiovascular diseases lead the ranking of lethal causalities worldwide, which has been largely attributed to the limited regenerative capacity of the human heart. This restricted myocardial renewing capacity and exacerbated fibrosis often result in heart failure. Currently, the only long-term efficient therapy for this condition is whole-organ transplantation, which is limited by the shortage of donors and physiological constraints. Hence, several cutting-edge strategies to improve cardiac function, namely, gene and cellular therapies, biomaterial design and delivery, either solely or combined, are under investigation. In parallel, studies on heart development and on regenerative mechanisms evolutionarily conserved amongst species have highlighted molecules that hold potential for future therapeutic purposes. This perspective gained further relevance with recent advances showing that murine hearts display regenerative potential yet restrict to a limited period after birth. This chapter will revisit the regenerative capacity of the heart across species and throughout the lifespan, while discussing current advances in therapeutic alternatives for heart failure.

What You Will Learn in This Chapter

At the moment, epidemiological studies show that cardiovascular diseases are responsible for 17 million deaths per year worldwide. Even though cardiovascular research is committed to reduce this number by developing novel and more efficient therapies, estimations suggest that the burden caused by these diseases will continue to increase. Hence, a fundamental challenge in cardiovascular research is to drive regeneration of a damaged heart that typically responds to injury by mounting a reparative response incapable of replacing the lost muscle.

The observation that lower vertebrates and neonatal mammalian hearts were able to regenerate after injury revolutionized this field of research and strengthened the idea of regenerating the adult mammalian heart. Through the contribution of several fields of expertise, cardiovascular research is now focused on extending the regenerative time window of mammalian hearts by promoting neomyogenesis, neovascularization and limiting fibrosis.

This chapter provides an overview of the known mechanisms governing cardiac regeneration and presents future perspectives on therapeutic approaches.

10.1 Epidemiology of Cardiovascular Disease

Cardiovascular diseases (CVD) are the leading cause of death worldwide. Epidemiological reports from 2013 showed that globally CVD were responsible for approximately 32% of total deaths (*i.e.* 17 million) [1]. An update showed that in 2016, 45% of deaths in Europe (*i.e.* 4 million) were attributed to CVD [2]. Although there is some genetic contribution, most risk factors associated with heart diseases are behavioural and environmental, such as age, gender, high blood pressure, hypercholesterolemia, smoking, obesity and alcohol consumption [3]. As a result, CVD are more prevalent in high-income countries. However, while high-income countries have seen declines in mortality rates over the past 20 years, an inverse trend is observed in lower- and middle-income countries [4].

Amongst CVD, ischemic heart diseases are the most predominant (◘ Fig. 10.1). Ischemia in the heart, as in the case of myocardial infarction (MI), generates a hypoxic environment that ultimately leads to cell death. Owing to the negligible regenerative capacity of the human heart, upon MI the damaged myocardium is replaced by a fibrotic scar, resulting in progressive deterioration of organ function and development of heart failure.

To prevent an ischemic event, clinical strategies mainly involve management of cardiovascular risk factors by lifestyle improvement, psychosocial support and prescription of cardioprotective drugs [5]. After an ischemic event, current clinical practice consists in the reperfusion of the ischemic myocardium by percutaneous coronary intervention to limit the deleterious effects of hypoxia to a small portion of myocardium [6]. Currently, heart failure drugs and interventions target disease progression, and thus, the only efficient treatment to re-establish the injured muscle is whole-organ transplantation, which is limited by the shortage of donors, organ rejection/failure and side effects related to immunosuppression [7].

The scarcity of alternative therapies prompted the cardiovascular community to search for novel regenerative mechanisms by dissecting the molecular machineries regulating heart development and response to injury. Because regeneration of the human adult myocardium is virtually inexistent, the heart has been classically considered a postmitotic organ. By the turn of the century, evidence of cardiomyocyte renewal throughout human life [8] and after MI [9] questioned this paradigm. In fact, the prospect of cardiomyocyte proliferation after birth, even at rates incompatible with regeneration, sustained the hypothesis that the heart is endowed with an intrinsic capacity for *de novo* muscle formation that could be exploited to promote a better outcome after injury. In the succeeding years, the discussion focused on the origin of these newly formed cardiomyocytes and two explanations gained strength: one claimed cardiomyocytes originated by differentiation of myocardium-resident cardiac progenitor cells (CPCs) [10, 11], and another postulated that pre-existing cardiomyocytes re-enter cell cycle and proliferate [8, 9, 12]. Despite the existence of empirical data supporting each hypothesis, after 15 years of research, the most consensual view is that new cardiomyocytes arise from the proliferation of pre-existing ones. Indeed, resorting to unbiased techniques such as single-cell mRNA sequencing and also genetic lineage tracing, Kretzschmar and colleagues reported that cycling cardiomyocytes were only robustly observed during early postnatal growth, appearing in insignificant numbers in the adult murine heart,

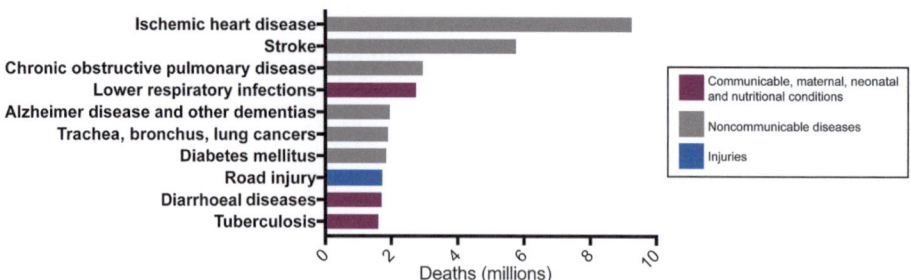

◘ Fig. 10.1 Top 10 global causes of deaths, 2016. (Adapted from World Health Organization [165])

both in homeostasis and after injury. Moreover, no evidence was found for the contribution of quiescent CPCs for the myocardial lineage upon damage [13]. In addition, Valente et al., by combining multiparametric surface marker analysis with single-cell transcriptional profiling, found that cardiomyocytes at different stages of differentiation coexist during development and identified heat-stable antigen (HSA)/CD24 as a marker for identification and prospective isolation of immature mononucleated cardiomyocytes [14].

10.2 Evolutionary Perspective on Cardiac Development, Physiology and Injury Response

The heart of vertebrates displays unique features distinguishable from invertebrate species as is the case of highly specialized chambers with internal endothelial lining (reviewed in [15]). However, substantial differences regarding cardiovascular anatomo-physiology are observed across vertebrates, most likely a result of independent processes of environmental adaptation.

Although fish detain the highest interclass diversity within vertebrates, exhibiting multiple anatomical adaptations to diverse habitats [16], cardiac knowledge in this class results largely from the use of zebrafish. The two-chambered, single-circulation zebrafish heart is subjected to lower pressure when compared to mammals and cardiomyocytes reside in a hypoxic environment [17]. Alike fish, amphibian cardiac studies rely on few model species, namely, axolotl and newt [18]. Amphibians and most reptiles have three-chambered hearts and incomplete separation of the pulmonary and systolic circulation, characterized by a combination of oxygenated and systemic blood in a single ventricle [19, 20] (◘ Fig. 10.2).

Some reptiles (*e.g.* crocodile), birds and mammals have four-chambered hearts, characterized by greater pressure levels and complete separation between oxygenated and deoxygenated blood, most likely an adaptation to high metabolic and oxygen demands.

Regeneration, herein defined as the ability to restore histo-functional integrity of a lost body part, is a phenomenon widely disseminated in metazoans. Regeneration often occurs by *de novo* synthesis of functional tissue through recapitulation to a varying extent of embryonic programs that gave rise to the original tissue [21]. Numerous efforts have been employed on the identification of conserved regenerative responses that could be activated in non-regenerative species for therapeutic proposes [22].

The first evidence of adult heart regeneration in lower vertebrates was obtained in zebrafish after surgical resection of the apex (10–20% of the ventricle). In this model, the cardiac tissue is progressively restored within 60 days [23]. Initially, a blood clot is formed which is gradually replaced by newly formed cardiomyocytes, restoring organ contractility to original levels. Genetic tracing of cardiomyocytes further confirmed that *de novo* cardiomyocyte formation encompasses dedifferentiation and proliferation of pre-existing cardiomyocytes [24]. The cardiac regenerative potential of zebrafish is presently well established and has been illustrated in different injury settings (*e.g.* cryoinjury, hypoxia/reoxygenation, cardiomyocyte genetic depletion) [25–29]. Also urodele amphibians, which include salamanders and newts, have been

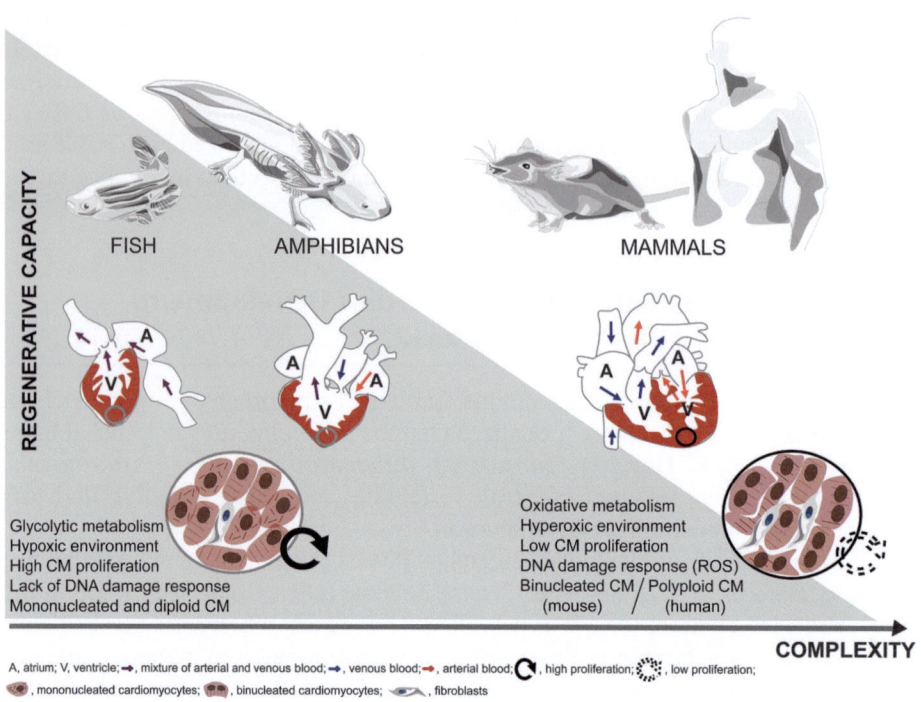

Fig. 10.2 The phylogenetic axis of cardiac complexity versus regenerative capacity

reported to regenerate their heart by re-activation of cardiomyocyte proliferation as reported in zebrafish [19, 20, 30, 31].

Recent evidences highlight similarities between cardiomyocytes of adult lower vertebrates and foetal mammalian cardiomyocytes, which are yet immature (Fig. 10.2). Whereas adult zebrafish and foetal mammalian cardiomyocytes display glucose-based metabolism and high proliferative capacity [32, 33], postnatal mammalian cardiomyocytes reside in an oxygen-rich environment, generate energy mainly through oxidative metabolism and have withdrawn from the cell cycle [32, 34]. Also, zebrafish cardiomyocytes alike foetal mammalian cardiomyocytes are mononucleated, diploid and lack markers of DNA damage response, which contrasts with the binucleation and polyploidy, high levels of mitochondrial reactive oxygen species and DNA damage response observed in adult mammalian cardiomyocytes [32, 35] (Fig. 10.2). As cardiac regeneration encompasses a balance between neomyogenesis/neovascularization and fibrosis, it is plausible that dissimilarities in non-myocytic cells and in the extracellular microenvironment also account for the differential regenerative capacity across vertebrates (reviewed in [36]). In line with this, the histological structure of the zebrafish heart has been suggested to resemble the trabeculated foetal mammalian heart in which endocardial cells surround cardiomyocytes and cardiac fibroblasts are scarce [36]. Globally, these comparative studies demonstrate that biochemical and anatomical complexity of the cardiovascular system across phylogeny is inversely correlated with the regenerative capacity of the heart (Fig. 10.2).

10.3 An Ontogenic Window for Cardiac Regeneration in Mammals

The first reference to heart regeneration in mammals was provided by Mario Robledo in 1956. In this study, myocardial lesions by cauterization were shown to induce cardiomyocyte cell division [37]. Remarkably, the idea of a developmental window for heart regeneration in mammals only became an established principle 55 years later, following the work by Sadek and colleagues [38]. Hearts of 1-day old mice (P1) after surgical apex resection progressively re-established the amputated region with negligible fibrosis. However, when the same procedure was performed at P7, hearts failed to regenerate. Hence, while the proliferation of pre-existing cardiomyocytes was the main regenerative mechanism at P1, the response at P7 resembled the adult heart, characterized by fibrosis without evidence of cardiomyocyte proliferation [38].

In addition to apex resection, other models showed the regenerative capacity of neonatal mice at various extents (*i.e.* cryoinjury, genetic ablation of cardiomyocytes, ligation of left anterior descending (LAD) coronary artery, ischemia/reperfusion (I/R) and pressure overload by transverse aortic constriction (TAC)). Interestingly, cryoinjury at P1 was typically associated with incomplete regeneration, yet depending on injury severity.

Aside from mice, the regenerative capacity of other mammalian species has been poorly investigated. In an attempt to extend this observation to other species with a greater body mass, Zogbi et al. subjected P1 and P7 rat neonates to ventricular resection. Younger rats showed neomyogenesis and long-term preservation of cardiac function in spite of reduced perfusion whereas older rats mainly exhibited scar tissue formation [39]. Similar observations were reported after clinical electrocution of rat hearts at 1, 2, 3 and 4 weeks after birth [40].

Regarding larger mammals, marked differences were found between foetal and adult sheep response to MI by ligation of the LAD coronary artery. While adult healing response was characterized by the formation of a fibrotic scar and loss of systolic function, foetal hearts recovered with minimal inflammatory response, increased proliferative activity in the infarct and border zone, marginal cardiac fibrosis and complete restoration of cardiac function [41]. Regenerative capacity at this early stage was also demonstrated in neonatal pigs by two independent studies. Porcine hearts were able to regenerate when MI was induced during the two first days of life. After this time frame, cardiomyocytes exit cell cycle precluding efficient regeneration [42, 43].

In humans, evidence suggests that the heart of children and infants may exhibit regenerative capacity. In the first half of the last century, analysis of *post-mortem* histological heart sections of children with diphtheria highlighted the presence of split myofibers and mitotic cells in the myocardium, which were interpreted as a result of cardiac regeneration [44, 45]. Moreover, a clinical report showed that young patients with anomalous left coronary artery from the pulmonary artery (ALCAPA) disease and displaying evident signs of left ventricle dysfunction and fibrosis healed after corrective surgery without or with minimal scarring. Following these results, authors speculated that the absence of a fibrotic scar was mediated by compensatory cardiomyocyte proliferation that either replaced the damaged tissue by newly functional myocardium or formed surrounding myocardium to an extent that scar became negligible [46]. Other studies showed that neonatal hearts were able to fully recover from spontaneous MI occurring after birth [47, 48]. These findings support that also human young hearts preserve the ability to regenerate. However, as opportunely highlighted in Vivien et al., further studies are required to substantiate these findings since the impossibility to perform mechanistic studies in these patients and confounding effects (*e.g.* pharmacotherapy) may be contributing to the observed functional recovery [49].

The work by Sadek and colleagues [38] has transformed the cardiovascular field and has been fully acknowledged by over 900 citations. However, in 2014, the controversy was settled when Andersen and colleagues, while attempting to establish the neonatal apex resected model, reported a complete absence of cardiac regeneration [50]. In opposition to the pioneer work, these authors documented a strong inflammatory response in the injury site, followed by massive collagen deposition. Additionally, resected hearts were shown to be smaller than the respective surgical controls, exhibiting impaired neovascularization and cardiomyocyte proliferation. On a later follow-up study, the same authors showed that the scarring tissue resulting from the apex resection was permanently deposited in the myocardium, and that these animals developed dilated cardiomyopathy [51].

The explanation for these disparate results is still not fully understood [52–54] but some progress was made in the subsequent years. Injury severity was shown to impact on the development of fibrosis both after apical resection [55] and cryoinjury [56]. Moreover, difficulties associated with the assessment of fibrosis deposition and cardiomyocyte proliferation can preclude an accurate interpretation of the biological events occurring after the injury.

In a recent report, Sampaio-Pinto et al. showed that neonatal apex resection triggers both regenerative (*i.e.* cardiomyocyte proliferation and neovascularization) and reparative mechanisms (*i.e.* cardiac fibrosis) which result in an adult heart with more cardiomyocytes, benign adaptive cardiac remodelling and restored systolic function [57]. One can speculate if the reported co-activation of regeneration and repair can explain the origin of disparate reports in the literature.

Hence, while some uncertainties concerning the ability of neonates to regenerate their hearts and on the reproducibility of the neonatal injury model emerged, the current view is that both apical resection and coronary ligation injury during the first 2 days of life result in, at least partial, heart regeneration in neonatal mammals [58].

10.4 Critical Synopsis on Neonatal Injury Models

Aiming to disclose the orchestrating mechanisms of cardiac regeneration, several types of neonatal cardiac injury have been established, namely, apex resection [38, 50, 57], cryoinjury [56, 59], genetic ablation [60], MI by LAD ligation [61, 62] and pressure overload by TAC [63]. Not surprisingly, the outcome varies amongst studies because distinct cellular events are differently activated according to the specific type and extension of damage.

Ventricular apex resection has been commonly used in lower vertebrates and was for the first time translated to mammals in 2011. This model has proven useful to dissect molecular mechanisms underlying heart regeneration in teleosts and newts. In mammals, it will certainly be valuable for mechanistic understanding of the different components of the injury response, namely, initial blood clot formation, infiltration of inflammatory cells, cell cycle re-entry of cardiomyocytes, scarring, neovascularization, epicardial cell activation and deposition of extracellular matrix (ECM) constituents. Despite the specific relevance for studying the aforementioned mechanisms, this model is artificial and has poor robustness for modelling pathophysiological alterations of the adult mammalian heart, a major goal of biomedical cardiovascular research.

With the purpose of mimicking ischemic injuries, two injury models, MI and ventricular cryoinjury, have been developed in the neonate. Contrarily to apex resection, which explores the ability of the neonate's heart to reconstitute lost tissue, the former models involve tissue damage and formation of necrotic areas that have to be removed prior to replacement by *de novo* formed tissue. Although cryoinjury mimics several pathologic aspects of MI [56, 59, 64], similar to apex resection, this form of injury is artificial and equally prone to surgical variability.

Neonatal MI was induced by permanent LAD ligation in neonatal mice. Although neonatal cardiomyocytes are particularly resistant to hypoxia, ischemia was efficiently attained, as demonstrated by the absence of troponin expressing cells, caspase 3-mediated programmed cell death at the injury site and severe dysfunction 24 h post MI. Impressively, at 7 days post ligation, cardiac regeneration was nearly complete by a process that encompasses the reactivation of neomyogenesis [61]. Although, amongst all the neonatal injury models, LAD ligation is seemingly best resembling adult heart pathophysiology, it is yet to determine whether the cardiac injury is indeed a result of ischemic damage or a consequence of local tissue destruction as a by-side effect of the procedure. Nonetheless, it is important to highlight that at this time point, cardiac vasculature is not completely developed [65]; therefore, coronary artery occlusion in neonatal mice will most likely have a less severe impact compared to the adult counterpart.

Recently, Malek Mohammadi and colleagues established the neonatal TAC injury model. In contrast to the previous model, this model evaluates the myocardial response to pressure overload, which had only been done in adult mice. Predictably, when TAC was performed in a non-regenerative stage (*i.e.* P7), hearts developed cardiac dysfunction, myocardial fibrosis and cardiomyocyte hypertrophy, recapitulating the well-described adult hypertrophic remodelling. Inversely, P1 hearts subjected to TAC did not develop these maladaptive responses showing instead increased cardiomyocyte proliferation and angiogenesis [63]. Hence, the development of this model revealed that the neonatal regenerative mechanisms are sufficient to revert/prevent the deleterious effects of pressure overload.

Finally, genetic cardiomyocyte ablation has been also employed to demonstrate that the ability to recover from cardiomyocyte ablation decreases with age [60, 66]. Although this method is important for elucidating cardiomyocyte regeneration and death-related processes, it bears disadvantages of only affecting one cell type and triggering diffuse and unspecific distribution of impaired cells throughout the myocardium, contrarily to what is observed post MI.

Although neonatal cardiac injury models currently available do not entirely reproduce the physiopathology of the adult heart, they are valid alternatives to elucidate specific mechanisms underlying mammalian response to cardiac insult, namely, myogenesis, angiogenesis and fibrosis.

10.5 From Regeneration to Repair – Ontogenic Changes During Neonatal Life

In mammals, the reactivation of neomyogenic mechanisms upon injury is quickly diminished after birth, although the mechanisms that govern this shift of potential are still incompletely understood. During intra- to extrauterine environmental transition, mammalian hearts undergo dramatic alterations that mark the end of cardiac morphogenesis and the beginning of myocardial maturation. A better understanding of tissue dynamics during this fundamental transition will certainly reveal cues for the development of innovative heart therapies.

10.5.1 Cardiomyocytes

The mammalian heart develops from a simple contractile tube to a four-chambered organ. At late developmental stages, the heart is still a hyperplastic organ composed of a highly proliferative population of mononucleated cardiomyocytes with partially organized sarcomeres that support the final stage of myocardium growth [15, 67] (◘ Figs. 10.2 and 10.3). At birth, in order to meet systemic requirements of extrauterine life, the mammalian heart undergoes profound physiological and metabolic remodelling. Blood oxygenation starts in the lungs, and closure of the extra- and intra-cardiac shunts suddenly increases the oxygen pressure and systemic vascular resistance [68]. In fact, arterial oxygen pressure rises from 30 to 100 mmHg, rendering cardiomyocytes to a hyperoxic environment and to subsequent drastic metabolic changes [32, 69]. At this stage, cardiomyocytes, which during development produce energy mainly by anaerobic glycolysis and lactate oxidation, undergo a permanent shift towards aerobic metabolism, while retaining for a short period the unique ability to resist ischemia and hypoxia [70]. More impressively, a correlation between this metabolic change and a cessation of cell cycle activity has been recently reported. Puente and colleagues precisely timed the metabolic switch in the mouse heart from

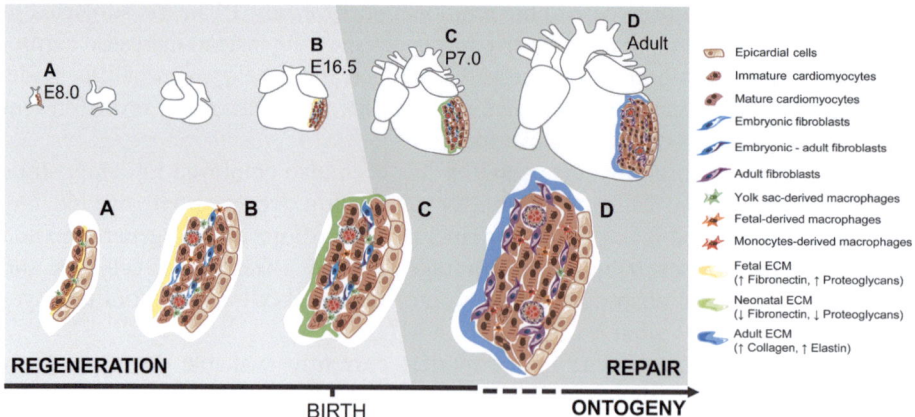

◘ Fig. 10.3 Schematic perspective of cardiac tissue cellular and extracellular dynamics and regenerative capacity throughout ontogeny

anaerobic glycolysis to mitochondrial oxidative phosphorylation to the first week after birth. Indeed, reactive oxygen species (ROS), a by-product of mitochondrial respiration formed during ATP synthesis, are responsible for the activation of DNA damage response pathways, triggering cardiomyocyte withdrawal from the cell cycle [32]. In line with this, an elegant study by Nakada et al. showed that exposure to gradual systemic hypoxaemia rescued infarcted hearts of more severe remodelling while leading to new myocardium at the expense of pre-existing cardiomyocyte proliferation [71].

Contrasting to rodents, in which cell cycle arrest occurs between karyokinesis and cytokinesis, prompting the emergence of binucleated cardiomyocytes [34, 72], in primates, a lack of karyokinesis renders polyploid cells as the predominant, although not exclusive, cardiomyocyte population [67]. In humans, Bergmann et al. took advantage of carbon-14 integration into DNA during Cold War nuclear tests to estimate a residual cardiac turnover of 1% at the age of 25 and 0.45% at the age of 75 [8]. Nevertheless, physiological heart growth in adulthood mostly relies on hypertrophic mechanisms that counterbalance pressure overload (e.g. pregnancy, chronic exercise) and are characterized by normal levels of fibrosis and cardiac function [73].

Apart from upstream environmental factors aforementioned, several molecular regulators of postnatal cardiomyocyte cell cycle arrest have been identified and can fall within one of three levels of regulation: (1) epigenetic regulation involving pre-transcriptional heterochromatin-mediated gene silencing of positive cell cycle regulators; (2) transcriptional activation of negative cell cycle regulators; and (3) post-transcriptional regulation through non-coding RNAs (ncRNAs) [74].

A cell-intrinsic timer mechanism was shown to limit the proliferation of rat neonatal cardiomyocytes by the activation of cyclin-dependent kinases (CDKs) inhibitors [75]. In fact, the activity of cyclin E- and cyclin A-associated CDK complexes is severely reduced around P5 [76]. Furthermore, knockdown of the CDK inhibitors p21 and p27 in adult cardiomyocytes induced cellular proliferation, highlighting a putative role for these elements in cell cycle exit [77]. In addition, by overexpressing CDK1, CDK4, cyclin B1 and cyclin D1, Mohamed and colleagues were able to induce stable cell division of postmitotic mouse, rat and human cardiomyocytes [78].

After birth, cardiomyocytes silence cardiac development-related genes (i.e. Notch and Wnt/b-catenin pathways) and cell cycle progression genes via hypermethylation [79]. In contrast, overall hypomethylation is detected in genes involved in heart contractility (i.e. myosin heavy chain, titins and troponin I) and mitochondrial activity [80]. Major epigenetic changes are set until P14 [79] although a partial reactivation of foetal development program was documented in response to pathological conditions [80, 81].

Identification of microRNAs (miRNAs/miRs) involved in cardiac disease has led to the study of their involvement also in heart development and cardiac regeneration (reviewed in [82]). Knockout of miRNA-133a induced proliferation of cardiomyocytes through derepression of cyclin D2 [83]. Similarly, the miR-15 family was found to target checkpoint kinase 1 (Check1), and thus, genetic deletion of this miRNA family increased cardiomyocyte proliferation [62, 84]. Inversely, miR-199a-3p and miR-590-3p induced cell cycle re-entry of cardiomyocytes by inhibiting Homer1 and Hopx, a regulator of calcium signalling in cardiac cells and a repressor of embryonic myocytes proliferation, respectively [85]. In fact, overexpression of miR-199a in pig hearts subjected to MI stimulated cardiomyocyte proliferation and improved cardiac repair [86]. Moreover, knockdown of antimitotic genes *meis1*, miR-548c-3p, miR-509-3p and miR-23b-3p promoted cardiomyocyte proliferation [87].

Recent reports also suggest that besides miRNAs, other ncRNAs can regulate cardiomyocyte proliferation. The long non-coding RNA (lncRNA) CAREL was found to be upregulated in non-mitotic cardiomyocytes, and its overexpression inhibited neonatal cardiac regeneration by suppressing cardiomyocyte proliferation [88].

10.5.2 Non-cardiomyocyte Heart Cell Subsets

The inflammatory response following injury, such as MI, has been extensively characterized. Necrotic tissues originated by ischemia release cytokines and chemokines that recruit neutrophils and monocytes/macrophages to the injury site, thereby promoting tissue repair through the removal of cellular debris [89]. Despite the well-recognized importance of macrophages on cardiac healing, the heterogeneity of subsets, developmental origin and activation upon injury were only recently scrutinized. Cardiac resident macrophages, mainly derived from the yolk sac (primitive haematopoiesis, embryonic day (E) 7.5–E11.5) and foetal liver (definitive haematopoiesis, E11.5–E16.5), were shown to persist until adult life (◘ Fig. 10.3). After the injury, these embryonic macrophages are permanently replaced by infiltrating monocyte-derived macrophages [90]. In addition, CX3C chemokine receptor 1 (CX3CR1)$^+$ embryonic cardiac macrophages, which constitute a major histocompatibility complex class II (MHCII)$^-$ homogenous population at birth, diversify into MHCII$^-$ and MHCII$^+$ subsets. Also, the initial high contribution of CX3CR1$^+$ embryonic macrophages to resident cardiac macrophages declines after birth as the proliferative capacity of CX3CR1$^+$ resident macrophages decreases upon diversification into subpopulations. As a compensatory mechanism, blood monocyte-derived macrophages replace embryonic macrophages in the myocardium, therefore preserving the pool of resident macrophages [91]. Aurora and colleagues reported that after MI P1 and P14, neonatal hearts display a distinct number and distribution of macrophages within the myocardium. Seven days post infarction, P1 hearts displayed a higher number of macrophages homogeneously distributed across the myocardium, while P14 hearts exhibited fewer cells, confined to the ischemic area [92]. Importantly, clodronate-based depletion of macrophages during the neonatal period abrogated heart regeneration and led to the development of fibrotic scars with impact on cardiac contractility and angiogenesis. An independent study, using inducible transgenic mice to enable cardiomyocyte depletion, further explored the role of distinct macrophage lineages in neonatal and adult response to injury. Whereas in the neonate following injury the number of resident cardiac macrophages increases with negligible monocyte recruitment, the adult heart engages on a vigorous immune response with higher levels of inflammatory cytokines and chemokines, strong monocyte recruitment and persistent monocyte-derived macrophages [66]. Importantly, inhibition of monocyte recruitment showed to be sufficient to reduce inflammation, protect the pool of resident embryonic-derived macrophages and, consequently, improve repair of the heart [66].

In the past, fibroblasts have been mostly regarded as a cell type with minor physiological and developmental importance, which were readily activated in pathological conditions to orchestrate a vigorous reparative response culminating in the formation of fibrosis. Currently, fibroblasts are gathering attention as bidirectional cardiomyocyte–fibroblast interactions became extensively reported [93]. In addition, fibroblasts display a vital role in the tight remodelling of cardiac ECM due to production/degradation of matrix molecules and synthesis of different signalling factors [94, 95]. Ieda et al. elegantly demonstrated that fibroblasts change with age, especially in the transition between intra- and extrauterine life, having thus distinctive roles over cardiomyocytes. Opposed to adult fibroblasts, embryonic fibroblasts predominantly secrete growth factors, such as heparin-binding endothelial growth factor (HBEGF), an important trigger of cardiomyocyte proliferation, and ECM components crucial for cardiac development (i.e. fibronectin, tenascin-C, periostin, hyaluronan and collagen) [96]. Adult fibroblasts exhibit a 58-fold increase in the expression of interleukin 6, a cytokine recognized to induce cardiomyocyte hypertrophy [96, 97] (◘ Fig. 10.3). Although several phenotypic modifications have been identified in fibroblast populations through ontogeny, its association with the loss of regenerative capacity of the mammalian heart after birth is yet to be addressed.

Endothelial cells were also shown to contribute to the early response of neonatal hearts to injury. Capillaries migrate into the apical thrombus as soon as 2 days after apical resection and undergo maturation into fully functional arteries within 5 days. This event precedes cardiomyocyte proliferation and is required for neomyogenesis [98]. Moreover, arterial endothelial cells were shown to migrate along pre-existing capillaries and form collateral arteries following neonatal MI. Such an event increases the reperfusion of the ischemic myocardium and allows efficient cardiac regeneration [99].

In addition, resorting to the pharmacological, genetic or mechanical inhibition of cardiac innervation both in zebrafish and neonatal mouse, Mahmoud et al. demonstrated that cardiac denervation negatively impacts on cardiomyocyte proliferation and inflammatory response hindering cardiac regeneration. Interestingly, administration of recombinant neuregulin 1 and nerve growth factor partially rescued the regenerative capacity of the heart [100].

10.5.3 Extracellular Cardiac Microenvironment

The cardiac ECM is a dynamic network composed of a diversity of molecules, including collagens, non-collagenous glycoproteins (e.g. fibronectin and laminin) and proteoglycans that are essential for tissue integrity but also in coordinating biological processes during heart morphogenesis, homeostasis and pathology [101]. Fibroblasts are the main producers of matrix proteins and, following stimulation, initiate cardiac ECM remodelling at the level of composition, stiffness and rearrangement [95].

During heart development, the ECM consists of a dynamic network enriched in fibronectin, proteoglycans and growth factors [102–104] (◘ Fig. 10.3). The early matrix influences the initial peristaltic pumping of the cardiac tube, the regulation of an epithelial–mesenchymal transition (EMT), which is responsible for the formation of cardiac valves and atrioventricular/outflow septa [105, 106] and myocardial growth [96]. The composition of the ECM undergoes a profound reduction postpartum, being progressively replaced by a collagen and elastin-rich meshwork [102, 107, 108]. In fact, collagen deposition and crosslink during cardiac tissue maturation are essential for normal cardiac performance by supporting the alignment of cardiac fibbers, cardiomyocyte–fibroblast bridging and electrical conductance [109, 110].

ECM derived from distinct ontogenic stages differently impacts on cardiac cells' behaviour. Through the use of ontogenic-tissue-specific ECM coatings, Williams and colleagues demonstrated that foetal cardiac ECM promotes cardiomyocyte expansion *in vitro* when compared with the adult counterpart [108]. Subsequently, decellularized foetal ECM showed greater potential to support and instruct cardiac cells in comparison to adult-derived ECM [111].

Recently, agrin was identified as an ECM molecule required for efficient regeneration of the murine neonatal heart. *In vitro*, agrin induced proliferation of cardiomyocytes derived from mouse- and human-induced pluripotent stem cells (iPSCs) via disassembly of dystrophin–glycoprotein complexes and activation of yes-associated protein (Yap) and extracellular signal-regulated kinase (ERK) pathways. *In vivo*, a single intramyocardial injection of agrin improved the response of adult mouse hearts to MI, and when conditionally knocked out, the regenerative capacity of neonatal hearts was significantly impaired [112].

Apart from biochemical alterations, mechanical properties of cardiac tissue are also altered with ageing. During the perinatal period, several sarcomeric proteins change their isoforms, which impacts on myocardial mechanical properties [80, 113]. For example, titin, a giant molecule that composes the stress-responsive machinery of sarcomeres, changes from foetal/neonatal to adult isoform in cardiomyocytes resulting in increased passive cardiac stiffness [113–115]. To discriminate between cellular and extracellular contribution to stiffness, Majkut *et al.* subjected E14 hearts to either a myosin inhibitor or collagenase resulting in 25% and 40% softer tissue, respectively, outlining the importance of ECM for organ structural maintenance since early developmental stages [116]. In this regard, mechanical testing of decellularized hearts demonstrated that neonatal and adult myocardium were twice as stiff as foetal hearts. This escalation in stiffness through heart maturation correlates with an increase in collagen I, collagen III and laminin, and a decrease in periostin and fibronectin [117] (Fig. 10.3).

In fact, side-by-side comparison of global heart transcriptomes of P1 and P2 mice revealed that most differentially expressed transcripts encode for ECM and cytoskeleton proteins, suggesting that cardiac regeneration is also determined by local microenvironment stiffness. Indeed, to test this hypothesis, the ECM of P3 mouse hearts was pharmacologically softened. The decreased stiffness rescued the ability of mice to regenerate the resected tissue and lengthen the time window for heart regeneration [118].

Corroborating this hypothesis, Yap, a transcriptional cofactor of the Hippo pathway, known to promote proliferation upon mechano-stimuli, was shown to be required for cardiac regeneration. Cardiac-specific deletion of Yap resulted in incomplete regeneration whereas forced overexpression stimulated cardiomyocyte proliferation and improved cardiac contractility after MI [119].

Stiffness of the adult heart (22–50 kPa) was shown to be optimal for cell morphology and function, with greater cell elongation, high contractile force development and well-developed striations, when compared to softer (3 kPa) and stiffer (144 kPa) substrates [120]. Contrarily, Jacot *et al.* showed that cardiomyocytes displayed enhanced functional properties and sarcomeric alignment within 10 kPa substrates [121]. Overall, and despite discrepancies on the most appropriate mechanical properties for the maintenance of cardiomyocytes, it has become well accepted that cardiomyocytes are extremely responsive to mechanical stimuli and phenotypically adapt to these.

Mechanistically, the Hippo pathway has recently been associated with the regulation of cardiomyocyte proliferation and transduction of mechanical cues responsible for phenotypical adaptation in heart cells [122, 123].

10.6 Past, Present and Future of Therapeutic Approaches

Besides the deteriorating impact on life expectancy and quality, CVD also impose a heavy socio-economic burden. Encouragingly, the cardiovascular community is strongly determined in developing innovative and more effective therapeutic strategies (◘ Fig. 10.4).

Neomyogenesis is a prime objective after MI. In an initial approach, the attempts resorted to skeletal myoblasts; however, transplantation of autologous skeletal myoblasts significantly increased the risk of developing ventricular arrhythmias [124]. Bone marrow mononuclear cells (BMMNCs) were also on the first-line for heart therapies [125–127]. However, likely due to poor engraftment and differentiation after transplantation, clinical trials using bone marrow-derived cells to treat MI proved disappointing [128]. In addition, mesenchymal stem cells (MSCs) isolated from the bone marrow and from sources such as adipose tissue and earlier ontogenic

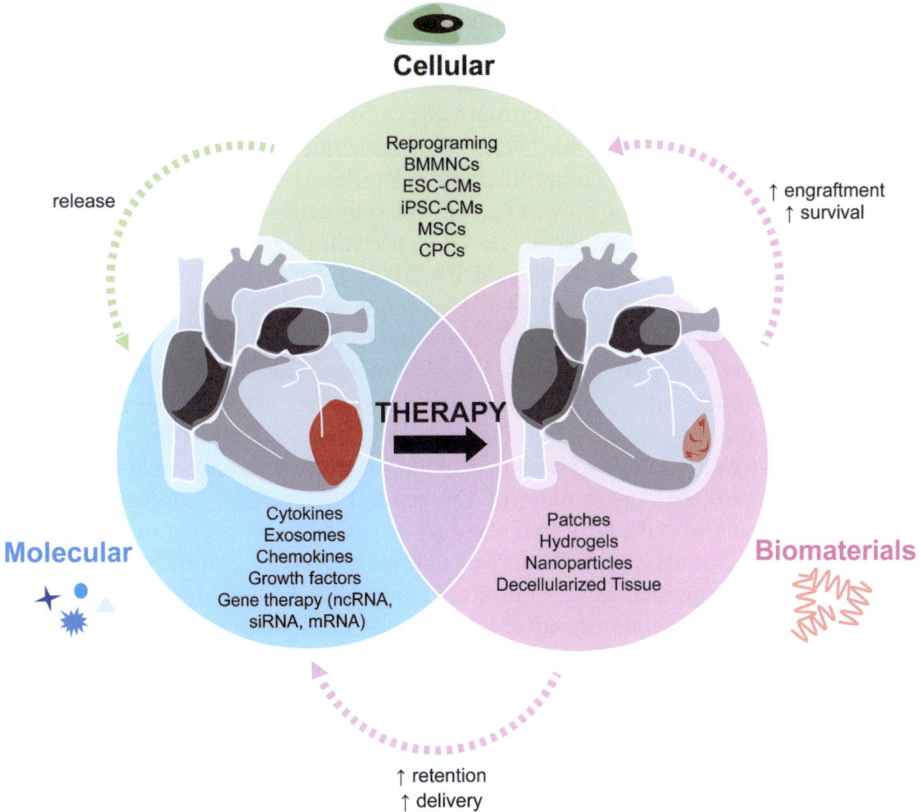

◘ Fig. 10.4 Therapeutic approaches to treat myocardial infarction

sources (*e.g.* Wharton's jelly) were pursued for the treatment of heart failure patients due to their potential to secrete growth factors and their immunomodulatory properties. These features hold therapeutic value and endorse mesenchymal cells for allogenic transplantation [129, 130]. Of note, although clinical reports are promising this far [131], larger and controlled trials are required to validate further the efficiency of MSCs. Remarkably, the cardiac function improvement has been accredited to a paracrine action rather than to direct differentiation into cardiomyocytes.

By the turn of the century, cells exhibiting stem cell-like properties were identified in the adult heart of rodents and humans. These cells, frequently referred to as "cardiac stem cells" (CSCs) or "cardiac progenitor cells" (CPCs), were identified by the expression of distinct cell surface markers (*i.e.* proto-oncogene receptor tyrosine kinase (c-Kit), stem cell antigen 1 (Sca-1), ATP-binding cassette superfamily G (Abcg-2) and platelet-derived growth factor receptor α (PDGFR-α)) and transcription factors (*e.g.* Isl-1, Nkx2.5, Gata-4 and Wilms' tumour 1 (Wt-1)) [132, 133]. However, the early excitement on the evidence for these cells to differentiate in cardiomyocytes and endothelial cells *in vitro* and *in vivo* was promptly challenged by a series of studies, some incorporating lineage tracing, which could neither replicate the findings nor show putative CSC/CPC contribution to *de novo* cardiomyocyte formation [134–136]. Hence, whether or not renewal on the adult myocardium is based on stem cells became a contentious topic in the cardiovascular area. Meanwhile, CPCs had gathered the biomedical community interest, and at least, two trials addressing different subsets of progenitor cells (c-Kit$^+$ for SCIPIO [137] and Sca-1$^+$ for CADUCEUS [138]) were launched to test the therapeutic potential on the patients' infarcted heart. Deplorably, ethical concerns on the research conducted by the main contributors to the endogenous c-Kit$^+$ CSCs [139] were raised over the last years overshadowing science [140, 141] and bringing the concept in the twilight.

Pluripotent stem cells, *i.e.* embryonic stem cells (ESC) and iPSCs, are, when stimulated by a defined cocktail of growth factors, the only cells that display robust differentiation into functional cardiomyocytes thus constituting the most promising cell source for cardiac regenerative therapy (◉ Fig. 10.4). Even so, ESC- and iPSC-derived cardiomyocytes display functional and morphological features of immature myocytes, thereby resembling foetal cardiomyocytes [142–145]. Promising increments emerge by the day as the efforts of the experimental cardiovascular community succeed on, for example, identifying transcriptional regulatory networks directing transition from the immature neonate to a mature cardiomyocyte stage [146]. In addition, human ESC-derived cardiomyocytes implanted into non-human primates after MI showed to increase the muscle mass of the heart. Interestingly, autologous muscle coupled electrically with the graft that was also irrigated by the vasculature of the host [147]. In a recent clinical case, ESC-derived cardiac progenitors, embedded in a fibrin scaffold, were transplanted into the infarcted area of a patient with advanced ischemic heart failure. Three months after the administration of the cells, the symptomatology of the patient improved while complications such as arrhythmias, tumour formation or immunosuppression-related adverse events were not described [148].

Nevertheless, despite the extensive enterprise to develop a cell-based therapy, myocardial cell delivery commonly results in poor engraftment, cell death or cell loss into the bloodstream and subsequent entrapment in the lungs and liver [149, 150]. Several distinct strategies have been considered to improve cell retention, survival and coupling in the heart [151], including the use of decellularized cardiac ECM

[111], injectable hydrogels [152], cardiac patches [153] and piezoelectric biomaterials that promote the alignment and stimulation of cardiomyocytes [154] (◘ Fig. 10.4).

Because functional improvement of cell therapy appears to rely on paracrine mechanisms, new strategies are focusing on replacing cells by soluble factors (*e.g.* chemokines, cytokines and growth factors). However, the administration of molecules with putative beneficial effects (*e.g.* vascular endothelial growth factor (VEGF), granulocyte-colony stimulating factor (G-CSF) and erythropoietin (Epo) produced none or small functional increment after MI [155–157], which may be related with inappropriate dosage or organ selectivity.

Owing to their ability to regulate gene expression, miRNAs and lncRNAs are also contributing to a new pipeline of gene therapies for CVD. Strategies involve the inhibition (*e.g.* via injection of antagomirs, locked nucleic acid (LNA) anti-miRs, small interfering RNAs (siRNAs) or short hairpin RNAs (shRNAs)) of ncRNAs promoting reparative mechanisms and/or upregulation (*e.g.* via adeno-associated virus or miRNA precursors) of ncRNAs supporting cardiac regeneration/CM proliferation [158] (◘ Fig. 10.4).

Alternatively, to replenish the myocardium with new cardiomyocytes, direct reprogramming of other cardiac cells into myocytes has been attempted (◘ Fig. 10.4). Due to their frequency and role in scar formation upon injury, fibroblasts are an obvious target. Initial reprogramming strategies were based on the use of cocktails of transcription factors and lately incorporated the use of miRNAs important for muscle development and differentiation [159]. In 2013, Nam and colleagues used a combination of four transcription factors (*i.e.* GATA binding protein 4, Hand2, T-box 5 and myocardin) and two miRNAs (*i.e.* miR-1 and miR-133) to reprogram neonatal and adult human fibroblasts. After 4–11 weeks in culture, fibroblasts revealed transcriptomic alterations, with the silencing of non-myocytic genes and upregulation of cardiac genes and more importantly, phenotypic changes, exhibiting sarcomere-like structures, calcium transients and contractility [160]. Albeit recent advances in this area, growing evidence suggests that cardiac fibroblasts, besides producing fibrotic tissue, attain other roles which would be lost upon reprogramming into cardiomyocytes. Furthermore, a recent study combining robust genetic models and cellular markers reported that cardiac fibroblasts compose a minor population in the murine heart [161]. Hence, direct reprogramming of fibroblasts into cardiomyocytes would still have the drawback of not generating enough cardiomyocytes to restore significantly heart function.

In the last decade, considerable attention was given to extracellular vesicles (EVs) and their role in promoting intercellular crosstalk [162]. All cardiac subsets release EVs and might interact with target cells in a non-selective or specific way [163]. In peripartum cardiomyopathy, upregulation of miR-146a in endothelial cells directs autocrine inhibition of angiogenesis, and miR-146a-loaded EVs released by endothelial cells lead to decreased metabolic activity and reduced expression of Erbb4, Notch1 and Irak1 in cardiomyocytes. Indeed, preventing paracrine signalling of miR-146a attenuated peripartum cardiomyopathy [164]. This biological process may therefore constitute a mean to modulate cell behaviour. Interestingly, EVs can be loaded with several different components (*e.g.* ncRNAs, siRNAs, mRNAs and/or proteins). Nevertheless, despite important advances in this field, a better understanding of EV biology is necessary for the development of new therapeutic approaches, including more efficient isolation methods and limiting potential off-targets.

Noteworthy, a combination of several of the previous strategies is at present contemplated towards a better therapeutic outcome; yet further investigation is required to understand possible synergistic or antagonistic interactions.

Take-Home Message

- Cardiovascular diseases are the leading cause of death due to the inability of mammalian hearts to regenerate efficiently.
- Distinct animal and injury models have been implemented to investigate cardiac regeneration and repair.
- Cardiac regeneration is conserved amongst lower vertebrates and in young mammals.
- Mammalian regenerative time window is determined by ontogenic alterations in the cellular and extracellular compartment of the heart.
- Cell cycle re-entry of pre-existing cardiomyocytes and not dedicated endogenous CPC differentiation seems to be responsible for the modest renewal of cardiac muscle.
- Investigation of cardiovascular biology endorses the development of new and more effective therapeutic alternatives.

Acknowledgments This work was funded by the European Regional Development Fund (ERDF) through COMPETE 2020, Portugal 2020 and by FCT (Fundação para a Ciência e Tecnologia, [POCI-01-0145-FEDER-030985] and [POCI-01-0145-FEDER-031120]; and by FCT/Ministério da Ciência, Tecnologia e Inovação in the framework of individual funding: [SFRH/BD/111799/2015] to V.S.-P. and [CEECINST/00091/2018] to DSN.

References

1. Naghavi M, Wang H, Lozano R, Davis A, Liang X, Zhou M, Vollset SE, Ozgoren AA, Abdalla S, Abd-Allah F, Abdel Aziz MI, et al. Global, regional, and national age-sex specific all-cause and cause-specific mortality for 240 causes of death, 1990–2013: a systematic analysis for the Global Burden of Disease Study 2013. Lancet. 2015;385(9963):117–71.
2. Townsend N, Wilson L, Bhatnagar P, Wickramasinghe K, Rayner M, Nichols M. Cardiovascular disease in Europe: epidemiological update 2016. Eur Heart J. 2016;37(42):3232–45.
3. Yusuf S, Hawken S, Ounpuu S, Dans T, Avezum A, Lanas F, et al. Effect of potentially modifiable risk factors associated with myocardial infarction in 52 countries (the INTERHEART study): case-control study. Lancet. 2004;364(9438):937–52.
4. Roth GA, Huffman MD, Moran AE, Feigin V, Mensah GA, Naghavi M, et al. Global and regional patterns in cardiovascular mortality from 1990 to 2013. Circulation. 2015;132(17):1667–78.
5. Piepoli MF, Hoes AW, Agewall S, Albus C, Brotons C, Catapano AL, et al. 2016 European Guidelines on cardiovascular disease prevention in clinical practice: The Sixth Joint Task Force of the European Society of Cardiology and Other Societies on Cardiovascular Disease Prevention in Clinical Practice (constituted by representatives of 10 societies and by invited experts)Developed with the special contribution of the European Association for Cardiovascular Prevention & Rehabilitation (EACPR). Eur Heart J. 2016;37(29):2315–81.
6. Ibanez B, James S, Agewall S, Antunes MJ, Bucciarelli-Ducci C, Bueno H, et al. 2017 ESC guidelines for the management of acute myocardial infarction in patients presenting with ST-segment elevation: The Task Force for the management of acute myocardial infarction in patients present-

ing with ST-segment elevation of the European Society of Cardiology (ESC). Eur Heart J. 2018;39(2):119–77.
7. John R, Rajasinghe HA, Chen JM, Weinberg AD, Sinha P, Mancini DM, et al. Long-term outcomes after cardiac transplantation: an experience based on different eras of immunosuppressive therapy. Ann Thorac Surg. 2001;72(2):440–9.
8. Bergmann O, Bhardwaj RD, Bernard S, Zdunek S, Barnabe-Heider F, Walsh S, et al. Evidence for cardiomyocyte renewal in humans. Science. 2009;324(5923):98–102.
9. Beltrami AP, Urbanek K, Kajstura J, Yan SM, Finato N, Bussani R, et al. Evidence that human cardiac myocytes divide after myocardial infarction. N Engl J Med. 2001;344(23):1750–7.
10. Ellison GM, Vicinanza C, Smith AJ, Aquila I, Leone A, Waring CD, et al. Adult c-kit(pos) cardiac stem cells are necessary and sufficient for functional cardiac regeneration and repair. Cell. 2013;154(4):827–42.
11. Hsieh PC, Segers VF, Davis ME, MacGillivray C, Gannon J, Molkentin JD, et al. Evidence from a genetic fate-mapping study that stem cells refresh adult mammalian cardiomyocytes after injury. Nat Med. 2007;13(8):970–4.
12. Senyo SE, Steinhauser ML, Pizzimenti CL, Yang VK, Cai L, Wang M, et al. Mammalian heart renewal by pre-existing cardiomyocytes. Nature. 2013;493(7432):433–6.
13. Kretzschmar K, Post Y, Bannier-Helaouet M, Mattiotti A, Drost J, Basak O, et al. Profiling proliferative cells and their progeny in damaged murine hearts. Proc Natl Acad Sci U S A. 2018;115(52): E12245–E54.
14. Valente M, Resende TP, Nascimento DS, Burlen-Defranoux O, Soares-da-Silva F, Dupont B, et al. Mouse HSA+ immature cardiomyocytes persist in the adult heart and expand after ischemic injury. PLoS Biol. 2019;17(6):e3000335.
15. Perez-Pomares JM, Gonzalez-Rosa JM, Munoz-Chapuli R. Building the vertebrate heart – an evolutionary approach to cardiac development. Int J Dev Biol. 2009;53(8–10):1427–43.
16. Burggren WW. Cardiac design in lower vertebrates: what can phylogeny reveal about ontogeny? Experientia. 1988;44(11–12):919–30.
17. Roesner A, Hankeln T, Burmester T. Hypoxia induces a complex response of globin expression in zebrafish (Danio rerio). J Exp Biol. 2006;209(Pt 11):2129–37.
18. Garcia-Gonzalez C, Morrison JI. Cardiac regeneration in non-mammalian vertebrates. Exp Cell Res. 2014;321(1):58–63.
19. Flink IL. Cell cycle reentry of ventricular and atrial cardiomyocytes and cells within the epicardium following amputation of the ventricular apex in the axolotl, Amblystoma mexicanum: confocal microscopic immunofluorescent image analysis of bromodeoxyuridine-labeled nuclei. Anat Embryol. 2002;205(3):235–44.
20. Laube F, Heister M, Scholz C, Borchardt T, Braun T. Re-programming of newt cardiomyocytes is induced by tissue regeneration. J Cell Sci. 2006;119(Pt 22):4719–29.
21. Stocum DL. Regenerative biology and medicine. J Musculoskelet Neuronal Interact. 2002;2(3): 270–3.
22. Sanchez Alvarado A, Tsonis PA. Bridging the regeneration gap: genetic insights from diverse animal models. Nat Rev Genet. 2006;7(11):873–84.
23. Poss KD, Wilson LG, Keating MT. Heart regeneration in zebrafish. Science. 2002;298(5601): 2188–90.
24. Jopling C, Sleep E, Raya M, Marti M, Raya A, Izpisua Belmonte JC. Zebrafish heart regeneration occurs by cardiomyocyte dedifferentiation and proliferation. Nature. 2010;464(7288):606–9.
25. Chablais F, Veit J, Rainer G, Jazwinska A. The zebrafish heart regenerates after cryoinjury-induced myocardial infarction. BMC Dev Biol. 2011;11(1):21.
26. Gonzalez-Rosa JM, Martin V, Peralta M, Torres M, Mercader N. Extensive scar formation and regression during heart regeneration after cryoinjury in zebrafish. Development. 2011;138(9): 1663–74.
27. Parente V, Balasso S, Pompilio G, Verduci L, Colombo GI, Milano G, et al. Hypoxia/reoxygenation cardiac injury and regeneration in zebrafish adult heart. PLoS One. 2013;8(1):e53748.
28. Schnabel K, Wu CC, Kurth T, Weidinger G. Regeneration of cryoinjury induced necrotic heart lesions in zebrafish is associated with epicardial activation and cardiomyocyte proliferation. PLoS One. 2011;6(4):e18503.
29. Wang J, Panakova D, Kikuchi K, Holdway JE, Gemberling M, Burris JS, et al. The regenerative capacity of zebrafish reverses cardiac failure caused by genetic cardiomyocyte depletion. Development. 2011;138(16):3421–30.

30. Witman N, Murtuza B, Davis B, Arner A, Morrison JI. Recapitulation of developmental cardiogenesis governs the morphological and functional regeneration of adult newt hearts following injury. Dev Biol. 2011;354(1):67–76.
31. Oberpriller JO, Oberpriller JC. Response of the adult newt ventricle to injury. J Exp Zool. 1974;187(2):249–53.
32. Puente BN, Kimura W, Muralidhar SA, Moon J, Amatruda JF, Phelps KL, et al. The oxygen-rich postnatal environment induces cardiomyocyte cell-cycle arrest through DNA damage response. Cell. 2014;157(3):565–79.
33. Sander V, Sune G, Jopling C, Morera C, Izpisua Belmonte JC. Isolation and in vitro culture of primary cardiomyocytes from adult zebrafish hearts. Nat Protoc. 2013;8(4):800–9.
34. Paradis AN, Gay MS, Zhang L. Binucleation of cardiomyocytes: the transition from a proliferative to a terminally differentiated state. Drug Discov Today. 2014;19(5):602–9.
35. Porrello ER, Olson EN. A neonatal blueprint for cardiac regeneration. Stem Cell Res. 2014;13(3 Pt B):556–70.
36. Ausoni S, Sartore S. From fish to amphibians to mammals: in search of novel strategies to optimize cardiac regeneration. J Cell Biol. 2009;184(3):357–64.
37. Robledo M. Myocardial regeneration in young rats. Am J Pathol. 1956;32(6):1215–39.
38. Porrello ER, Mahmoud AI, Simpson E, Hill JA, Richardson JA, Olson EN, et al. Transient regenerative potential of the neonatal mouse heart. Science. 2011;331(6020):1078–80.
39. Zogbi C, Saturi de Carvalho AE, Nakamuta JS, Caceres Vde M, Prando S, Giorgi MC, et al. Early postnatal rat ventricle resection leads to long-term preserved cardiac function despite tissue hypoperfusion. Physiol Rep. 2014;2(8):e12115.
40. Nag AC, Carey TR, Cheng M. DNA synthesis in rat heart cells after injury and the regeneration of myocardia. Tissue Cell. 1983;15(4):597–613.
41. Herdrich BJ, Danzer E, Davey MG, Allukian M, Englefield V, Gorman JH 3rd, et al. Regenerative healing following foetal myocardial infarction. Eur J Cardiothorac Surg. 2010;38(6):691–8.
42. Ye L, D'Agostino G, Loo SJ, Wang CX, Su LP, Tan SH, et al. Early regenerative capacity in the porcine heart. Circulation. 2018;138(24):2798–808.
43. Zhu W, Zhang E, Zhao M, Chong Z, Fan C, Tang Y, et al. Regenerative potential of neonatal porcine hearts. Circulation. 2018;138(24):2809–16.
44. Macmahon HE. Hyperplasia and regeneration of the myocardium in infants and in children. Am J Pathol. 1937;13(5):845–54.5.
45. Warthin AS. The myocardial lesions of diphtheria. J Infect Dis. 1924;35(1):32–66.
46. Fratz S, Hager A, Schreiber C, Schwaiger M, Hess J, Stern HC. Long-term myocardial scarring after operation for anomalous left coronary artery from the pulmonary artery. Ann Thorac Surg. 2011;92(5):1761–5.
47. Haubner BJ, Schneider J, Schweigmann U, Schuetz T, Dichtl W, Velik-Salchner C, et al. Functional recovery of a human neonatal heart after severe myocardial infarction. Circ Res. 2016;118(2):216–21.
48. Saker DM, Walsh-Sukys M, Spector M, Zahka KG. Cardiac recovery and survival after neonatal myocardial infarction. Pediatr Cardiol. 1997;18(2):139–42.
49. Vivien CJ, Hudson JE, Porrello ER. Evolution, comparative biology and ontogeny of vertebrate heart regeneration. NPJ Regen Med. 2016;1:16012.
50. Andersen DC, Ganesalingam S, Jensen CH, Sheikh SP. Do neonatal mouse hearts regenerate following heart apex resection? Stem Cell Reports. 2014;2(4):406–13.
51. Andersen DC, Jensen CH, Baun C, Hvidsten S, Zebrowski DC, Engel FB, et al. Persistent scarring and dilated cardiomyopathy suggest incomplete regeneration of the apex resected neonatal mouse myocardium – a 180 days follow up study. J Mol Cell Cardiol. 2016;90:47–52.
52. Andersen DC, Jensen CH, Sheikh SP. Response to Sadek et al. and Kotlikoff et al. Stem Cell Reports. 2014;3(1):3–4.
53. Kotlikoff MI, Hesse M, Fleischmann BK. Comment on "Do neonatal mouse hearts regenerate following heart apex resection"? Stem Cell Reports. 2014;3(1):2.
54. Sadek HA, Martin JF, Takeuchi JK, Leor J, Nie Y, Giacca M, et al. Multi-investigator letter on reproducibility of neonatal heart regeneration following apical resection. Stem Cell Reports. 2014;3(1):1.
55. Bryant DM, O'Meara CC, Ho NN, Gannon J, Cai L, Lee RT. A systematic analysis of neonatal mouse heart regeneration after apical resection. J Mol Cell Cardiol. 2015;79:315–8.

56. Darehzereshki A, Rubin N, Gamba L, Kim J, Fraser J, Huang Y, et al. Differential regenerative capacity of neonatal mouse hearts after cryoinjury. Dev Biol. 2015;399(1):91–9.
57. Sampaio-Pinto V, Rodrigues SC, Laundos TL, Silva ED, Vasques-Novoa F, Silva AC, et al. Neonatal apex resection triggers cardiomyocyte proliferation, neovascularization and functional recovery despite local fibrosis. Stem Cell Reports. 2018;10(3):860–74.
58. Lam NT, Sadek HA. Neonatal heart regeneration. Circulation. 2018;138(4):412–23.
59. Jesty SA, Steffey MA, Lee FK, Breitbach M, Hesse M, Reining S, et al. c-kit+ precursors support postinfarction myogenesis in the neonatal, but not adult, heart. Proc Natl Acad Sci U S A. 2012;109(33):13380–5.
60. Sturzu AC, Rajarajan K, Passer D, Plonowska K, Riley A, Tan TC, et al. Fetal mammalian heart generates a robust compensatory response to cell loss. Circulation. 2015;132(2):109–21.
61. Haubner BJ, Adamowicz-Brice M, Khadayate S, Tiefenthaler V, Metzler B, Aitman T, et al. Complete cardiac regeneration in a mouse model of myocardial infarction. Aging. 2012;4(12):966–77.
62. Porrello ER, Mahmoud AI, Simpson E, Johnson BA, Grinsfelder D, Canseco D, et al. Regulation of neonatal and adult mammalian heart regeneration by the miR-15 family. Proc Natl Acad Sci U S A. 2013;110(1):187–92.
63. Malek Mohammadi M, Abouissa A, Isyatul A, Xie Y, Cordero J, Shirvani A, et al. Induction of cardiomyocyte proliferation and angiogenesis protects neonatal mice from pressure overload-associated maladaptation. JCI Insight. 2019;5:128336.
64. Strungs EG, Ongstad EL, O'Quinn MP, Palatinus JA, Jourdan LJ, Gourdie RG. Cryoinjury models of the adult and neonatal mouse heart for studies of scarring and regeneration. Methods Mol Biol. 2013;1037:343–53.
65. Jain RK. Molecular regulation of vessel maturation. Nat Med. 2003;9(6):685–93.
66. Lavine KJ, Epelman S, Uchida K, Weber KJ, Nichols CG, Schilling JD, et al. Distinct macrophage lineages contribute to disparate patterns of cardiac recovery and remodeling in the neonatal and adult heart. Proc Natl Acad Sci U S A. 2014;111(45):16029–34.
67. Laflamme MA, Murry CE. Heart regeneration. Nature. 2011;473(7347):326–35.
68. Blackburn S. Maternal, fetal, & neonatal physiology. St. Louis: Elsevier Health Sciences; 2014.
69. Breckenridge R. Molecular control of cardiac fetal/neonatal remodeling. J Cardiovasc Dev Dis. 2014;1(1):29–36.
70. Lopaschuk GD, Collins-Nakai RL, Itoi T. Developmental changes in energy substrate use by the heart. Cardiovasc Res. 1992;26(12):1172–80.
71. Nakada Y, Canseco DC, Thet S, Abdisalaam S, Asaithamby A, Santos CX, et al. Hypoxia induces heart regeneration in adult mice. Nature. 2017;541(7636):222–7.
72. Soonpaa MH, Kim KK, Pajak L, Franklin M, Field LJ. Cardiomyocyte DNA synthesis and binucleation during murine development. Am J Physiol Heart Circ Physiol. 1996;271(5):H2183–H9.
73. Bernardo BC, Weeks KL, Pretorius L, McMullen JR. Molecular distinction between physiological and pathological cardiac hypertrophy: experimental findings and therapeutic strategies. Pharmacol Ther. 2010;128(1):191–227.
74. Locatelli P, Gimenez CS, Vega MU, Crottogini A, Belaich MN. Targeting the cardiomyocyte cell cycle for heart regeneration. Curr Drug Targets. 2019;20(2):241–54.
75. Burton PB, Raff MC, Kerr P, Yacoub MH, Barton PJ. An intrinsic timer that controls cell-cycle withdrawal in cultured cardiac myocytes. Dev Biol. 1999;216(2):659–70.
76. Ikenishi A, Okayama H, Iwamoto N, Yoshitome S, Tane S, Nakamura K, et al. Cell cycle regulation in mouse heart during embryonic and postnatal stages. Develop Growth Differ. 2012;54(8):731–8.
77. Tane S, Ikenishi A, Okayama H, Iwamoto N, Nakayama KI, Takeuchi T. CDK inhibitors, p21(Cip1) and p27(Kip1), participate in cell cycle exit of mammalian cardiomyocytes. Biochem Biophys Res Commun. 2014;443(3):1105–9.
78. Mohamed TMA, Ang YS, Radzinsky E, Zhou P, Huang Y, Elfenbein A, et al. Regulation of cell cycle to stimulate adult cardiomyocyte proliferation and cardiac regeneration. Cell. 2018;173(1):104–16.e12.
79. Sim CB, Ziemann M, Kaspi A, Harikrishnan KN, Ooi J, Khurana I, et al. Dynamic changes in the cardiac methylome during postnatal development. FASEB J. 2015;29(4):1329–43.
80. Gilsbach R, Preissl S, Gruning BA, Schnick T, Burger L, Benes V, et al. Dynamic DNA methylation orchestrates cardiomyocyte development, maturation and disease. Nat Commun. 2014;5:5288.

81. Dirkx E, da Costa Martins PA, De Windt LJ. Regulation of fetal gene expression in heart failure. Biochim Biophys Acta. 2013;1832(12):2414–24.
82. Peters MMC, Sampaio-Pinto V, da Costa Martins PA. Non-coding RNAs in endothelial cell signalling and hypoxia during cardiac regeneration. Biochim Biophys Acta, Mol Cell Res. 2020;1867:118515.
83. Liu N, Bezprozvannaya S, Williams AH, Qi X, Richardson JA, Bassel-Duby R, et al. microRNA-133a regulates cardiomyocyte proliferation and suppresses smooth muscle gene expression in the heart. Genes Dev. 2008;22(23):3242–54.
84. Porrello ER, Johnson BA, Aurora AB, Simpson E, Nam YJ, Matkovich SJ, et al. MiR-15 family regulates postnatal mitotic arrest of cardiomyocytes. Circ Res. 2011;109(6):670–9.
85. Eulalio A, Mano M, Dal Ferro M, Zentilin L, Sinagra G, Zacchigna S, et al. Functional screening identifies miRNAs inducing cardiac regeneration. Nature. 2012;492(7429):376–81.
86. Gabisonia K, Prosdocimo G, Aquaro GD, Carlucci L, Zentilin L, Secco I, et al. MicroRNA therapy stimulates uncontrolled cardiac repair after myocardial infarction in pigs. Nature. 2019;569(7756):418–22.
87. Pandey R, Yang Y, Jackson L, Ahmed RP. MicroRNAs regulating meis1 expression and inducing cardiomyocyte proliferation. Cardiovasc Regen Med. 2016;3:e1468.
88. Cai B, Ma W, Ding F, Zhang L, Huang Q, Wang X, et al. The long noncoding RNA CAREL controls cardiac regeneration. J Am Coll Cardiol. 2018;72(5):534–50.
89. Jiang B, Liao R. The paradoxical role of inflammation in cardiac repair and regeneration. J Cardiovasc Transl Res. 2010;3(4):410–6.
90. Epelman S, Lavine Kory J, Beaudin Anna E, Sojka Dorothy K, Carrero Javier A, Calderon B, et al. Embryonic and adult-derived resident cardiac macrophages are maintained through distinct mechanisms at steady state and during inflammation. Immunity. 2014;40(1):91–104.
91. Molawi K, Wolf Y, Kandalla PK, Favret J, Hagemeyer N, Frenzel K, et al. Progressive replacement of embryo-derived cardiac macrophages with age. J Exp Med. 2014;211(11):2151–8.
92. Aurora AB, Porrello ER, Tan W, Mahmoud AI, Hill JA, Bassel-Duby R, et al. Macrophages are required for neonatal heart regeneration. J Clin Invest. 2014;124(3):1382–92.
93. Howard CM, Baudino TA. Dynamic cell-cell and cell-ECM interactions in the heart. J Mol Cell Cardiol. 2014;70:19–26.
94. Krenning G, Zeisberg EM, Kalluri R. The origin of fibroblasts and mechanism of cardiac fibrosis. J Cell Physiol. 2010;225(3):631–7.
95. Souders CA, Bowers SL, Baudino TA. Cardiac fibroblast: the renaissance cell. Circ Res. 2009; 105(12):1164–76.
96. Ieda M, Tsuchihashi T, Ivey KN, Ross RS, Hong TT, Shaw RM, et al. Cardiac fibroblasts regulate myocardial proliferation through beta1 integrin signaling. Dev Cell. 2009;16(2):233–44.
97. Fredj S, Bescond J, Louault C, Delwail A, Lecron JC, Potreau D. Role of interleukin-6 in cardiomyocyte/cardiac fibroblast interactions during myocyte hypertrophy and fibroblast proliferation. J Cell Physiol. 2005;204(2):428–36.
98. Ingason AB, Goldstone AB, Paulsen MJ, Thakore AD, Truong VN, Edwards BB, et al. Angiogenesis precedes cardiomyocyte migration in regenerating mammalian hearts. J Thorac Cardiovasc Surg. 2018;155(3):1118–27.. e1
99. Das S, Goldstone AB, Wang H, Farry J, D'Amato G, Paulsen MJ, et al. A unique collateral artery development program promotes neonatal heart regeneration. Cell. 2019;176(5):1128–42.. e18
100. Mahmoud AI, O'Meara CC, Gemberling M, Zhao L, Bryant DM, Zheng R, et al. Nerves regulate cardiomyocyte proliferation and heart regeneration. Dev Cell. 2015;34(4):387–99.
101. Goldsmith EC, Hoffman A, Morales MO, Potts JD, Price RL, McFadden A, et al. Organization of fibroblasts in the heart. Dev Dyn. 2004;230(4):787–94.
102. Hanson KP, Jung JP, Tran QA, Hsu SP, Iida R, Ajeti V, et al. Spatial and temporal analysis of extracellular matrix proteins in the developing murine heart: a blueprint for regeneration. Tissue Eng Part A. 2013;19:1132–43.
103. Hurle JM, Icardo JM, Ojeda JL. Compositional and structural heterogenicity of the cardiac jelly of the chick embryo tubular heart: a TEM, SEM and histochemical study. J Embryol Exp Morphol. 1980;56:211–23.
104. Kim H, Yoon CS, Kim H, Rah B. Expression of extracellular matrix components fibronectin and laminin in the human fetal heart. Cell Struct Funct. 1999;24(1):19–26.

105. Barry A. The functional significance of the cardiac jelly in the tubular heart of the chick embryo. Anat Rec. 1948;102(3):289–98.
106. Lockhart M, Wirrig E, Phelps A, Wessels A. Extracellular matrix and heart development. Birth Defects Res A Clin Mol Teratol. 2011;91(6):535–50.
107. Gazoti Debessa CR, Mesiano Maifrino LB, Rodrigues de Souza R. Age related changes of the collagen network of the human heart. Mech Ageing Dev. 2001;122(10):1049–58.
108. Williams C, Quinn KP, Georgakoudi I, Black LD 3rd. Young developmental age cardiac extracellular matrix promotes the expansion of neonatal cardiomyocytes in vitro. Acta Biomater. 2014;10(1):194–204.
109. Bowers SL, Banerjee I, Baudino TA. The extracellular matrix: at the center of it all. J Mol Cell Cardiol. 2010;48(3):474–82.
110. Camelliti P, Borg TK, Kohl P. Structural and functional characterisation of cardiac fibroblasts. Cardiovasc Res. 2005;65(1):40–51.
111. Silva AC, Rodrigues SC, Caldeira J, Nunes AM, Sampaio-Pinto V, Resende TP, et al. Three-dimensional scaffolds of fetal decellularized hearts exhibit enhanced potential to support cardiac cells in comparison to the adult. Biomaterials. 2016;104:52–64.
112. Bassat E, Mutlak YE, Genzelinakh A, Shadrin IY, Baruch Umansky K, Yifa O, et al. The extracellular matrix protein agrin promotes heart regeneration in mice. Nature. 2017;547(7662):179–84.
113. Kruger M, Kohl T, Linke WA. Developmental changes in passive stiffness and myofilament Ca2+ sensitivity due to titin and troponin-I isoform switching are not critically triggered by birth. Am J Physiol Heart Circ Physiol. 2006;291(2):H496–506.
114. Lahmers S, Wu Y, Call DR, Labeit S, Granzier H. Developmental control of titin isoform expression and passive stiffness in fetal and neonatal myocardium. Circ Res. 2004;94(4):505–13.
115. Linke WA. Sense and stretchability: the role of titin and titin-associated proteins in myocardial stress-sensing and mechanical dysfunction. Cardiovasc Res. 2008;77(4):637–48.
116. Majkut S, Idema T, Swift J, Krieger C, Liu A, Discher DE. Heart-specific stiffening in early embryos parallels matrix and myosin expression to optimize beating. Curr Biol. 2013;23(23):2434–9.
117. Gershlak JR, Resnikoff JI, Sullivan KE, Williams C, Wang RM, Black LD 3rd. Mesenchymal stem cells ability to generate traction stress in response to substrate stiffness is modulated by the changing extracellular matrix composition of the heart during development. Biochem Biophys Res Commun. 2013;439(2):161–6.
118. Notari M, Ventura-Rubio A, Bedford-Guaus SJ, Jorba I, Mulero L, Navajas D, et al. The local microenvironment limits the regenerative potential of the mouse neonatal heart. Sci Adv. 2018;4(5):eaao5553.
119. Xin M, Kim Y, Sutherland LB, Murakami M, Qi X, McAnally J, et al. Hippo pathway effector Yap promotes cardiac regeneration. Proc Natl Acad Sci U S A. 2013;110(34):13839–44.
120. Bhana B, Iyer RK, Chen WL, Zhao R, Sider KL, Likhitpanichkul M, et al. Influence of substrate stiffness on the phenotype of heart cells. Biotechnol Bioeng. 2010;105(6):1148–60.
121. Jacot JG, McCulloch AD, Omens JH. Substrate stiffness affects the functional maturation of neonatal rat ventricular myocytes. Biophys J. 2008;95(7):3479–87.
122. Heallen T, Morikawa Y, Leach J, Tao G, Willerson JT, Johnson RL, et al. Hippo signaling impedes adult heart regeneration. Development. 2013;140(23):4683–90.
123. Mosqueira D, Pagliari S, Uto K, Ebara M, Romanazzo S, Escobedo-Lucea C, et al. Hippo pathway effectors control cardiac progenitor cell fate by acting as dynamic sensors of substrate mechanics and nanostructure. ACS Nano. 2014;8(3):2033–47.
124. Menasche P, Alfieri O, Janssens S, McKenna W, Reichenspurner H, Trinquart L, et al. The Myoblast Autologous Grafting in Ischemic Cardiomyopathy (MAGIC) trial: first randomized placebo-controlled study of myoblast transplantation. Circulation. 2008;117(9):1189–200.
125. Gnecchi M, Zhang Z, Ni A, Dzau VJ. Paracrine mechanisms in adult stem cell signaling and therapy. Circ Res. 2008;103(11):1204–19.
126. Murry CE, Soonpaa MH, Reinecke H, Nakajima H, Nakajima HO, Rubart M, et al. Haematopoietic stem cells do not transdifferentiate into cardiac myocytes in myocardial infarcts. Nature. 2004;428(6983):664–8.
127. Orlic D, Kajstura J, Chimenti S, Jakoniuk I, Anderson SM, Li B, et al. Bone marrow cells regenerate infarcted myocardium. Nature. 2001;410(6829):701–5.

128. Traverse JH, Henry TD, Pepine CJ, Willerson JT, Zhao DX, Ellis SG, et al. Effect of the use and timing of bone marrow mononuclear cell delivery on left ventricular function after acute myocardial infarction: the TIME randomized trial. JAMA. 2012;308(22):2380–9.
129. Musialek P, Mazurek A, Jarocha D, Tekieli L, Szot W, Kostkiewicz M, et al. Myocardial regeneration strategy using Wharton's jelly mesenchymal stem cells as an off-the-shelf "unlimited" therapeutic agent: results from the Acute Myocardial Infarction First-in-Man Study. Postepy Kardiol Interwencyjnej = Adv Interv Cardiol. 2015;11(2):100–7.
130. Santos Nascimento D, Mosqueira D, Sousa LM, Teixeira M, Filipe M, Resende TP, et al. Human umbilical cord tissue-derived mesenchymal stromal cells attenuate remodeling after myocardial infarction by proangiogenic, antiapoptotic, and endogenous cell-activation mechanisms. Stem Cell Res Ther. 2014;5(1):5.
131. Mathiasen AB, Qayyum AA, Jorgensen E, Helqvist S, Fischer-Nielsen A, Kofoed KF, et al. Bone marrow-derived mesenchymal stromal cell treatment in patients with severe ischaemic heart failure: a randomized placebo-controlled trial (MSC-HF trial). Eur Heart J. 2015;36(27):1744–53.
132. Uchida S, De Gaspari P, Kostin S, Jenniches K, Kilic A, Izumiya Y, et al. Sca1-derived cells are a source of myocardial renewal in the murine adult heart. Stem Cell Reports. 2013;1(5):397–410.
133. Valente M, Nascimento DS, Cumano A, Pinto-do OP. Sca-1+ cardiac progenitor cells and heart-making: a critical synopsis. Stem Cells Dev. 2014;23(19):2263–73.
134. Li Y, He L, Huang X, Bhaloo SI, Zhao H, Zhang S, et al. Genetic lineage tracing of nonmyocyte population by dual recombinases. Circulation. 2018;138(8):793–805.
135. Sultana N, Zhang L, Yan J, Chen J, Cai W, Razzaque S, et al. Resident c-kit(+) cells in the heart are not cardiac stem cells. Nat Commun. 2015;6:8701.
136. van Berlo JH, Kanisicak O, Maillet M, Vagnozzi RJ, Karch J, Lin SC, et al. c-kit+ cells minimally contribute cardiomyocytes to the heart. Nature. 2014;509(7500):337–41.
137. Bolli R, Chugh AR, D'Amario D, Loughran JH, Stoddard MF, Ikram S, et al. Cardiac stem cells in patients with ischaemic cardiomyopathy (SCIPIO): initial results of a randomised phase 1 trial. Lancet. 2011;378(9806):1847–57.
138. Makkar RR, Smith RR, Cheng K, Malliaras K, Thomson LE, Berman D, et al. Intracoronary cardiosphere-derived cells for heart regeneration after myocardial infarction (CADUCEUS): a prospective, randomised phase 1 trial. Lancet. 2012;379(9819):895–904.
139. Beltrami AP, Barlucchi L, Torella D, Baker M, Limana F, Chimenti S, et al. Adult cardiac stem cells are multipotent and support myocardial regeneration. Cell. 2003;114(6):763–76.
140. Dyer O. NEJM retracts article from former researcher once hailed as heart stem cell pioneer. BMJ. 2018;363:k4432.
141. The Lancet Editors. Expression of concern: the SCIPIO trial. Lancet. 2014;383(9925):1279.
142. Lundy SD, Gantz JA, Pagan CM, Filice D, Laflamme MA. Pluripotent stem cell derived cardiomyocytes for cardiac repair. Curr Treat Options Cardiovasc Med. 2014;16(7):319.
143. Lundy SD, Zhu WZ, Regnier M, Laflamme MA. Structural and functional maturation of cardiomyocytes derived from human pluripotent stem cells. Stem Cells Dev. 2013;22(14):1991–2002.
144. Madonna R, Van Laake L, Davidson SM, Engel FB, Hausenloy DJ, Lecour S, et al. Position Paper of the European Society of Cardiology Working Group Cellular Biology of the Heart: cell-based therapies for myocardial repair and regeneration in ischemic heart disease and heart failure. Eur Heart J. 2016;37(23):1789–98d.
145. Tu C, Chao BS, Wu JC. Strategies for improving the maturity of human induced pluripotent stem cell-derived cardiomyocytes. Circ Res. 2018;123(5):512–4.
146. Nicks Amy M, Humphreys David T, Holman Sara R, Chan Andrea Y, Djordjevic D, Naqvi N, et al. Abstract 15573: Transcription factors driving postnatal cardiomyocyte maturation. Circulation. 2018;138(Suppl_1):A15573-A.
147. Chong JJ, Yang X, Don CW, Minami E, Liu YW, Weyers JJ, et al. Human embryonic-stem-cell-derived cardiomyocytes regenerate non-human primate hearts. Nature. 2014;510(7504):273–7.
148. Menasche P, Vanneaux V, Hagege A, Bel A, Cholley B, Cacciapuoti I, et al. Human embryonic stem cell-derived cardiac progenitors for severe heart failure treatment: first clinical case report. Eur Heart J. 2015;36(30):2011–7.
149. Bui QT, Gertz ZM, Wilensky RL. Intracoronary delivery of bone-marrow-derived stem cells. Stem Cell Res Ther. 2010;1(4):29.
150. Muller-Ehmsen J, Krausgrill B, Burst V, Schenk K, Neisen UC, Fries JW, et al. Effective engraftment but poor mid-term persistence of mononuclear and mesenchymal bone marrow cells in acute and chronic rat myocardial infarction. J Mol Cell Cardiol. 2006;41(5):876–84.

151. Cambria E, Pasqualini FS, Wolint P, Gunter J, Steiger J, Bopp A, et al. Translational cardiac stem cell therapy: advancing from first-generation to next-generation cell types. NPJ Regen Med. 2017;2(1):17.
152. Song Y, Zhang C, Zhang J, Sun N, Huang K, Li H, et al. An injectable silk sericin hydrogel promotes cardiac functional recovery after ischemic myocardial infarction. Acta Biomater. 2016;41:210–23.
153. Georgiadis V, Knight RA, Jayasinghe SN, Stephanou A. Cardiac tissue engineering: renewing the arsenal for the battle against heart disease. Integr Biol. 2014;6(2):111–26.
154. Gouveia PJ, Rosa S, Ricotti L, Abecasis B, Almeida HV, Monteiro L, et al. Flexible nanofilms coated with aligned piezoelectric microfibers preserve the contractility of cardiomyocytes. Biomaterials. 2017;139:213–28.
155. Stewart DJ, Kutryk MJ, Fitchett D, Freeman M, Camack N, Su Y, et al. VEGF gene therapy fails to improve perfusion of ischemic myocardium in patients with advanced coronary disease: results of the NORTHERN trial. Mol Ther. 2009;17(6):1109–15.
156. Ripa RS, Jorgensen E, Wang Y, Thune JJ, Nilsson JC, Sondergaard L, et al. Stem cell mobilization induced by subcutaneous granulocyte-colony stimulating factor to improve cardiac regeneration after acute ST-elevation myocardial infarction: result of the double-blind, randomized, placebo-controlled stem cells in myocardial infarction (STEMMI) trial. Circulation. 2006;113(16):1983–92.
157. Ott I, Schulz S, Mehilli J, Fichtner S, Hadamitzky M, Hoppe K, et al. Erythropoietin in patients with acute ST-segment elevation myocardial infarction undergoing primary percutaneous coronary intervention: a randomized, double-blind trial. Circ Cardiovasc Interv. 2010;3(5):408–13.
158. Lucas T, Bonauer A, Dimmeler S. RNA therapeutics in cardiovascular disease. Circ Res. 2018;123(2):205–20.
159. Raso A, Dirkx E. Cardiac regenerative medicine: at the crossroad of microRNA function and biotechnology. Noncoding RNA Res. 2017;2(1):27–37.
160. Nam YJ, Song K, Luo X, Daniel E, Lambeth K, West K, et al. Reprogramming of human fibroblasts toward a cardiac fate. Proc Natl Acad Sci U S A. 2013;110(14):5588–93.
161. Pinto AR, Ilinykh A, Ivey MJ, Kuwabara JT, D'Antoni ML, Debuque R, et al. Revisiting cardiac cellular composition. Circ Res. 2016;118(3):400–9.
162. Ottaviani L, Sansonetti M, da Costa Martins PA. Myocardial cell-to-cell communication via microRNAs. Noncoding RNA Res. 2018;3(3):144–53.
163. Corrado C, Raimondo S, Chiesi A, Ciccia F, De Leo G, Alessandro R. Exosomes as intercellular signaling organelles involved in health and disease: basic science and clinical applications. Int J Mol Sci. 2013;14(3):5338–66.
164. Halkein J, Tabruyn SP, Ricke-Hoch M, Haghikia A, Nguyen NQ, Scherr M, et al. MicroRNA-146a is a therapeutic target and biomarker for peripartum cardiomyopathy. J Clin Invest. 2013;123(5):2143–54.
165. World Health Organization. Global health estimates 2016: deaths by cause, age, sex, by Country and by Region, 2000–2016. Geneva: World Health Organization; 2018.

Reproducing Human Brain Development In Vitro: Generating Cerebellar Neurons for Modelling Cerebellar Ataxias

Evguenia Bekman, Teresa P. Silva, João P. Cotovio, and Rita Mendes de Almeida

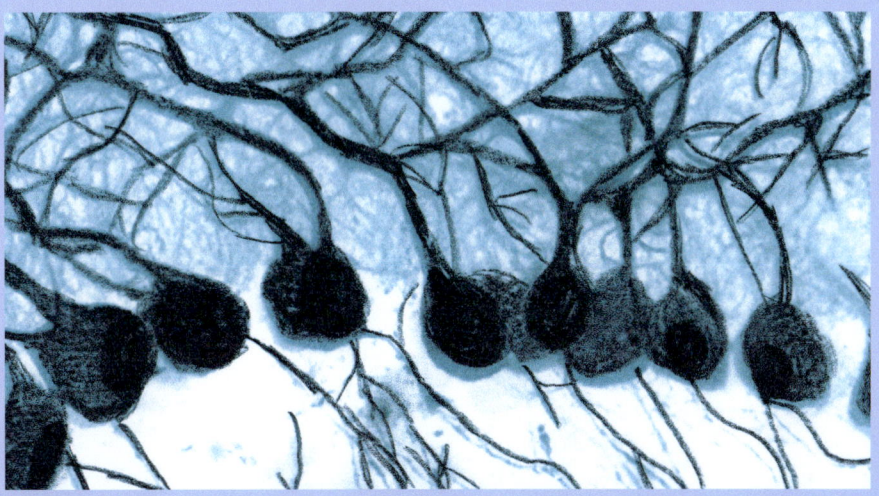

© Springer Nature Switzerland AG 2020
G. Rodrigues, B. A. J. Roelen (eds.), *Concepts and Applications of Stem Cell Biology*,
Learning Materials in Biosciences, https://doi.org/10.1007/978-3-030-43939-2_11

Contents

11.1 Human Brain Development – 215
11.1.1 How Billions of Neurons Are Generated – 217
11.1.2 How Is Neuronal Diversity Obtained? – 218
11.1.3 Cerebellar Development – 222

11.2 Case Study: Generating Cerebellar Neurons for Modelling Cerebellar Ataxias – 223

References – 225

What You Will Learn in This Chapter

The main purpose of this chapter is to introduce the reader to the complexity of nervous system development, to better understand the huge challenge to faithfully reproduce this system in vitro for the purpose of neurodevelopmental and neurodegenerative disease modelling. The emphasis in this chapter is on the aspects of embryonic development that are needed to be mimicked in vitro to generate functionally validated neuronal cells. As an example of the latter, a case study of the generation of functionally mature cerebellar neurons from human-induced pluripotent stem cells (hiPSCs) is given.

A brief outline of crucial steps in laying out of the central nervous system (CNS) is a compilation of up-to-date knowledge based on experimental data collected mainly from animal studies, due to difficulty of studying the development of the human brain. Though the same main principles govern neural development in higher vertebrates, some important differences between murine and human cerebellar commitment have been reported. For the sake of clarity, we do not discuss these differences and focus entirely on the human cerebellar model. After reading this chapter, the reader will have acquired information on the most important features of neural development for successful in vitro modelling.

11.1 Human Brain Development

Have you ever thought that there is a galaxy of neurons inside your skull? If we compare the number of stars in the Milky Way, estimated between 100 and 400 billion ($2.5 \times 10^{11} \pm 1.5 \times 10^{11}$), with the number of neurons just in the human cerebellum (101 billion) plus some 20–25 billion neurons in the neocortex, and even without other brain areas and glial cells (which in some brain areas outnumber neurons tenfold), we will realise that we have a galaxy of neurons inside our brain.

And then we can add some extra complexity to this, with 164,000 billion synapses (~7000 synapses/neuron), ~12,000,000 km of dendrites and ~100,000 km of axons…

Thinking of these numbers the following question arises: how is this complexity generated? And maintained? And of course, how is it translated into our thoughts, memories, feelings, communication, imagination? We still do not know the answers for all these questions, but now we are able to outline, with ever-growing precision in detail, the process of the building of the brain structure. So, how does it start?

Everything starts with the egg, a wonder of nature and the origin of all the cells in the body including those that produce other eggs, upon fertilisation. The successive divisions of the fertilised egg sequentially give rise to the morula, the blastula and the primitive epiblast undergoing gastrulation to originate the three definitive germ layers, ecto-, meso- and endoderm. The formation of the ectoderm is the essential initial step in laying out the neural tissue that arises from its dorsal (axial) part. Dorsal ectoderm acquires neural identity in response to a signal from the underlying mesoderm in the process called neural induction. The inductive signal consists in the concerted action of bone morphogenetic protein (BMP) antagonists, fibroblast growth factors (FGFs) and wingless proteins (Wnts) that together efficiently inhibit ongoing transforming growth factor beta (TGFβ) signalling and induce the switch from non-neural to neural ectoderm identity. The immediate consequences induced by this switch are the onset of the expression of neural identity genes, such as *SOX1* [1], and the replacement of E-Cadherin by N-Cadherin on the cell surface. N-Cadherin localises at the subapical membrane domain, where its subcellular domain contacts with the PAR3 complex to recruit β-catenin and rearranging local actin cytoskeleton [2]. As a result, elongation and apical constriction of epithelial cells occur, producing a thickened neuroepithelial sheet, called a neural plate that, due to much reduced apical surface area, starts to bend along the anteroposterior axis and forms the neural tube (◘ Fig. 11.1). The bending and closing of the neural tube occur more easily and faster in its middle part (future spinal cord), while it is less thick than the anterior part which will give rise to brain vesicles and takes more time to close. The most posterior part of the neural tube is the last to close, and the neurogenesis here occurs in a different manner than in the rest of the neural tube and is, therefore, called secondary neurogenesis (reviewed in [3]), to distinguish from the primary neurogenesis that generates the more anterior nervous system. At the end of neurulation, three different regions are formed: neural ectoderm that originates the central nervous system (CNS); non-neural ectoderm, which will form the epidermis; and a region between neural and non-neural ectoderm, which will give rise to the neural crest [4, 5].

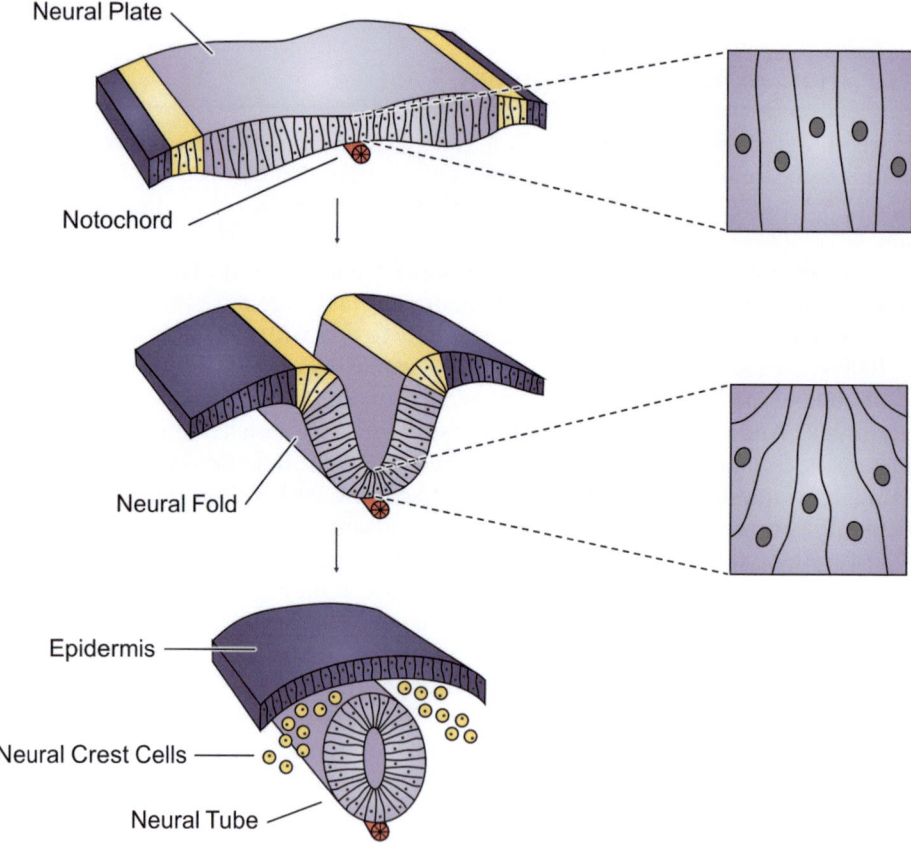

◘ Fig. 11.1 Schematic view of vertebrate neural tube formation. Neural plate is formed by elongation of epithelial cells upon inductive signal from underlying mesoderm (notochord). Apical constriction of neuroepithelial cells facilitates the bending of neural plate and subsequent neural tube formation. As a result, three different regions are formed: neural ectoderm that originates the CNS, non-neural ectoderm (epidermis) and neural crest forming in between neural and non-neural ectoderm

11.1.1 How Billions of Neurons Are Generated

The closed neural tube is built up of a pseudostratified epithelium composed by fusiform progenitors, each of them trespassing the neuroepithelium from its apical (inner, or luminal) side to basal (outer) side (◘ Fig. 11.2). Unlike cell nuclei of other epithelia, the nuclei of neuroepithelial cells do not lie at the same level; in fact, they are dispersed along the apico-basal axis, being in constant movement from apical side to basal and back. In this clever way, the neuroepithelium can fit a great number of progenitors in a very compact space. One of the amazing features of this peculiar organisation is that the movement of nuclei, called interkinetic nuclei migration, INM, is coupled to cell cycle, so that mitoses always occur at the apical side and S phase takes place at the basal side ([6]; reviewed in [7]). From the very beginning, and until the late embryonic stages, the neural tube is the place of intense proliferation and differentiation. For example, during the first half of pregnancy the rate of production of newborn neurons is over 200,000 neurons per minute. To keep up with such a high rate of neuronal production, neuroepithelium needs to have an efficient mechanism of the maintenance of progenitor pool. This is assured by Notch signalling, where a newborn neuron starts to express Notch ligand, Delta, that binds its receptor Notch on the surface of adjacent progenitor cell and this binding exposes the cleavage site releasing the intracellular domain of Notch (NICD) [8]. This domain goes to the nucleus where it forms a complex with CBF1 to activate target genes that will maintain the progenitor state of the cell and inhibit proneural genes necessary for the exit to differentiation (◘ Fig. 11.3). Thus, a newborn neuron signals to neighbouring cells to prevent them from exiting for differentiation at the same time, preserving the progenitors for differentiation at later stages [9]. This mechanism of cell–cell interaction is called lateral inhibition and is widely used during development whenever a binary decision between two cell fates must be made. The importance of lateral inhibition in neural development is demonstrated by studies of Notch path-

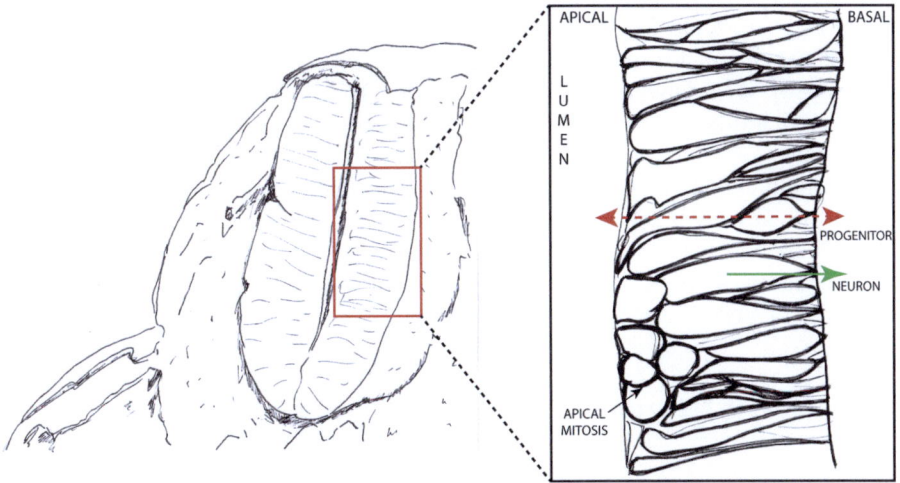

◘ **Fig. 11.2** The structure of neuroepithelium. Closed neural tube takes form of a pseudostratified epithelium with basal (outer) side and apical (inner, or luminal) side where mitoses occur. While progenitor cells stretch from apical to basal surface, neurons lose apical endfeet and accumulate at the basal side

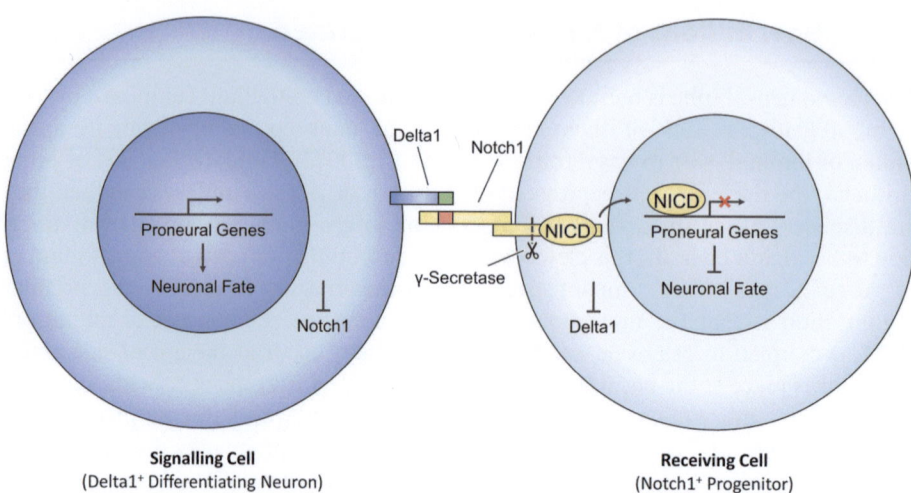

Fig. 11.3 Notch signalling pathway. Newborn neuron expresses Notch ligand, Delta1, which binds its receptor Notch on the surface of adjacent progenitor cell. Cleavage releases the intracellular domain of Notch, NICD, which goes to the nucleus and inhibits proneural genes necessary for the exit to differentiation. As a result of this inhibition, cells adjacent to the newborn neuron remain as progenitors until the next cell cycle, in which, due to interkinetic nuclear migration and to the movement of neurons towards mantle layer, a new combination of neighbouring cells is generated and another cell is singled out for differentiation and lateral inhibition exerted on surrounding progenitors

way mutants, where excessive signalling results in overproliferation of neural progenitors and dramatic decrease in neuronal production, while the lack of signalling causes massive premature neuronal differentiation and reduction of neural tube thickness due to exhaustion of progenitor pool [8, 9].

With the progress of neurogenesis in the neural tube its thickness increases, so the later stage progenitors, called radial glia, are obliged to stretch out and form a very long basal endfeet that plays an important role in guiding neuronal migration [10]. Radial glia is thought to divide asymmetrically, each division giving rise to another radial glial cell and a transit-amplifying cell that will divide several times to produce neurons [10]. With time, the neural tube subdivides into the apical ventricular zone, where progenitors persist, and basal mantle layer, where differentiating neurons accumulate.

The maintenance of the progenitor pool throughout embryonic development is one of the key mechanisms underlying the diversity of neural cell types in the CNS. Once the constant supply of progenitors is assured, the diversity is generated by conjugation of spatial and temporal cues combined with gradual changes in progenitor competence, as will be discussed below.

11.1.2 How Is Neuronal Diversity Obtained?

The great diversity of neuronal subtypes is obtained by a combination of intrinsic and extrinsic cues that together will determine which type of neuron will be born at a specific time and place, meaning that both positional and temporal cues are in charge of this process.

The most important and probably the earliest positional cue is the location of the progenitor along the rostro-caudal (R-C) axis. The regional patterning of neural progenitor cells starts with the most rostral identity, the "primitive identity" [11–14]. While the forebrain territory is specified in the absence of all major signalling molecules, more caudal fates require the action of some morphogens (see below), including retinoic acid (RA), WNT and FGF [15–17]. For the midbrain/hindbrain identity, FGF signalling is essential, while RA confers spinal cord identity [18]. As a result of the concerted action of different morphogens, four major regions are created along the R-C axis of the neural tube: forebrain (prosencephalon), midbrain (mesencephalon), hindbrain (rhombencephalon) and spinal cord (◘ Fig. 11.4). Within these four regions, the same morphogen gradients induce overlapping expression of Hox homeodomain proteins that generate a segmented pattern of positional identities determining neuronal fates [19]. *Positional cues along the R-C axis define the functional specificity of neural cells with respect to different body segments.*

In addition to the R-C axis, dorsal–ventral (D-V) direction is determined by different concentrations of morphogens provided by different organising centres. Morphogens are diffusible molecules that are able to establish a graded concentration distribution to elicit distinct cellular responses in a dose-dependent manner.

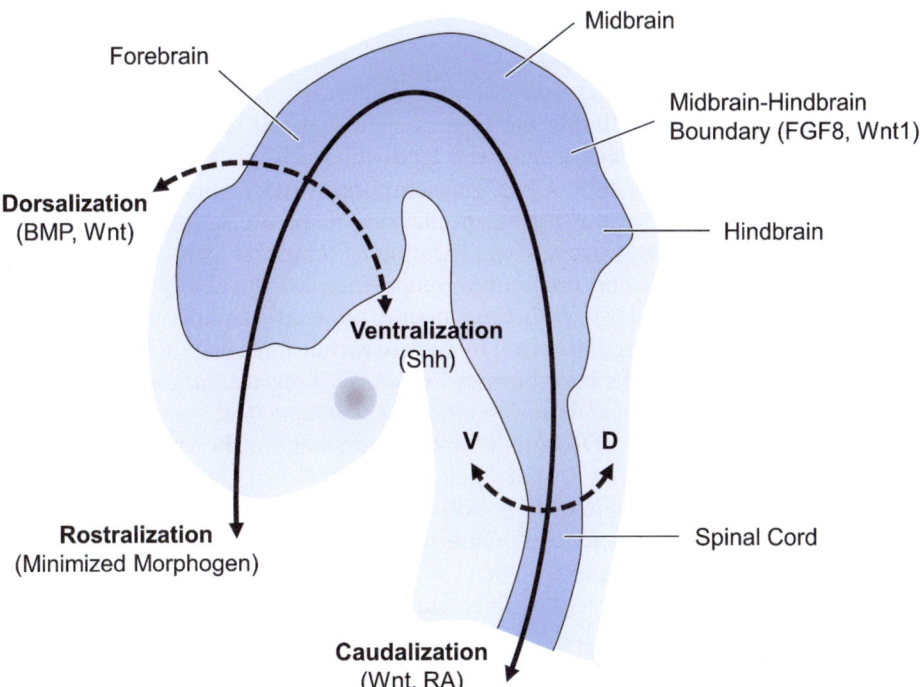

◘ **Fig. 11.4** Regional patterning of neural tube. Rostral–caudal (R-C) and dorsal–ventral (D-V) axis are determined by the action of various morphogens from different organising centres. Forebrain identity is established in the absence of major signalling molecules, while midbrain requires FGF activity, where FGF8 in particular is essential for the positioning of the midbrain–hindbrain boundary. Spinal cord identity is conferred by concerted action of Wnts and RA. In the dorsoventral plane, BMPs produced by overlying epidermis and Wnts coming from the roofplate oppose ventrally produced Shh establishing a dorsoventral gradient according to which different types of neurons will be generated

Gradients of these signalling molecules direct tissue patterning during embryogenesis [20, 21]. For instance, while sonic hedgehog (SHH) is produced at the ventral side of the neural tube [22, 23], at the opposite side BMPs and WNTs constitute dorsal signals [24–26] (◘ Fig. 11.4). In this way, the neural tube is organised into different zones along the D-V axis: roof plate (dorsal-most), alar plate (dorsal), basal plate (ventral) and floor plate (ventral-most), where at each given point of the D-V axis a combination of opposing ventralising and dorsalising signals specifies a unique type of progenitor [27]. Subsequently, different types of neural progenitors are formed with the capacity to originate specific types of neurons and glial cells. For example, in the spinal cord, several classes of interneurons are produced by progenitors in dorsal and intermediate domains while motor neurons arise from the ventral motor neuron (MN) domain. The interneurons of each longitudinal spinal cord segment integrate circuits that will orchestrate the coordinated action of body muscles by regulating motor neuron activity. Examples of these are discrete circuits commanding trot and gallop, i.e. simultaneous or alternate leg movements, composed by different interneuron types produced in different D-V domains of the same R-C segment of the spinal cord. Thus, *the diversity of neuronal types produced at the level of each segment of the neural tube is essential for the formation of local neuronal circuits and provides functional specificity in each segment.*

11.1.2.1 Timing and Competence

Individual neural progenitors possess spatial identity determined by their position within the neural tissue, defining the type of neuronal cell they can originate. In addition to this, they are able to give rise to distinct cell types over time, further increasing neural diversity in the CNS. This temporal switch of the progenitor identity is determined by the expression of specific subsets of transcription factors and results from two different processes: specification of temporal identity by changing intrinsic or extrinsic cues and progenitor competence, i.e. the ability of progenitor to respond to these cues [28]. With time, neural progenitors undergo competence restriction, gradually losing the ability to specify earlier-born cell fates and acquiring the competence to make later-born cell types. *This means that every neural cell type has a restricted time window during which it can be specified.*

Multiple studies both in Drosophila and mammals have shown that early progenitors are able to give rise to later neuronal fates when transplanted to later embryonic stages, but the opposite is not always true. This happens because the switch of progenitor competence is reinforced through gene silencing, either by repositioning of a given genomic locus into a gene-silencing hub such as the nuclear lamina or by recruitment of Polycomb repressive complexes (PRCs) which promote heritable gene silencing (reviewed in [28]).

Temporal patterning is best understood in Drosophila melanogaster neuroblasts, where sequential expression of transcription factors Hunchback (Hb), Kruppel (Kp), POU domain protein (Pdm) and Castor (Cas) determines the transition from early to late neuronal fates [29]. After the first two neuroblast divisions, Hb expression is downregulated and by the fifth division the Hb locus is relocated to the nuclear lamina and permanently silenced [30]. This relocation coincides with the time window of expression of Distal antenna (Dan), a member of the Centromere protein B (CENP-B)/transposase family of proteins. Although the exact role of Dan in this relocation is still unknown, this is one of the few examples where the mechanism of the temporal switch of neuroblast competence has been elucidated.

In mammals, orthologues of *Drosophila* transcription factors (TFs) have also been shown to define the temporal identity of progenitors in several contexts. In the developing retina, a timely succession of seven cell types has been described [31]. The Hb orthologue Ikaros is expressed in retinal progenitors where it specifies early-born neuronal fates [32]. Interestingly, while *Ikaros* mRNA is expressed throughout entire retinal development, the protein is detected only in early progenitors, suggesting that temporal restriction of progenitor competence occurs via post-translational regulation [32]. Misexpression of Ikaros in the older retina is able to restore some but not all early neuronal types blocking the differentiation of the latest cell type, Muller glia [32]. Temporal regulation of different retinal cell type production acts in parallel with stochastic mechanisms generating progenitor heterogeneity to which Notch-mediated lateral inhibition is thought to contribute [33]. By the concerted action of these mechanisms, several cell types are produced in retina simultaneously, allowing the proper laying out of complex neuronal circuitry.

The structural and functional complexity of the mammalian cortex is also generated by several mechanisms. Cortical excitatory neurons are generated from radial glial progenitors in the ventricular zone (VZ) of the dorsal telencephalon, often with an intermediate amplification step via the proliferation of basal progenitors in the subventricular zone (SVZ) [10]. Inhibitory neurons, in contrast, originate from the ventral telencephalon and populate cortical layers by concerted radial migration that is tightly coupled with their birthdate, with early-born neurons populating deep layers and later-born ones settling in the outer layers. Pioneering heterochronic transplantation studies of Susan McConnell demonstrated that young cortical progenitors generate late-born neuronal types when transplanted into the old cortical environment [34, 35] suggesting that early neural progenitors can respond to late extrinsic cues by generating temporally matched neuronal types. Older cortical progenitors, in contrast, were not able to produce younger, deep-layer neuronal types even when they had undergone cell divisions in a younger cortical environment [36]. However, when progenitors of layer VI (late-born) neurons were transplanted into layer IV (earlier-born), they were able to give rise to layer V (but not layer IV) neurons despite that at the time of transplantation the production of layer V neurons already ceased. This demonstrates an important property of neural progenitors: the interval of their competence to specify temporal identity spreads beyond the time of a given cell fate transition [28].

11.1.2.2 Neuron-to-Glia Switch

The most common switch in progenitor competence is the transition from neuronal to glial production that occurs in the different brain and spinal cord regions [37]. Neuronal identity of the cell is assured by the expression of proneural basic helix–loop–helix (bHLH) transcription factors known to promote neurogenesis and inhibit gliogenesis [38]. Glial identity is promoted by the gliogenic factor SOX9, and cytokines such as leukemia inhibitory factor (LIF), and Notch and BMP signalling [39]. Chromatin regulators also play a role in this switch, by demethylating the *GFAP* (glial fibrillary acidic protein) promoter and silencing proneural genes at the end of neurogenesis [40].

11.1.3 Cerebellar Development

Cerebellar specification starts early in human embryonic development, at 6 weeks, while its final cytoarchitecture is only achieved postnatally [41, 42]. During early development, when the neural tube is being regionalised, the hindbrain or rhombencephalon is presented as a segmental structure, containing 11 different rhombomeres [43, 44]. The cerebellum primordium, called "cerebellar anlage", originates from one of the hindbrain segments, the rhombomere 1 – r1 [45], which comprises the most anterior zone of the hindbrain caudally to the mid-hindbrain boundary (MHB), the isthmic organizer (IsO) [46]. This boundary appears to be maintained by the differential expression of transcription factors OTX2 and GBX2, which are important for the development of forebrain/midbrain and anterior hindbrain respectively [47–50]. The organising activity of the IsO is essentially mediated by the secretion of FGF8, which is strongly expressed in the MHB and its confined localisation is induced by the interaction of different transcription factors [50–54]. The organising action of IsO plays an important role in the formation of the cerebellum, because it regulates expression of different transcription factors involved in r1 patterning, including EN2, PAX2 and WNT1 [55]. The limits of the cerebellar territory are determined by the rostral expression of OTX2 and caudal expression of HOX genes, particularly HOXA2, in the hindbrain region, also in response to FGF8 signalling from MHB [55, 56].

After cerebellar territory formation, the cerebellar anlage is divided into two germinal centres that originate all GABAergic and glutamatergic cerebellar neurons, the ventricular zone (VZ) and the rhombic lip (RL) [57, 58]. The VZ is characterised by the expression of pancreas-specific transcription factor 1a (PTF1a) and gives rise to all inhibitory GABAergic neurons (Purkinje cells, Golgi, Lugaro, Stellate, Basket, Candelabrum, mid-sized GABAergic inhibitory projection neurons and small GABAergic interneurons) present in the adult cerebellum [59]. In a similar way to described above for the cerebral cortex, cerebellar excitatory neurons have a separate origin. Thus, in the cerebellum, the RL is a source of all excitatory glutamatergic neurons (Granule cells, unipolar brush cells and large glutamatergic projection neurons) and is essentially generated by atonal homolog 1 (ATOH1, also known as MATH1)-expressing progenitors [60, 61]. The cerebellar regionalisation is achieved by radial and tangential migration of post-mitotic neurons from the different germinal zones that will contribute to the final shape and size of the cerebellum [43, 46]. The appearance of a temporary layer containing ATOH1$^+$ proliferative progenitors derived from RL, at the surface of the developing cerebellum [62, 63], the external germinal layer (EGL), is a key feature in cerebellar development [57]. Already at a postnatal stage, the EGL-derived granule cells differentiate and migrate radially from the molecular layer across the Purkinje cell layer to their final destination, the granular cell layer. When granule cell migration is completed, the final stage of cerebellum foliation is achieved [57, 58].

The adult cerebellum is anatomically arranged into the cerebellar cortex that surrounds the white matter and the cerebellar nuclei. The cerebellar cortex is composed of different cell layers containing several types of neurons with an organised arrangement. This includes the Purkinje cell layer, containing a monolayer of the Purkinje cell bodies, Bergmann glial cells and a lower number of Candelabrum cells; between the innermost dense layer of Granule cells and Interneurons (Golgi cells, Unipolar

Brush cells, and Lugaro cells), constituting the Granular layer; and the outermost layer with the inhibitory Interneurons (Stellate cells and Basket cells), which is the molecular layer. Cerebellar nuclei are constituted by three major different neuronal types: large glutamatergic projection neurons, mid-sized GABAergic inhibitory projection neurons and small GABAergic interneurons [57, 58, 64]. The involvement of this brain structure in motor functions is well established, comprising the maintenance of balance and posture and the coordination of voluntary movements [65–67]. More recently, the cerebellum has also been associated with non-motor functions, including auditory processing tasks [68], reward expectation [69] and other forms of emotional processing [70].

The dysfunction of the cerebellum is translated into ataxia, a symptom detected in different neurodegenerative disorders consisting of motor dysfunction, balance problems, as well as limb movement and gait abnormalities. Thus, there is a phenotypically and genotypically heterogeneous group of disorders called cerebellar ataxias characterised by neurodegeneration of the cerebellum [71]. Up to date, there is no effective cure available for ataxias, and the majority of recently performed trials have failed, mostly because the assessed drugs did not target a specific deleterious pathway [72]. The identification of the molecular and cellular mechanisms involved in disease pathogenesis is necessary for the development of therapies aimed to target relevant pathogenic pathways.

11.2 Case Study: Generating Cerebellar Neurons for Modelling Cerebellar Ataxias

Different model systems used until now have been important for providing information about the function of the cerebellum, the pathogenesis of cerebellar disorders and have also given some clues about therapies for cerebellar ataxias. For neurodegenerative disorders in general, most of the current knowledge about disease-related neuronal phenotypes is based on post mortem studies hampering the understanding of disease progression and development [73]. Besides that, the current pre-clinical models used to test the potential positive effects of some drugs and to study the molecular and cellular pathways of cerebellar ataxias include animal models and immortalised human cell lines [74, 75]. Although these models help in understanding the various mechanisms of cerebellar neurodegeneration, differences in anatomy, metabolism and behaviour between animals and humans make it difficult to fully recapitulate the human disease [76]. Furthermore, many candidate drugs that presented significant effects in these models have failed to show relevant positive effects in clinical trials [74]. On the other hand, human pluripotent stem cells (PSCs) provide a human cell source that has demonstrated great potential for disease modelling, drug screening and toxicology, since they have unlimited in vitro expansion potential and differentiation capacity [77, 78]. The knowledge about the signalling pathways involved in the maintenance of pluripotency as well as the generation of different germ layer derivatives has allowed the manipulation and control of PSC commitment to different lineages and further differentiation into specific cell types, including brain cells. In recent years, advances in our understanding of cerebellar development and differentiation have fostered the generation of techniques for obtaining different types of cerebellar neurons from PSCs [79–81].

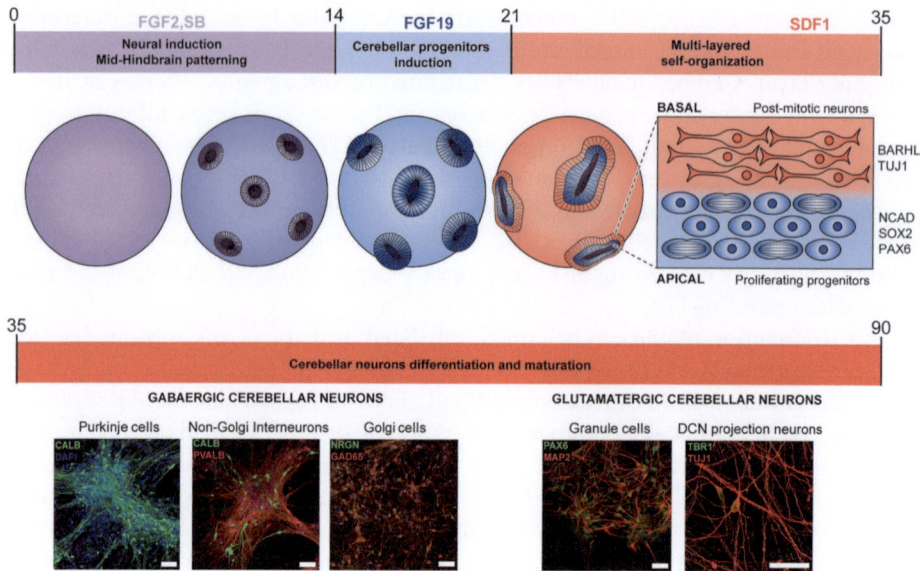

◘ **Fig. 11.5** Generation of cerebellar neurons from human pluripotent stem cells. Upper panel: schematic representation of cerebellar differentiation from human pluripotent stem cells. Neural tube-like structures form within floating aggregates, with apical domains delineated by NCAD, Sox2 and PAX6 staining and basal layer of post-mitotic neurons (Tuj1+/BARHL1+). Lower panel: immunostaining analysis for the indicated markers, supporting the presence of different mature cerebellar neurons. Scale bars 50 μm

For cerebellar commitment, an efficient neural induction is required as the first step of differentiation. To achieve this, the Nodal/Activin signalling inhibition by a chemical antagonist SB431542 (SB) of TGFβ signalling is used to prevent the meso-endodermal differentiation and drive neural commitment of PSC [82, 83]. After this step, the PSC-derived neural progenitors are ready to acquire their regional identity, and different regions of the human neural tube can be mimicked in vitro by adding specific morphogens to the culture medium. For cerebellar patterning, the sequential addition of defined morphogens, including FGF2, FGF19 and SDF1, can reproduce the sequential progression of human cerebellar development (◘ Fig. 11.5). FGF2 has an inductive role in cerebellar commitment, acting as a moderate caudalising factor and leading to an efficient generation of mid-hindbrain progenitors. After the establishment of cerebellar territory, FGF19 signalling promotes the spontaneous generation of rostral hindbrain-like structures with apico-basal polarity, which are reorganised into different layers after the SDF1 addition as seen at the developmental stage when cerebellar neurogenesis occurs. By initiating differentiation using PSC-derived aggregates, after 14 days of neural induction, aggregates are mostly composed of neural progenitors expressing typical neural markers NESTIN and PAX6, that organise into small neural rosettes structures, similar to the embryonic neural tube (◘ Fig. 11.5). After 21 days in culture and upon the action of FGF19 signalling, these neural rosettes reorganise into larger neuroepithelium with apico-basal polarity, strongly expressing the apical marker N-Cadherin (NCAD) on the apical side of the neural rosette. Going onwards until day 35 of differentiation, neural rosettes reorganise into polarised neuroepithelial structures with different layers, with proliferating cerebellar progenitors expressing PAX6 and SOX2 on the apical (luminal) side, and more mature post-mitotic neurons on the basal side, expressing TUJ1 and BARHL1 (◘ Fig. 11.5). By promoting further maturation of the PSC-derived cerebellar progenitors, different types of functional cerebellar neurons can be obtained, including Purkinje cells (Calbindin (CALB)$^+$), Non-Golgi interneurons (Parvalbumin (PVALB$^+$; CALB$^-$), Golgi cells (Neurogranin (NRGN)$^+$), Granule cells (PAX6$^+$ and MAP 2$^+$) and large glutamatergic projection neurons (TBR1$^+$; ◘ Fig. 11.5). This procedure represents a differentiation strategy to generate different types of cerebellar cells in a well-organised structure that can form functional cerebellar neurons. This strategy gives the opportunity to study cerebellar development together with the possibility to efficiently generate cerebellar neurons from patient-derived iPSCs for the purpose of drug screening and for the study of specific pathways involved in cerebellar dysfunctions.

Take Home Message

- Billions of neurons composing our CNS are generated in an orderly fashion in accordance with spatial and temporal cues.
- To sustain the continuous generation of neural cells during CNS development, the progenitor pool must be maintained by asymmetric progenitor divisions and by lateral inhibition via Notch signalling.
- Neural development can be efficiently reproduced in vitro using pluripotent cell differentiation in a controlled environment.
- Many neurodevelopmental and neurodegenerative disorders have cerebellar ataxia as one of the major symptoms, and there is still no effective cure for ataxia.
- Functionally mature cerebellar neurons can be efficiently produced in vitro using human iPSCs as a source, being an excellent model for studying diseases affecting the cerebellar function.

References

1. Pevny LH, Sockanathan S, Placzek M, Lovell-Badge R. A role for SOX1 in neural determination. Development. 1998;125(10):1967–78.
2. Afonso C, Henrique D. PAR3 acts as a molecular organizer to define the apical domain of chick neuroepithelial cells. J Cell Sci. 2006;119(20):4293–304. https://doi.org/10.1242/jcs.03170.
3. Henrique D, Abranches E, Verrier L, Storey KG. Neuromesodermal progenitors and the making of the spinal cord. Development. 2015;142(17):2864–75. https://doi.org/10.1242/dev.119768.
4. Gammill LS, Bronner-Fraser M. Neural crest specification: migrating into genomics. Nat Rev Neurosci. 2003;4(10):795–805.
5. Sadler TW. Embryology of neural tube development. Am J Med Genet C Semin Med Genet. 2005;135C(1):2–8.
6. Sauer FC. Mitosis in the neural tube. J Comp Neurol. 1935;62(2):377–405.
7. Spear PC, Erickson CA. Interkinetic nuclear migration: a mysterious process in search of a function. Develop Growth Differ. 2012;54(3):306–16.
8. Louvi A, Artavanis-Tsakonas S. Notch signalling in vertebrate neural development. Nat Rev Neurosci. 2006;7(2):93–102.
9. Henrique D, Hirsinger E, Adam J, Le Roux I, Pourquié O, Ish-Horowicz D, et al. Maintenance of neuroepithelial progenitor cells by Delta-Notch signalling in the embryonic chick retina. Curr Biol. 1997;7(9):661–70.

10. Kriegstein A, Alvarez-Buylla A. The glial nature of embryonic and adult neural stem cells. Annu Rev Neurosci. 2009;32(1):149–84. https://doi.org/10.1146/annurev.neuro.051508.135600.
11. Levine AJ, Brivanlou AH. Proposal of a model of mammalian neural induction. Dev Biol. 2007;308:247–56.
12. Stern CD. Initial patterning of the central nervous system: how many organizers? Nat Rev Neurosci. 2001;2(2):92–8.
13. Suzuki IK, Vanderhaeghen P. Is this a brain which I see before me? Modeling human neural development with pluripotent stem cells. Development. 2015;142(18):3138–50. https://doi.org/10.1242/dev.120568.
14. Wilson SW, Houart C. Early steps in the development of the forebrain. Dev Cell. 2004;6:167–81.
15. Imaizumi K, Sone T, Ibata K, Fujimori K, Yuzaki M, Akamatsu W, et al. Controlling the regional identity of hPSC-derived neurons to uncover neuronal subtype specificity of neurological disease phenotypes. Stem Cell Reports. 2015;5(6):1010–22.
16. Irioka T, Watanabe K, Mizusawa H, Mizuseki K, Sasai Y. Distinct effects of caudalizing factors on regional specification of embryonic stem cell-derived neural precursors. Dev Brain Res. 2005;154(1):63–70.
17. Kirkeby A, Grealish S, Wolf DA, Nelander J, Wood J, Lundblad M, et al. Generation of regionally specified neural progenitors and functional neurons from human embryonic stem cells under defined conditions. Cell Rep. 2012;1(6):703–14.
18. Petros TJ, Tyson JA, Anderson SA. Pluripotent stem cells for the study of CNS development. Front Mol Neurosci. 2011;4:30.
19. Lumsden A, Krumlauf R. Patterning the vertebrate neuraxis. Science. 1996;274(5290):1109–15.
20. Crick F. Diffusion in embryogenesis. Nature. 1970;225(5231):420–2.
21. Wolpert L. Positional information and the spatial pattern of cellular differentiation. J Theor Biol. 1969;25(1):1–47.
22. Chiang C, Litingtung Y, Lee E, Young KE, Corden JL, Westphal H, et al. Cyclopia and defective axial patterning in mice lacking Sonic hedgehog gene function. Nature. 1996;383(6599):407–13.
23. Martínez S, Puelles E, Puelles L, Echevarria D. Molecular regionalization of the developing neural tube. In: The mouse nervous system. London/Waltham: Elsevier/Academic Press; 2012. p. 2–18.
24. Basler K, Edlund T, Jessell TM, Yamada T. Control of cell pattern in the neural tube: regulation of cell differentiation by dorsalin-1, a novel TGFβ family member. Cell. 1993;73(4):687–702.
25. Dickinson ME, Selleck MA, McMahon AP, Bronner-Fraser M. Dorsalization of the neural tube by the non-neural ectoderm. Development. 1995;121(7):2099–106.
26. Ulloa F, Martí E. Wnt won the war: antagonistic role of Wnt over Shh controls dorso-ventral patterning of the vertebrate neural tube. Dev Dyn. 2010;239:69–76.
27. Jessell TM. Neuronal specification in the spinal cord: inductive signals and transcriptional codes. Nat Rev Genet. 2000;1(1):20–9.
28. Kohwi M, Doe CQ. Temporal fate specification and neural progenitor competence during development. Nat Rev Neurosci. 2013;14(12):823–38.
29. Isshiki T, Pearson B, Holbrook S, Doe CQ. Drosophila neuroblasts sequentially express transcription factors which specify the temporal identity of their neuronal progeny. Cell. 2001;106(4):511–21.
30. Kohwi M, Lupton JR, Lai SL, Miller MR, Doe CQ. Developmentally regulated subnuclear genome reorganization restricts neural progenitor competence in Drosophila. Cell. 2013;152(1–2):97–108.
31. Turner DL, Snyder EY, Cepko CL. Lineage-independent determination of cell type in the embryonic mouse retina. Neuron. 1990;4(6):833–45.
32. Elliott J, Jolicoeur C, Ramamurthy V, Cayouette M. Ikaros confers early temporal competence to mouse retinal progenitor cells. Neuron. 2008;60(1):26–39.
33. Rocha SF, Lopes SS, Gossler A, Henrique D. Dll1 and Dll4 function sequentially in the retina and pV2 domain of the spinal cord to regulate neurogenesis and create cell diversity. Dev Biol. 2009;328(1):54–65.
34. McConnell SK, Kaznowski CE. Cell cycle dependence of laminar determination in developing neocortex. Science. 1991;254(5029):282–5.
35. Desai AR, McConnell SK. Progressive restriction in fate potential by neural progenitors during cerebral cortical development. Development. 2000;127(13):2863–72.
36. Frantz GD, McConnell SK. Restriction of late cerebral cortical progenitors to an upper-layer fate. Neuron. 1996;17(1):55–61.

37. Okano H, Temple S. Cell types to order: temporal specification of CNS stem cells. Curr Opin Neurobiol. 2009;19(2):112–9.
38. Ross S, Greenberg M, Stiles C. Basic helix-loop-helix factors in cortical development. Neuron. 2003;39(1):13–25.
39. Rowitch DH, Kriegstein AR. Developmental genetics of vertebrate glial-cell specification. Nature. 2010;468:214–22.
40. Hirabayashi Y, Suzki N, Tsuboi M, Endo TA, Toyoda T, Shinga J, et al. Polycomb limits the neurogenic competence of neural precursor cells to promote astrogenic fate transition. Neuron. 2009;63(5):600–13.
41. Cho KH, Rodríguez-Vázquez JF, Kim JH, Abe H, Murakami G, Cho BH. Early fetal development of the human cerebellum. Surg Radiol Anat. 2011;33(6):523–30.
42. Millen KJ, Wurst W, Herrup K, Joyner AL. Abnormal embryonic cerebellar development and patterning of postnatal foliation in two mouse Engrailed-2 mutants. Development. 1994;120(3):695–706.
43. Andreu A, Crespo-Enriquez I, Echevarria D. Molecular events directing the patterning and specification of the cerebellum. Eur J Anat. 2014;18(4):245–52.
44. Morales D, Hatten ME. Molecular markers of neuronal progenitors in the embryonic cerebellar anlage. J Neurosci. 2006;26(47):12226–36.
45. Zervas M, Millet S, Ahn S, Joyner AL. Cell behaviors and genetic lineages of the mesencephalon and rhombomere 1. Neuron. 2004;43(3):345–57.
46. Butts T, Green MJ, Wingate RJT. Development of the cerebellum: simple steps to make a "little brain". Development. 2014;141(21):4031–41. https://doi.org/10.1242/dev.106559.
47. Broccoli V, Boncinelli E, Wurst W. The caudal limit of Otx2 expression positions the isthmic organizer. Nature. 1999;401(6749):164–8.
48. Wassarman KM, Lewandoski M, Campbell K, Joyner AL, Rubenstein JL, Martinez S, et al. Specification of the anterior hindbrain and establishment of a normal mid/hindbrain organizer is dependent on Gbx2 gene function. Development. 1997;124(15):2923–34.
49. Millet S, Campbell K, Epstein DJ, Losos K, Harris E, Joyner AL. A role for Gbx2 in repression of Otx2 and positioning the mid/hindbrain organizer. Nature. 1999;401(6749):161–4.
50. Wurst W, Bally-Cuif L, Bally-Cuif L. Neural plate patterning: upstream and downstream of the isthmic organizer. Nat Rev Neurosci. 2001;2(2):99–108.
51. Heikinheimo M, Lawshé A, Shackleford GM, Wilson DB, MacArthur CA. Fgf-8 expression in the post-gastrulation mouse suggests roles in the development of the face, limbs and central nervous system. Mech Dev. 1994;48(2):129–38.
52. Crossley PH, Martin GR. The mouse Fgf8 gene encodes a family of polypeptides and is expressed in regions that direct outgrowth and patterning in the developing embryo. Development. 1995;121(2):439–51.
53. Joyner AL, Liu A, Millet S. Otx2, Gbx2 and Fgf8 interact to position and maintain a mid-hindbrain organizer. Curr Opin Cell Biol. 2000;12:736–41.
54. Chizhikov VV, Millen KJ. Neurogenesis in the cerebellum. In: Comprehensive developmental neuroscience: patterning and cell type specification in the developing CNS and PNS. Amsterdam: Academic Press; 2013. p. 417–34.
55. Irving C, Mason I. Signalling by FGF8 from the isthmus patterns anterior hindbrain and establishes the anterior limit of Hox gene expression. Development. 2000;127(1):177–86.
56. Gavalas A, Davenne M, Lumsden A, Chambon P, Rijli FM. Role of Hoxa-2 in axon pathfinding and rostral hindbrain patterning. Development. 1997;124(19):3693–702.
57. White JJ, Sillitoe RV. Development of the cerebellum: from gene expression patterns to circuit maps. Wiley Interdiscip Rev Dev Biol. 2013;2(1):149–64.
58. Marzban H, Del Bigio MR, Alizadeh J, Ghavami S, Zachariah RM, Rastegar M. Cellular commitment in the developing cerebellum. Front Cell Neurosci. 2015;8(January):1–26.
59. Hoshino M, Nakamura S, Mori K, Kawauchi T, Terao M, Nishimura YV, et al. Ptf1a, a bHLH transcriptional gene, defines GABAergic neuronal fates in cerebellum. Neuron. 2005;47(2):201–13.
60. Wang VY, Rose MF, Zoghbi HY. Math1 expression redefines the rhombic lip derivatives and reveals novel lineages within the brainstem and cerebellum. Neuron. 2005;48(1):31–43.
61. MacHold R, Fishell G. Math1 is expressed in temporally discrete pools of cerebellar rhombic-lip neural progenitors. Neuron. 2005;48(1):17–24.

62. Ben-Arie N, McCall AE, Berkman S, Eichele G, Bellen HJ, Zoghbi HY. Evolutionary conservation of sequence and expression of the bHLH protein Atonal suggests a conserved role in neurogenesis. Hum Mol Genet. 1996;5(9):1207–16.
63. Ben-Arie N, Bellen HJ, Armstrong DL, McCall AE, Gordadze PR, Guo Q, et al. Math1 is essential for genesis of cerebellar granule neurons. Nature. 1997;390(6656):169–72.
64. Hoshino M. Neuronal subtype specification in the cerebellum and dorsal hindbrain. Dev Growth Diff. 2012;54:317–26.
65. Ito M. The modifiable neuronal network of the cerebellum. Jpn J Physiol. 1984;34:781–92.
66. Glickstein M. Motor skills but not cognitive tasks. Trends Neurosci. 1993;16(11):450–1.
67. Schmahmann JD. Rediscovery of an early concept. Int Rev Neurobiol. 1997;41:3–27.
68. McLachlan NM, Wilson SJ. The contribution of brainstem and cerebellar pathways to auditory recognition. Front Psychol. 2017;8:265. https://doi.org/10.3389/fpsyg.2017.00265.
69. Wagner MJ, Kim TH, Savall J, Schnitzer MJ, Luo L. Cerebellar granule cells encode the expectation of reward. Nature. 2017;544(7648):96–100.
70. Adamaszek M, D'Agata F, Ferrucci R, Habas C, Keulen S, Kirkby KC, et al. Consensus paper: cerebellum and emotion. Cerebellum. 2017;16:552–76.
71. Klockgether T. Update on degenerative ataxias. Curr Opin Neurol. 2011;24(4):339–45.
72. Marmolino D, Manto M. Past, present and future therapeutics for cerebellar ataxias. Curr Neuropharmacol. 2010;8(1):41–61.
73. Durnaoglu S, Genc S, Genc K. Patient-specific pluripotent stem cells in neurological diseases. Stem Cells Int. 2011;2011:1–17.
74. Manto M, Marmolino D. Animal models of human cerebellar ataxias: a cornerstone for the therapies of the twenty-first century. Cerebellum. 2009;8:137–54.
75. Ziv Y, Danieli T, Amiel A, Ravia Y, Shiloh Y, Etkin S, et al. Cellular and molecular characteristics of an immortalized ataxia-telangiectasia (group AB) cell line. Cancer Res. 1989;49(9):2495–501.
76. Cendelin J, Schmahmann J, Sherman J, Manto M, Manto M, Marmolino D, et al. From mice to men: lessons from mutant ataxic mice. Cerebellum Ataxias. 2014;1(1):4.
77. Kropp C, Massai D, Zweigerdt R. Progress and challenges in large-scale expansion of human pluripotent stem cells. Process Biochem. 2017;59:244–54.
78. Yamamoto T, Takenaka C, Yoda Y, Oshima Y, Kagawa K, Miyajima H, et al. Differentiation potential of Pluripotent Stem Cells correlates to the level of CHD7. Sci Rep. 2018;8(1):241.
79. Muguruma K, Nishiyama A, Kawakami H, Hashimoto K, Sasai Y. Self-organization of polarized cerebellar tissue in 3D culture of human pluripotent stem cells. Cell Rep. 2015;10(4):537–50.
80. Salero E, Hatten ME. Differentiation of ES cells into cerebellar neurons. Proc Natl Acad Sci U S A. 2007;104(8):2997–3002.
81. Wang S, Wang B, Pan N, Fu L, Wang C, Song G, et al. Differentiation of human induced pluripotent stem cells to mature functional Purkinje neurons. Sci Rep. 2015;5:9232.
82. Laping NJ, Grygielko E, Mathur A, Butter S, Bomberger J, Tweed C, et al. Inhibition of transforming growth factor (TGF)-beta1-induced extracellular matrix with a novel inhibitor of the TGF-beta type I receptor kinase activity: SB-431542. Mol Pharmacol. 2002;62(1):58–64.
83. Chambers SM, Fasano CA, Papapetrou EP, Tomishima M, Sadelain M, Studer L. Highly efficient neural conversion of human ES and iPS cells by dual inhibition of SMAD signaling. Nat Biotechnol. 2009;27(3):275–80.

Neurotoxicology and Disease Modelling

Carolina Nunes and Marie-Gabrielle Zurich

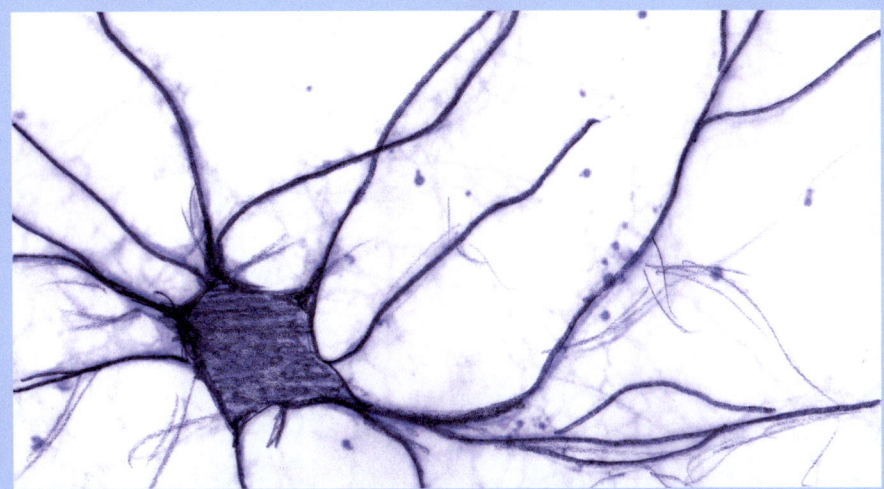

Contents

12.1   Emergence of Toxicology – 231

12.2   Birth of Experimental Toxicology: In Vivo Testing – 233

12.3   The Way Forward in Toxicology: *In Vitro* Testing – 234

12.4   Stem Cells and Neurotoxicology – 235

12.5   Stem Cells and Disease Modelling – 237

© Springer Nature Switzerland AG 2020
G. Rodrigues, B. A. J. Roelen (eds.), *Concepts and Applications of Stem Cell Biology*,
Learning Materials in Biosciences, https://doi.org/10.1007/978-3-030-43939-2_12

12.6 Practical Class—iPSC-Derived Neural Progenitor Cells (NPCs) – 238
12.6.1 Materials – 238
12.6.2 Media – 239

References – 243

12

Neurotoxicology and Disease Modelling

What You Will Learn in This Chapter

In our everyday life, we are exposed to a continuously increasing number of chemical compounds, such as pesticides, industrial chemicals, metals, solvents, nanomaterials, natural toxins, pharmaceutical drugs and drugs of abuse that are potentially harmful. We begin this chapter with a short history of toxicology, where we give examples of catastrophic events that raised awareness and led to the creation of a system of regulation of chemicals. We next look at the ways to assess the potential toxicity of chemicals and drugs *in vivo*, and then *in vitro* taking advantage of the breakthrough discovery of human-induced pluripotent stem cells. Furthermore, since aging of the population pushes the scientific community to better understand and cure diseases, in the next chapter, we quickly go through the possibilities given by stem cells to model diseases. We conclude this chapter by a practical class describing the way to derive neuroprogenitors from human-induced pluripotent stem cells. As a whole, this chapter takes the reader through the history of toxicology from hunting venoms to the use of human stem cell-based models to detect hazardous effects of chemicals, in particular on the brain, and to decipher the mechanisms of neurodegenerative diseases. In the end, the reader will have a broad view on the use of human stem cells for chemical safety evaluation and human disease modelling.

12.1 Emergence of Toxicology

Back in time, when the other animal's hunting leftovers were not enough to feed an increasing human population, humans started to hunt with sharp objects, traps and venoms extracted from animals and plants. Sources indicate that the Greeks, Scythians, and Nubians used poisoned dipped arrows, javelins, and other weapons in battle, indicating that there was, at least in some parts of the ancient world, a vast knowledge of animal and plant poisons and some understanding of their effects on living organisms [1]. Soon in society, murder became a way to get higher in the hierarchy, and venoms a clean way to reach this goal. At the same time, humans understood that plants and animals could also be the source of substances helping them to overcome some health issues.

The need to categorize substances started early in human history. It is safe to assume that even prehistoric humans categorized some plants as harmful and other as safe [1]. One of the most ancient documentation on substances is *Ebers Papyrus* (1500 B.C.), an Egyptian record containing mentions on more than 700 substances and over 300 recipes detailing incantations and mixtures using poisonous or medicinal minerals and plants. We can also see references of venoms through several tales like *The Odyssey and The Elliad* (850 B.C.), where Homer describes the use of poisoned arrows. Most of the publications were based on experimentations performed by the authors, who would test hundreds of substances on themselves, what sometimes ended up in their own death.

With the increasing number of listed substances, the classification became more complex. In ancient Greek literature, numerous poisons and their use are described [2]. Interestingly, Greeks defined all drugs or potions as "pharmaka" or "pharmakon" without making any distinction between harmful and safe compounds, or compounds used for the treatment of diseases [3, 4]. Dioscorides (40–90 A.D.), a Greek physician in the court of the Roman emperor Nero, attempted to classify poisons, and accompanied the descriptions by drawings. The classification of about 600 substances he published in *De Materia Medica* remained used till 1600s and became the basis for the modern pharmacopoeia [2, 3].

Already around 81 B.C., Lucius Cornelius Sulla (138–78 B.C.) a Roman general, issued the law Lex Cornelia de Sicariis et Veneficiis to punish the act of poisoning. The law clearly stated that any person who makes, sells, possesses, or purchases a "venenum malum" (poisonous or noxious substance) with the intention of committing murder may be sent for trial [3]. However, despite these efforts, poisoning kept being common and socially accepted. In the Medieval and Renaissance periods, there are several reports of influent people, like popes, being poisoned by their opponents. One could even learn the dark art of poisoning in specialized schools. Payment for poisoning people was quite common [3], as testified by the price scale emitted for this type of services by the Venetian Council of Ten (1310–1797), the most important governing body in the state of Venice.

Philippus Aureolus Theophrastus Bombastus von Hohenheim (1493–1541), best known as Paracelsus, was a Swiss physician, alchemist, and astrologer. Paracelsus defined for the first time the importance of the dose by saying, '*All substances are poisons; there is none which is not a poison. The right dose differentiates a poison and a remedy*', paraphrased as the dose makes the poison. Paracelsus also believed that diseases were primarily associated with a specific organ in the body, giving rise to the idea of "target organ toxicity". He questioned the long-standing ideas of his time and tried to rationalize the treatment of diseases [3].

In the eighteenth and nineteenth century, following the birth of the synthetic chemical industry the field of toxicology emerged in response to the need to understand how chemical substances might affect the health of workers and consumers. Bernardino Ramazzini (1633–1714), an Italian physician, published *De Morbis Artificum Diatriba* (*Diseases of Workers*) where he outlined the health hazards of chemicals in more than 50 occupations. Although he was not the first to correlate occupation and diseases, he gave rise to occupational medicine, advising physicians to relate their patients' occupation to their symptoms. In 1775, Percival Potts (1714–1788), an English surgeon became the first scientist to find an association between environmental chemicals and cancer. Indeed, he established the occupational link between soot and scrotal cancer and can, therefore, be considered as the founder of occupational toxicology [5]. At the same time, there was very little, if any, legislation to govern the production or distribution of chemicals. Furthermore, poisons such as arsenic were still easily accessible, being sold at drugstores, and medicines for children were containing opium against hyperactivity or for teething or colicky babies, heroin sold as sedative for cough, and cocaine against toothache.

In 1906, one of the first laws regulating the marketing of drugs was released in USA. It was the first effort to obtain from the manufacturers an accurate labelling of dosage and contents of their products. Prior to this, many drugs advertised benefits from secret ingredients. In 1937, more than 100 deaths by liver failure were linked to the consumption of a new formulation of the antibiotic sulfanilamide produced in USA, due to the toxicity of the diethylene glycol used as an excipient. This disaster showed that labelling of dosage is not sufficient to protect the population, and emphasized the importance to regulate the commerce of chemicals and their safety. In consequence, in USA, the Federal Food, Drug and Cosmetic Act, which required

Food and Drug Administration (FDA) approval before the release of a new drug, was accepted in 1938. In Europe, the conscience of the importance of this regulation only came after another tragic episode. Thalidomide released as a sedative in 1957 and prescribed to alleviate morning sickness in pregnant women produced malformations in more than 10,000 children born in 46 countries. This drug was withdrawn from the market in 1961. Nowadays, despite this tragedy, thalidomide is back for the treatment of skin lesions caused by leprosy and for multiple myeloma [6]. In most Western countries, access to the drug is strictly controlled and its use is restricted to specific cases. However, in many developing countries, drug control is deficient and babies continue to be born with birth defects, while it could be avoided. This disaster highlighted the importance of the evaluation of the toxicity of chemicals before their release on the market.

12.2 Birth of Experimental Toxicology: In Vivo Testing

Nowadays, toxicology is defined by the Society of Toxicology as the study of the adverse effects of chemical, physical, or biological agents on living organisms and the ecosystem, including the prevention and amelioration of such adverse effects. Although there have been numerous notorious poisonings throughout the ages and rather astute descriptions of toxic agents as seen above, the scientific study of toxicology did not commence until the nineteenth century. There was a rapid development of analytical methods in the late nineteenth century and then an acceleration of both method and scientific development in the latter half of the twentieth century [5]. The first scientific experimental determinations of the toxic effects of chemicals were performed on animals.

In 1927, the British pharmacologist J.W Trevan proposed to determine the median lethal dose (LD_{50}) of chemicals to evaluate their toxicity. LD_{50} is the single dose of a given chemical required to kill half of the members of a given tested population after a given exposure time. This measurement appealed to governments and regulators since for the first time a value was used to rank the chemicals according to their toxicity. In spite of clear limitations, ranging from high inter-laboratories variability to species differences to assess human safety, LD_{50} remains the most prevalent animal test used for human risk assessment to this day.

Other animal tests were established during the twentieth century. The standardized Draize test was developed by John Draize in 1942 to assess eye and skin irritation caused by chemicals. It consists in applying the test substance in the eye or skin of a restrained, conscious animal, usually albino rabbits, and to assess the effects until 14 days after exposure. Animals that did not exhibit any permanent effects could be used for other tests. In the 1950s decade, a test was developed to identify carcinogenic chemicals through the daily dosing of rats and mice for 2 years [7]. After the thalidomide incident, reproductive and developmental toxicity studies were introduced, where the animals are exposed to chemicals at different steps of reproduction to assess the effects on the development of the offspring.

Animal experimentation allowed building the principles of toxicology such as the importance of establishing dose-response relationships, the importance of the duration of the exposure and of the route of exposure. Soon it was recognized that industrial chemicals and even drugs could cause a wide range of diseases and/or disabilities, such as, but not restricted to, birth defects, impairment of development, cancer, and impairment of learning and memory. Regulatory toxicologists rely almost exclusively on animal *in vivo* experimentation for human safety assessment and in the absence of other techniques and with very limited understanding of toxicological mechanisms it was probably the only way to best ensure human safety [8]. However, a paradigm shift in toxicology was inevitable because the descriptive high-dose animal-based toxicity testing used was not based on sound science and we understand much more now about biology, species differences in xenobiotic handling and molecular toxicology. Furthermore, the huge financial cost of animal testing and the burden on non-human mammals speak also in favour of a radical change in our way to assess human safety. This shift was proposed by the US National Research Council (NRC) in 2007, in a report entitled "Toxicity Testing in the 21st Century: A Vision and a Strategy" [9]. The basic proposal was to reorient testing to the molecular level, rather than observing phenotypic responses at the level of whole organisms.

12.3 The Way Forward in Toxicology: *In Vitro* Testing

The panel of experts that generated the Toxicity Testing in the *twenty-first Century: A Vision and a Strategy* highlighted the concept of "toxicity pathways" within cells, as the way forward to understand and to interpret the toxicant-induced mode of action [9] (NRC, 2007). Pathway responses are dose-dependent. At some low dose, a pathway may begin to be disrupted by a toxicant exposure, but the pathway will continue to function, due to a homeostatic response, also called an "adaptive" behaviour. At a higher dose, the adaptive response is overwhelmed, and an adverse effect takes place. While some degree of toxicant-induced adversity can be repaired, ultimately a dose will be reached that causes an irreversible change in pathway function with severe consequences for the cell. Molecular pathways are better studied *in vitro* than *in vivo*, therefore, the implementation of this new strategy required the development of new *in vitro* tests, preferably based on human cells.

Years before, in 1959, Russell and Burch have delineated the 3Rs principle in their seminal book entitled, "*The principles of humane experimental technique*" [10]. 3Rs stands for the Replacement, Reduction and Refinement methods for animals used in experimental sciences. The aim of Russel and Burch was to remove inhumanity in the way we treat animals in experimentation. The 3Rs principle can be explained as follows: "Replace" the animal experimentation by methods which permit a given purpose to be achieved without conducting experiment or other scientific procedures on animals; "Reduce" by methods for obtaining comparable levels of information from the use of fewer animals in scientific procedures, or for obtaining more information from the same number of animals; and "Refine" by using methods which alleviate or minimise potential pain, suffering and distress, and which enhance animal well-being.

The first attempt to culture animal cells was made in 1885 by the zoologist Wilhelm Roux who maintained embryonic chicken cells in a warm saline solution for 13 days. His work on tissue cultures was pursued by Ross Granville Harrison (1870–1959) and Paul Alfred Weiss (1898–1989). Later on, in 1955, Eagle published the recipe for a basal nutrient formulation, an 'isosmotic, pH-balanced mixture of salts, amino acids, sugars, vitamins, and other necessary nutrients' able to support cells in culture [11]. All this work paved the way for Björn Ekwall (1940–2000), an outstanding Swedish cell scientist who made pioneering contributions to the field of *in vitro* toxicology. Indeed, during his Ph.D. studies, he proposed to predict human acute systemic toxicity of chemicals by the use of cell culture tests, instead of the use of animal LD_{50} determinations. He further formulated the so called "basal cytotoxicity concept" [12–14], and for further details, see [15]. Therefore, Ekwall established the foundations for *in vitro* toxicology. He classified the effect of chemicals into three categories: (1) basal cytotoxicity resulting from interference with structures and/or properties essential for cell survival, proliferation and/or functions; (2) organ-specific cytotoxicity, which affects organ-specific structures or functions; and (3) toxicity at the organizational level [12, 16].

From that time the field of *in vitro* toxicology progressed tremendously, with the development of very performant organ-specific cell culture systems and of high-throughput and high-content analytical techniques. However, until recently, most of the cell cultures were derived from animals, and as already pointed out by Russel and Burch (1959), similar species do not necessarily represent good models for each other. They called this the "high fidelity fallacy", which is well-illustrated in the high rate of drug candidate failures during clinical testing, due to both a lack of efficacy in humans and the identification of unacceptable toxicities not previously identified in pre-clinical animal testing [8]. The ultimate progress in the field of *in vitro* toxicology was made possible by the truly ground-breaking discovery by Yamanaka and colleagues in 2008 of the technique allowing the derivation of induced pluripotent stem cells (iPSCs) from human somatic cells [17]. The iPSCs can then be specifically differentiated in almost all human cell types which can be used, among other potential applications, for toxicity testing and disease modelling. In the next two paragraphs, we will give an overview of these applications, with an emphasis on the use of human iPSCs brain-derived cultures for neurotoxicology testing and brain disease modelling.

12.4 Stem Cells and Neurotoxicology

Neurotoxicity or a neurotoxic effect is an adverse change in the structure or function of the nervous system following exposure to a chemical agent [18].

Nowadays, guidelines for neurotoxicity testing (NT) and developmental neurotoxicity testing (DNT) are based exclusively on animal experimentation [19, 20]. The described *in vivo* tests are raising ethical, financial and scientific concerns. Indeed, while such guideline studies are currently necessary for consumer safety, it is already known that these animal tests might have limited prediction for human neurotoxicity due to inter-species variation [21]. Therefore, although decades of work on neuronal and glial cellular systems derived from several animal species has delivered a range of reliable *in vitro* assays highly valuable for neurotoxicity testing [22–28], there is a growing need for human-based *in vitro* models.

Human brain development consists of a series of complex spatiotemporal processes that if disturbed by chemical exposure causes irreversible impairments of the nervous system. To evaluate a chemical adverse effect, it is thought that the complex procedure of brain development can be disassembled into several neurodevelopmental endpoints, such as proliferation, migration, differentiation, network formation and function, apoptosis, which can be assessed by a combination of different assays. It is assumed that developmental neurotoxicants exert their toxicity by disturbing at least one of these processes (for reviews, see [29, 30]). To test the effect of chemicals on these different neurodevelopmental processes, various human neuronal-like immortalized cell lines are available. However, as the expression of tumour growth-related genes expressed in these cell lines may affect the cellular responses to a chemical exposure, the use of human stem cells and their progeny is strongly advised [31, 32]. The exposure of human neuroprogenitor cells (NPCs) to chemicals allows to study the effects of chemicals at very early stages of the nervous system differentiation. Recently, it was shown that NPCs are more sensitive to chemical-induced apoptosis than differentiated neurons [33]. NPCs are commercially available or can be derived either from embryonic stem cells (ESCs) or iPSCs. Adverse effects of chemicals on neuronal or glial differentiation can be studied by exposing differentiating NPCs to chemicals, and fully differentiated neurons and glial cells derived from the NPCs constitute the perfect tools for the evaluation of toxic effects on mature cells [34]. Human iPSC-derived neurons were, for example, already used to assess neurite outgrowth inhibition after exposure to chemicals [35, 36].

Due to their few ethical constraints, human-induced pluripotent stem cells (hiPSC) derived neuronal and glial models are currently gaining increasing scientific interest. Not surprisingly, it has been shown in the past that a given brain cell type reacts differently to a toxic substance when grown in a single cell-type culture than in a mixed-cell types culture [37–39]. Therefore, complex 3D cell culture systems containing many brain cell types, allowing a maximum of cell-to-cell interactions and recapitulating the main neurodevelopmental processes are also of the utmost importance for the evaluation of the adverse effects of chemicals. Pamies et al. (2017) recently described such a model derived from hiPSCs (◘ Fig. 12.1), and used it to demonstrate the developmental neurotoxicity exerted by rotenone [40]. Another 3D model containing several brain cell types derived from human embryonic stem cells was also shown to be able to detect the toxicity of trimethyltin and paraquat, two known environmental neurotoxicants [41]. Lancaster and colleagues (2014) reached an even higher level of biological complexity, by developing a human iPSC-derived 3D organoid culture system; however, this model has not been yet used for neurotoxicity testing [42]. It is important to stress that growing stem cells and delivering reliable and well-characterized cultures for toxicity assessment require a high level of standardization of both undifferentiated and differentiated cell cultures, in order to ensure the establishment of robust test systems. It is, therefore, of pivotal importance to define a set of quality control parameters suitable to properly characterize stem cell-derived models before using them for toxicity testing, especially those derived from PSCs [43–45].

Based on the current knowledge, it can be stated that *in vitro* models of human brain cells, such as those derived from hiPSCs, can recapitulate a sequence of neurodevelopmental processes starting from NPC proliferation until an advanced stage of neuronal and glial differentiation and maturation. Quantitative evaluation of the impairment of these processes due to chemical exposure can serve as reliable readouts for *in vitro* neurotoxicology evaluation. Nevertheless, further efforts should be made to upscale the throughput applicability of some measured endpoints, particularly when 3D systems are required.

Neurotoxicology and Disease Modelling

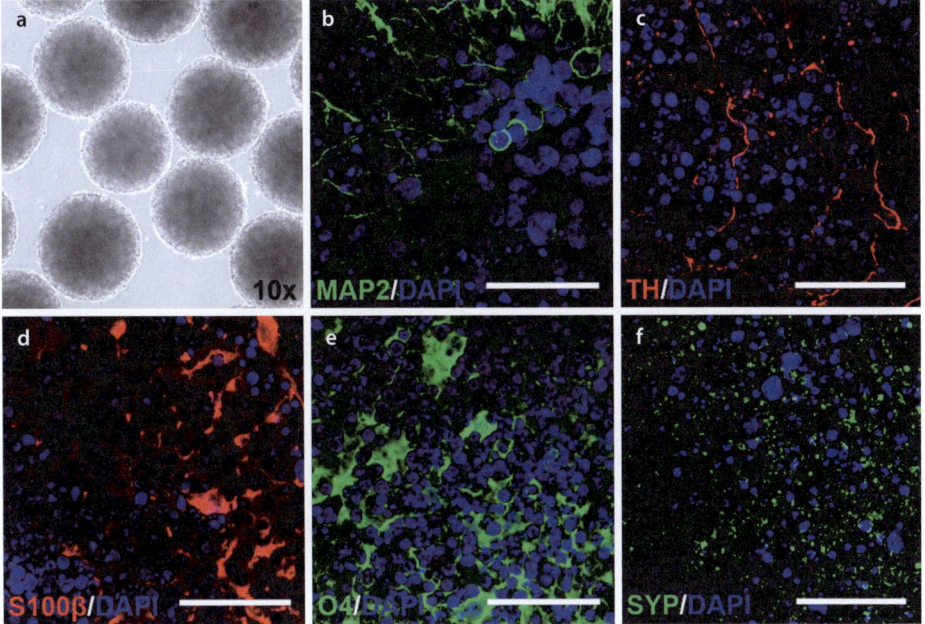

Fig. 12.1 hiPSC-derived BrainSpheres prepared according to Pamies et al. [52]. **a** Bright-field picture showing BrainSpheres at day 12 (day 0 = NPCs dissociation and starting day for 3D cultures). **b–f** Immunolabelling for cell type-specific markers of 8-week-old BrainSpheres. **b** Microtubule-associated protein 2 (MAP2), specific for neurons. **c** Tyrosine hydroxylase (TH), specific for dopaminergic neurons. **d** Calcium-binding protein β (S100 β), specific for astrocytes. **e** Oligodendrocyte 4 (O4), specific for oligodendrocytes. **f** Synaptophysin (SYP), a pre-synaptic protein. All procedures are described in Pamies et al. [52]. Scale bar: 50 μm

12.5 Stem Cells and Disease Modelling

Disease models are an essential tool for the elucidation of the molecular basis of pathologies and to allow the development of novel therapies. Historically, model organisms such as *Drosophila melanogaster, Caenorhabditis elegans* and zebrafish have been very helpful. However, more complex models are required to decipher the molecular mechanisms underlying human diseases. The most popular animal models are based on genetically engineered mice that can further be "humanized" by the engraftments of human cells. However, the recent availability of human pluripotent stem cells which have the potential to differentiate virtually into any cell type can now help to overcome the limitations of animal models for certain disorders (for review, see [46–48]). Several types of stem cells are currently used to model human diseases: human adult stem cells, human ESCs, hiPSCs. Stem cells and gene edition allow to model diseases as diverse as cancer, haemophilia, sickle cell disease, Parkinson's disease, schizophrenia (for review, [46]).

Although many diseases can be modelled using a single cell-type *in vitro*, some diseases require more complex systems. For example, amyotrophic lateral sclerosis (ALS) was better mimicked by using co-cultures of neurons and glial cells than by using neurons only [49, 50]. In that perspective, the development of 3D cell culture systems allowing intense cell-to-cell interactions between several cell types, constitute a very promising tool for the field [42, 51, 52]. Cancer can be better modelled in genetically modified organoids prepared from human embryonic and adult stem cells [53]. Furthermore, organoids can be grown from patient-derived healthy and tumour tissues, potentially enabling patient-specific drug testing and the development of individualized treatment [53]. 3D organoid-derived human glomeruli have been shown to represent an accessible approach to the *in vitro* modelling of human podocytopathies and screening for podocyte toxicity [54]. Neurodegenerative diseases are also modelled by 3D cultures derived from iPSCs; however, none of these models has managed yet to fully mimic the pathological hallmarks observed in the diseased human brain [55–57].

In spite of some limitations not exposed here (see [53]), 3D cultures have emerged as relevant *in vitro* models to study diseases. Although further efforts have to be made to improve these models, the recent advances in the reprogrammation of somatic cells into iPSCs, followed by the derivation into various cell types, associated to the progress in *in vitro* 3D culture technologies create exciting new possibilities for the establishment of more accurate, relevant and comprehensive models to understand disease mechanisms and to identify new therapeutic targets.

12.6 Practical Class—iPSC-Derived Neural Progenitor Cells (NPCs)

This protocol describes the generation of NPCs from high-quality hiPSC based on a previously published protocol by Gibco (▶ https://tools.thermofisher.com/content/sfs/manuals/MAN0008031.pdf). These NPCs have then the ability to give rise to neurons, astrocytes and oligodendrocytes and can be used to generate 3D BrainSpheres for human-based *in vitro* toxicity assays.

12.6.1 Materials

- Advanced™ DMEM/F-12 Medium (Cat. no. 12634).
- Neurobasal® Medium (Gibco A1647801).
- Neural Induction Supplement (Gibco A1647701).
- StemPro® Accutase® Cell Dissociation Reagent (Cat. no. A11105-01).
- DPBS [–/–] without $CaCl_2$ and $MgCl_2$ (Cat. no. 14190) (DPBS [–/–]).
- 6-well plates (Costar® 3516)
- Versene (Lonza, BE17-711E).
- ROCK Inhibitor Y27632 (Sigma-Aldrich, Cat. no. Y0503).
- 15-mL and 50-mL sterile polypropylene conical tubes
- Nalgene® Mr. FrostyR Freezing Container (Fisher Scientific, Cat. no. 15-350-50).
- 37 °C humidified cell culture incubator with 5% CO_2
- Liquid nitrogen storage.
- Centrifuge.
- 37 °C water bath
- 0.7 µL sterile Eppendorf tubes

12.6.2 Media

- **Neural Differentiation Medium (NDM) 100 mL**
- 100 mL Neurobasal® Medium
- 2 mL Neural Induction Supplement

- **Neural Expansion Medium (NEM) 100 mL**
- 49 mL Neurobasal® Medium
- 49 mL Advanced™ DMEM/F-12 Medium
- 2 mL Neural Induction Supplement.

- *Note*: Media can be stored at 2 – 8 °C for up to 2 weeks. Before use, pre-warm the required volume of complete medium for that day in a sterile bottle at 37 °C in a water bath.

Step-by-Step Protocol
A. *Preparation of hiPSC Culture for Neural Induction*
    A.1 Start with high-quality hiPSC (with minimal or no differentiated colonies) preferably cultured in feeder-free conditions.
    A.2 Coat a 6-well plate with the appropriate substrate on which to culture your hiPSC (e.g. Vitronectin, Geltrex® matrix) and place the plate in the incubator at 37 °C for 1 h.
    A.3 When hiPSC reach ~70–80% confluence, remove any differentiated and partially differentiated colonies with the help of an aspiration pump.
    A.4 Dissociate hiPSCs with Versene to generate cell clumps for passaging by following the protocol provided by the supplier.
    Note: When passaging hiPSCs, cells should be plated as small clumps and not as a single-cell suspension to avoid cell death since hiPSCs do not like to be isolated.
    A.5 Estimate the cell number of the hiPSC clumps suspension as follows:
        A.5.1 Transfer 100 µL of the cell suspension to a 0.7 µL Eppendorf.
        A.5.2 Centrifuge at $200 \times g$ for 3 minutes and discard the supernatant.
        A.5.3 Add 100 µL of pre-warmed (37 °C) StemPro® Accutase® and incubate for 5 minutes at 37 °C.
        A.5.4 Vigorously pipet the cells up and down with a 200 µL-pipette 5 times to dissociate the cells into a single-cell suspension.
        A.5.5 Determine the total cell number using your preferred method.
    A.6 Remove the supernatant from the freshly coated 6-well plate (from step 1.2), and add 2.5 mL of hiPSCs culture medium into each well.
    A.7 Gently shake the 15 mL tube containing the hiPSC cell suspension (from step 1.4) and plate the hiPSCs into the coated 6-well plate at several seeding densities: 1.5×10^5, 2.5×10^5 and 3×10^5 hiPSCs/well, in duplicates.

A.8 Add 10 µM ROCK inhibitor Y27632 to the hiPSCs culture media overnight, to prevent cell death.

A.9 Move the plate in several quick back-and-forth and side-to-side motions to disperse the cells across the surface and place it gently in a 37 °C incubator with a humidified atmosphere of 5% CO_2.

Note: If the day after the cells are not ready to undergo differentiation (less than 15–25% confluence), change the medium to normal hiPSC medium without ROCK inhibitor. Keep on changing the medium with hiPSC medium without ROCK inhibitor till they reach 15–25% confluence.

B. *Neural Induction* (Fig.12.2)

B.1 Warm 3 mL NDM at 37 °C in a water bath.

B.2 *Day 0*: (about 24 hours after hiPSC splitting), hiPSCs should be at 15–25% confluence. Remove entirely the spent medium and add 2.5 mL/well of NDM.

B.3 *Day 2*: The morphology of cell colonies should be uniform. Remove entirely the spent medium and add 2.5 mL of NDM/well.

B.4 *Day 4*: Cells will be reaching confluency. Remove entirely the spent medium and add 5 mL of NDM/well.

B.5 *Day 5*: Medium exchange idem 2.4.

B.6 *Day 6*: Cells should be close to maximal confluence. Medium exchange idem 2.4.

Note: A brownish colour of cells with many floating cells between days 4 and 7 of neural induction indicates that the starting density of hiPSCs was too high.

Note: At any point, remove any non-neural differentiated colonies (Fig. 12.2e–g)

B.7 *Day 7*: NPCs are ready to be harvested and expanded. Choose the well where cells reached confluency on day 6 and:

   B.7.1 Pre-warm 15 mL NEM to 37 °C.

   B.7.2 Coat one T25 flask with 3 mL of Geltrex® diluted in DMEM/F12 medium (1:20) and incubate for 1 h at 37 °C.

   Note: The 1:20 Geltrex® solution should cover the bottom of the well.

   B.7.3 Remove the spent NDM from the chosen well on step 2.7.

   B.7.4 Wash by gently adding 2 mL of DPBS [−/−] to the well. Remove the DPBS [−/−].

   Note: Add DPBS [−/−] towards the wall of the well to avoid cell detachment.

   B.7.5 Add 1.5 mL of pre-warmed (37 °C) StemPro® Accutase® to the well and incubate for 5–8 minutes at 37 °C until most cells are detached (check by observing with a phase contrast microscope).

   B.7.6 When the cells are fully retracted and round, add 1.5 mL of DPBS [−/−] to the well.

   B.7.7 Transfer the cells to a 15 mL conical tube.

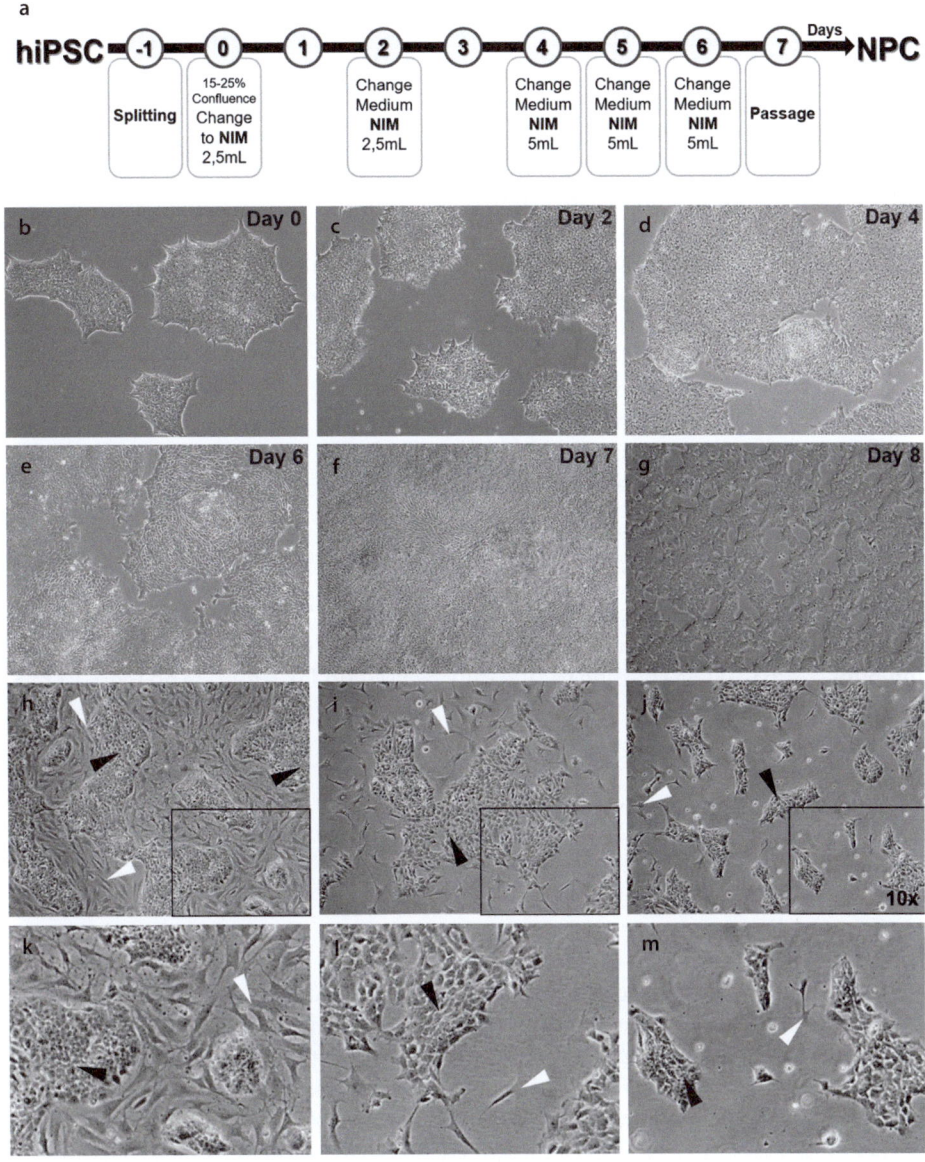

Fig. 12.2 **a** NPC differentiation from hiPSC timeline. **b** hiPSC cultured in feeder-free conditions, 15–25% confluent, ready to start be differentiated into NPCs (day 0). **c–f** Differentiating cells at day 2, 4, 6 and 7 of neural induction. At day 7, cells are 100% confluent. **g** P0 NPC cultured in Geltrex, 1 day after passage. **h–j** Example of selective passage. **h** P1 NSCs at 100% confluency showing a heterogeneous morphology, with flat non-neural cells (white arrowheads) and compact NSCs (black arrowheads). **i** P2 NSCs after one selective passage. **j** P3 NSCs after another passage with selective passage. In **h, i, j**: black boxes indicate the area enlarged in **k, l, m**, respectively

B.7.8 Add 3 mL of DPBS [−/−] to the well to collect residual cells and then transfer the cell suspension to the 15 mL conical tube from step 2.7.7.
B.7.9 Count the cells.
B.7.10 Centrifuge at 300 × g for 4 minutes.
B.7.11 Discard the supernatant and re-suspend the cells with NEM containing 10 µM ROCK inhibitor Y27632 to a final concentration of 2.5×10^5 cells/mL.
B.7.12 Discard the supernatant from the T25 flask (step 2.7.2).
B.7.13 Seed 6 mL of step 2.7.11 solution (1.5×10^6 cells/T25). This corresponds to passage 0 (P0).
B.7.14 After overnight incubation, replace NEM to eliminate the ROCK inhibitor Y27632.
B.7.15 Exchange the NEM every other day thereafter. Do not add ROCK inhibitor Y27632.
Note: NSCs should reach confluence on day 4–6 after plating.

C. *Passage*

After step 2.7.15 when NPC reach ~80–90% confluence you can passage the cells for further expansion by the following steps 2.7.1 to 2.7.15 adapting the volumes and amount of cells to the surface of the flasks used. The treatment with 10 µM ROCK inhibitor Y27632 when splitting should only be used till P4. From P5 no more ROCK inhibitor is needed.

$$T150\,cm^2 - 14 \times 10^6 \,cells\,(28\,mL\,NEM)$$
$$T75\,cm^2 - 3.5 \times 10^6 \,cells\,(18\,mL\,NEM)$$
$$T25\,cm^2 - 1.5 \times 10^6 \,cells\,(6\,mL\,NEM)$$

D. *Selective Passage*

In case of heterogeneous cell morphology observed during NSC expansion (see ◘ Fig. 12.2 h–j) indicating a contamination with non-neural cells (◘ Fig. 12.2e–j), proceed to a selective passage to select the NPCs. Repeat the process until reaching a pure population of NSCs, as judged by eye. This can take 4–5 selective passages.
D.1 Discard spent medium when NPCs reach 90–100% cell confluence.
D.2 Wash by gently adding 5 mL of DPBS [−/−] to the T25 flask. Discard DPBS [−/−].
D.3 Add 3 mL of pre-warmed StemPro® Accutase.
D.4 Observe the cells with a phase contrast microscope.
D.5 When the non-neural cells detach from the bottom, discard the supernatant.
D.6 Collect the remaining cells with 5 mL DPBS [−/−] in a 50 mL tube.
D.7 Proceed with steps 2.7.9 to 2.7.16 adapting the volumes to the vessel in use.

The next day NPCs should look as seen in ◘ Fig. 12.2j.

> **Take-Home Message**
>
> Chemicals we are exposed to need to be evaluated for their harmful potential. The use of animals for this evaluation should be avoided for scientific and ethical reasons. Cell cultures have shown their ability to predict toxicity. However, they are still mostly derived from animals. The breaking through the discovery of hiPSCs allows now to move completely away from the use of animals, although further work is needed to develop the most predictive methods. The fields of disease modelling and drug development also benefit strongly from the discovery of hiPSCs.

References

1. Mayor A. Greek fire, poison arrows, and scorpion bombs : biological and chemical warfare in the ancient world. Overlook Duckworth, New York, NY; 2003. 319 p.
2. Gallo MA. History and scope of toxicology. In: Casarett and Doull's Toxicology: The Basic Science of Poisons [Internet]. 8th ed. McGraw-Hill Education/Medical; 2013 [cited 2019 Jan 14]. Available from: https://accesspharmacy.mhmedical.com/content.aspx?bookid=958&sectionid=53483720.
3. Hayes AN, Gilbert SG. Historical milestones and discoveries that shaped the toxicology sciences. In 2009 [cited 2019 Jan 14]. p. 1–35. Available from: http://www.springerlink.com/index/10.1007/978-3-7643-8336-7_1.
4. Bailey MD. Magic and superstition in Europe: a concise history from antiquity to the present: Rowman & Littlefield Pub, Lanham, MD; 2007. 275 p.
5. Pappas AA, Massoll NA, Cannon DJ. Toxicology: past, present, and future. Ann Clin Lab Sci. 1999;29(4):253–62.
6. Rehman W, Arfons LM, Lazarus HM. The rise, fall and subsequent triumph of thalidomide: lessons learned in drug development. Ther Adv Hematol [Internet]. 2011 [cited 2019 Oct 3];2(5):291–308. Available from: http://www.ncbi.nlm.nih.gov/pubmed/23556097.
7. Parasuraman S. Toxicological screening. J Pharmacol Pharmacother [Internet]. 2011 [cited 2018 Nov 22];2(2):74. Available from: http://www.jpharmacol.com/text.asp?2011/2/2/74/81895.
8. Jennings P. The future of in vitro toxicology. Toxicol Vitr [Internet]. 2015 [cited 2018 Nov 22];29(6):1217–21. Available from: https://linkinghub.elsevier.com/retrieve/pii/S0887233314001891.
9. NRC. Toxicity Testing in the 21st Century [Internet]. Washington, D.C.: National Academies Press; 2007 [cited 2019 Jan 11]. Available from: http://www.nap.edu/catalog/11970
10. Russell WMS, Burch RL. The principles of humane experimental technique. Princ Hum Exp Tech [Internet]. 1959 [cited 2019 Jan 14]; Available from: https://www.cabdirect.org/cabdirect/abstract/19601401516.
11. Eagle H. Nutrition needs of mammalian cells in tissue Cult Sci (80-) [Internet]. 1955;122(3168):501–4. Available from: http://www.jstor.org/stable/1751011.
12. Ekwall B. Screening of toxic compounds in mammalian cell cultures. Ann N Y Acad Sci [Internet]. 1983 [cited 2019 Jan 14];407(1 Cellular Syst):64–77. Available from: http://doi.wiley.com/10.1111/j.1749-6632.1983.tb47814.x.
13. Ekwall B, Johansson A. Preliminary studies on the validity of in vitro measurement of drug toxicity using HeLa cells I. Comparative in vitro cytotoxicity of 27 drugs. Toxicol Lett [Internet]. 1980 [cited 2019 Jan 14];5(5):299–307. Available from: https://www.sciencedirect.com/science/article/pii/0378427480900314.
14. Ekwall B. Screening of toxic compounds in tissue culture. Toxicology [Internet]. 1980 [cited 2019 Jan 14];17(2):127–42. Available from: https://www.sciencedirect.com/science/article/pii/0300483X80900852.
15. Kolman A, Walum E. Björn Ekwall, an outstanding Swedish cell toxicologist. Toxicol Vitr [Internet]. 2010 [cited 2019 Jan 14];24(8):2060–2. Available from: https://www.sciencedirect.com/science/article/pii/S0887233310000871.

16. Bernson V, Clausen J, Ekwall B, Holme J, Hogberg J, Niemi M, et al. Trends in Scandinavian cell toxicology. ATLA-ALTERNATIVES TO Lab Anim. 1986;13(3):162–79.
17. Nakagawa M, Koyanagi M, Tanabe K, Takahashi K, Ichisaka T, Aoi T, et al. Generation of induced pluripotent stem cells without Myc from mouse and human fibroblasts. Nat Biotechnol [Internet]. 2008 [cited 2019 Jan 14];26(1):101–6. Available from: http://www.nature.com/articles/nbt1374.
18. Congress Office of Technology. Neurotoxicity: identifying and controlling poisons of the nervous system. 1990 [cited 2019 Jan 14]; Available from: https://digital.library.unt.edu/ark:/67531/metadc39975/.
19. Masjosthusmann S, Barenys M, El-gamal M, Geerts L, Gorreja A, Kühne B, et al. Literature review and appraisal on alternative neurotoxicity testing methods. 2018;(April). https://doi.org/102903/sp.efsa2018EN-1410.
20. Fritsche E, Alm H, Baumann J, Geerts L, Håkansson H, Witters H. Literature review on in vitro and alternative Developmental Neurotoxicity. 2015;1–186. https://doi.org/102903/sp.efsa2015EN-778
21. Leist M, Hartung T. Inflammatory findings on species extrapolations: humans are definitely no 70-kg mice. Arch Toxicol [Internet]. 2013 [cited 2019 Jan 10];87(4):563–7. Available from: http://link.springer.com/10.1007/s00204-013-1038-0.
22. Veronesi B. In vitro screening batteries for neurotoxicants. Neurotoxicology [Internet]. 1992 [cited 2019 Jan 14];13(1):185–95. Available from: http://www.ncbi.nlm.nih.gov/pubmed/1508418.
23. Tiffany-Castiglioni E. Cell culture models for lead toxicity in neuronal and glial cells. Neurotoxicology [Internet]. 1993 [cited 2019 Jan 14];14(4):513–36. Available from: http://www.ncbi.nlm.nih.gov/pubmed/8164894.
24. Zurich MG, Honegger P, Schilter B, Costa LG, Monnet-Tschudi F. Use of aggregating brain cell cultures to study developmental effects of organophosphorus insecticides. Neurotoxicology [Internet]. 2000 [cited 2019 Jan 14];21(4):599–605. Available from: http://www.ncbi.nlm.nih.gov/pubmed/11022867.
25. Forsby A, Bal-Price AK, Camins A, Coecke S, Fabre N, Gustafsson H, et al. Neuronal in vitro models for the estimation of acute systemic toxicity. Toxicol Vitr [Internet]. 2009 [cited 2019 Jan 14];23(8):1564–9. Available from: https://www.sciencedirect.com/science/article/pii/S0887233309001891.
26. Zurich M-G, Stanzel S, Kopp-Schneider A, Prieto P, Honegger P. Evaluation of aggregating brain cell cultures for the detection of acute organ-specific toxicity. Toxicol Vitr [Internet]. 2013 [cited 2019 Jan 14];27(4):1416–24. Available from: https://www.sciencedirect.com/science/article/pii/S0887233312002007.
27. Gabriel E, Ramani A, Karow U, Gottardo M, Natarajan K, Gooi LM, et al. Recent Zika virus isolates induce premature differentiation of neural progenitors in human brain organoids. Cell Stem Cell. 2017;20(3):397–406.e5.
28. Schmidt BZ, Lehmann M, Gutbier S, Nembo E, Noel S, Smirnova L, et al. In vitro acute and developmental neurotoxicity screening: an overview of cellular platforms and high-throughput technical possibilities. Arch Toxicol [Internet]. 2017 [cited 2019 Jan 16];91(1):1–33. Available from: http://link.springer.com/10.1007/s00204-016-1805-9.
29. Fritsche E, Barenys M, Klose J, Masjosthusmann S, Nimtz L, Schmuck M, et al. Development of the concept for stem cell-based developmental neurotoxicity evaluation. Toxicol Sci [Internet]. 2018 [cited 2019 Jan 14];165(1):14–20. Available from: https://academic.oup.com/toxsci/article/165/1/14/5046970.
30. Bal-Price A, Hogberg HT, Crofton KM, Daneshian M, Fitzgerald RE, Fritsche E, et al. Recommendation on test readiness criteria for new approach methods in toxicology: exemplified for developmental neurotoxicity. ALTEX [Internet]. 2018 [cited 2019 Jan 14];35(3):306. Available from: https://tampub.uta.fi/bitstream/handle/10024/104131/recommendation_on_test_readiness_2018.pdf?sequence=1&isAllowed=y.
31. Bal-Price A, Pistollato F, Sachana M, Bopp SK, Munn S, Worth A. Strategies to improve the regulatory assessment of developmental neurotoxicity (DNT) using in vitro methods. Toxicol Appl Pharmacol [Internet]. 2018 [cited 2019 Jan 10];354:7–18. Available from: https://linkinghub.elsevier.com/retrieve/pii/S0041008X18300541.
32. Bal-Price C, Coecke S, Costa L, Crofton KM, Fritsche E, Goldberg A, et al. Conference report: advancing the science of Developmental Neurotoxicity (DNT): testing for better safety evaluation

[Internet]. 2012 [cited 2019 Jan 14]. Available from: http://nrs.harvard.edu/urn-3:HUL.InstRepos:12605451.
33. Druwe I, Freudenrich TM, Wallace K, Shafer TJ, Mundy WR. Sensitivity of neuroprogenitor cells to chemical-induced apoptosis using a multiplexed assay suitable for high-throughput screening. Toxicology [Internet]. 2015 [cited 2019 Jan 14];333:14–24. Available from: https://www.sciencedirect.com/science/article/pii/S0300483X15000621.
34. Pei Y, Peng J, Behl M, Sipes NS, Shockley KR, Rao MS, et al. Comparative neurotoxicity screening in human iPSC-derived neural stem cells, neurons and astrocytes. Brain Res [Internet]. 2016 [cited 2019 Jan 14];1638:57–73. Available from: https://www.sciencedirect.com/science/article/pii/S0006899315005934.
35. Harrill JA, Freudenrich TM, Robinette BL, Mundy WR. Comparative sensitivity of human and rat neural cultures to chemical-induced inhibition of neurite outgrowth. Toxicol Appl Pharmacol [Internet]. 2011 [cited 2019 Jan 14];256(3):268–80. Available from: https://www.sciencedirect.com/science/article/pii/S0041008X11000676.
36. Ryan KR, Sirenko O, Parham F, Hsieh J-H, Cromwell EF, Tice RR, et al. Neurite outgrowth in human induced pluripotent stem cell-derived neurons as a high-throughput screen for developmental neurotoxicity or neurotoxicity. Neurotoxicology [Internet]. 2016 [cited 2019 Jan 14];53:271–81. Available from: https://www.sciencedirect.com/science/article/pii/S0161813X16300134.
37. Zurich MG, Monnet-Tschudi F, Bérode M, Honegger P. Lead acetate toxicity in vitro: dependence on the cell composition of the cultures. Toxicol Vitr [Internet]. 1998 [cited 2019 Jan 14];12(2):191–6. Available from: https://www.sciencedirect.com/science/article/pii/S0887233397000891.
38. Eskes C, Honegger P, Juillerat-Jeanneret L, Monnet-Tschudi F. Microglial reaction induced by noncytotoxic methylmercury treatment leads to neuroprotection via interactions with astrocytes and IL-6 release. Glia [Internet]. 2002 [cited 2019 Jan 14];37(1):43–52. Available from: http://doi.wiley.com/10.1002/glia.10019.
39. Eskes C, Juillerat-Jeanneret L, Leuba G, Honegger P, Monnet-Tschudi F. Involvement of microglia-neuron interactions in the tumor necrosis factor-? Release, microglial activation, and neurodegeneration induced by trimethyltin. J Neurosci Res [Internet]. 2003 [cited 2019 Jan 14];71(4):583–90. Available from: http://doi.wiley.com/10.1002/jnr.10508.
40. Pamies D, Block K, Lau P, Gribaldo L, Pardo CA, Barreras P, et al. Rotenone exerts developmental neurotoxicity in a human brain spheroid model. Toxicol Appl Pharmacol [Internet]. 2018 [cited 2019 Jan 14];354:101–14. Available from: https://www.sciencedirect.com/science/article/pii/S0041008X18300425.
41. Sandström J, Eggermann E, Charvet I, Roux A, Toni N, Greggio C, et al. Development and characterization of a human embryonic stem cell-derived 3D neural tissue model for neurotoxicity testing. Toxicol Vitr [Internet]. 2017 [cited 2019 Jan 14];38:124–35. Available from: https://www.sciencedirect.com/science/article/pii/S0887233316302089.
42. Lancaster MA, Knoblich JA. Generation of cerebral organoids from human pluripotent stem cells. Nat Protoc [Internet] 2014;9(10):2329–40. Available from: http://www.nature.com/doifinder/10.1038/nprot.2014.158.
43. Coecke S, Balls M, Bowe G, Davis J, Gstraunthaler G, Hartung T, et al. Guidance on good cell culture practice. A report of the second ECVAM task force on good cell culture practice. Altern Lab Anim [Internet]. 2005 [cited 2019 Jan 14];33(3):261–87. Available from: http://www.ncbi.nlm.nih.gov/pubmed/16180980.
44. Pistollato F, Bremer-Hoffmann S, Healy L, Young L, Stacey G. Standardization of pluripotent stem cell cultures for toxicity testing. Expert Opin Drug Metab Toxicol [Internet]. 2012 [cited 2019 Jan 14];8(2):239–57. Available from: http://www.tandfonline.com/doi/full/10.1517/17425255.2012.639763.
45. Pamies D, Bal-Price A, Chesné C, Coecke S, Dinnyes A, Eskes C, et al. Advanced Good Cell Culture Practice for human primary, stem cell-derived and organoid models as well as microphysiological systems. ALTEX [Internet]. 2018 [cited 2019 Jan 14];35(3):353–78. Available from: https://www.altex.org/index.php/altex/article/view/1000.
46. Martín F, Sánchez-Gilabert A, Tristán-Manzano M, Benabdellah K. Stem cells for modeling human disease. In: Pluripotent stem cells – from the bench to the clinic [Internet]. InTech; 2016 [cited 2019 Jan 14]. Available from: http://www.intechopen.com/books/pluripotent-stem-cells-from-the-bench-to-the-clinic/stem-cells-for-modeling-human-disease.

47. Avior Y, Sagi I, Benvenisty N. Pluripotent stem cells in disease modelling and drug discovery. Nat Rev Mol Cell Biol [Internet]. 2016 [cited 2019 Jan 14];17(3):170–82. Available from: http://www.nature.com/articles/nrm.2015.27.
48. Merkle FT, Eggan K. Modeling human disease with pluripotent stem cells: from genome association to function. Cell Stem Cell [Internet]. 2013 [cited 2019 Jan 14];12(6):656–68. Available from: https://www.sciencedirect.com/science/article/pii/S1934590913002099.
49. Di Giorgio FP, Carrasco MA, Siao MC, Maniatis T, Eggan K. Non–cell autonomous effect of glia on motor neurons in an embryonic stem cell–based ALS model. Nat Neurosci [Internet]. 2007 [cited 2019 Jan 14];10(5):608–14. Available from: http://www.nature.com/articles/nn1885.
50. Di Giorgio FP, Boulting GL, Bobrowicz S, Eggan KC. Human embryonic stem cell-derived motor neurons are sensitive to the toxic effect of glial cells carrying an ALS-causing mutation. Cell Stem Cell [Internet]. 2008 [cited 2019 Jan 14];3(6):637–48. Available from: https://www.sciencedirect.com/science/article/pii/S1934590908005225
51. Huch M, Koo B-K. Modeling mouse and human development using organoid cultures. Development [Internet]. 2015 [cited 2019 Jan 14];142(18):3113–25. Available from: http://www.ncbi.nlm.nih.gov/pubmed/26395140
52. Pamies D, Barreras P, Block K, Makri G, Kumar A, Wiersma D, et al. A human brain microphysiological system derived from induced pluripotent stem cells to study neurological diseases and toxicity. ALTEX. 2017;34(3):362–76.
53. Drost J, Clevers H. Organoids in cancer research. Nat Rev Cancer [Internet]. 2018 [cited 2019 Jan 14];18(7):407–18. Available from: http://www.nature.com/articles/s41568-018-0007-6
54. Hale LJ, Howden SE, Phipson B, Lonsdale A, Er PX, Ghobrial I, et al. 3D organoid-derived human glomeruli for personalised podocyte disease modelling and drug screening. Nat Commun [Internet]. 2018 [cited 2019 Jan 14];9(1):5167. Available from: http://www.nature.com/articles/s41467-018-07594-z
55. Bordoni M, Fantini V, Pansarasa O, Cereda C. From Neuronal Differentiation of iPSCs to 3D Neural-Organoids: Modeling of Neurodegenerative Diseases. 2018 [cited 2019 Jan 14]; Available from: https://www.intechopen.com/online-first/from-neuronal-differentiation-of-ipscs-to-3d-neural-organoids-modeling-of-neurodegenerative-diseases/.
56. Ranjan VD, Qiu L, Tan EK, Zeng L, Zhang Y. Modelling Alzheimer's disease: insights from *in vivo* to *in vitro* three-dimensional culture platforms. J Tissue Eng Regen Med [Internet]. 2018 [cited 2019 Jan 14];12(9):1944–58. Available from: http://doi.wiley.com/10.1002/term.2728.
57. Rowland HA, Hooper NM, Kellett KAB. Modelling sporadic Alzheimer's disease using induced pluripotent stem cells. Neurochem Res [Internet]. 2018 [cited 2019 Jan 14];43(12):2179–98. Available from: http://link.springer.com/10.1007/s11064-018-2663-z.

Mesenchymal Stem Cells for Cutaneous Wound Healing

Sérgio P. Camões, Jorge M. Santos, Félix Carvalho, and Joana P. Miranda

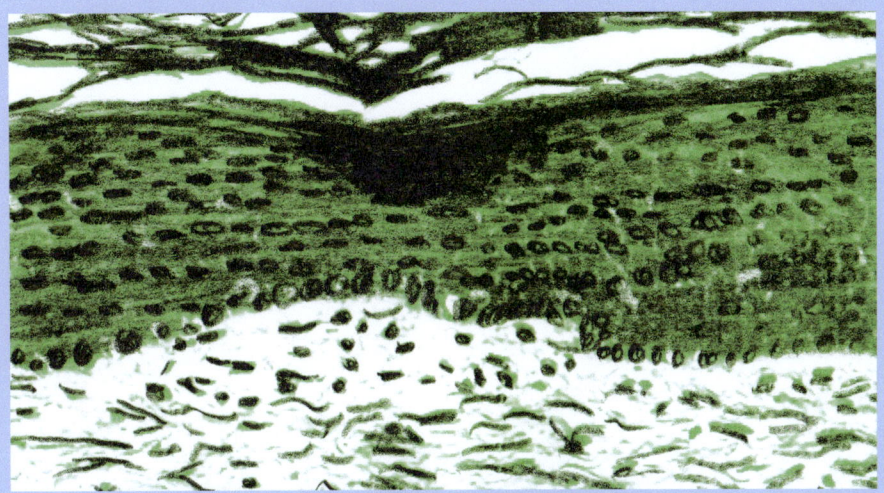

Contents

13.1 Introduction – 249

13.2 Mesenchymal Stem Cells in Wound Healing: Sources and Mechanisms – 251
13.2.1 Homing of Mesenchymal Stem Cells – 252
13.2.2 Mesenchymal Stem Cell Paracrine Mechanisms – 254

13.3		Mesenchymal Stem Cells in Skin Regeneration/ Cutaneous Wound Healing – 255
13.3.1		Endogenous MSCs – 255
13.3.2		Exogenous MSCs – 255
13.3.3		The Role of Exosomes Derived from MSCs – 258
13.4		Priming MSCs: Strategies for Improving Cutaneous Wound Healing – 260
13.4.1		Preconditioning Strategies – 260
13.4.2		Tissue Engineering: Three-Dimensional Cultures – 261
13.4.3		Genetic Manipulation – 262
13.5		Conclusion – 262
		References – 263

Mesenchymal Stem Cells for Cutaneous Wound Healing

What You Will Learn in This Chapter

The success of wound healing is impaired in several medical conditions resulting in increased morbidity. The current available therapeutic options frequently fail to promote full tissue regeneration, and stem cell-based therapies have emerged as promising alternatives. Mesenchymal stem/stromal cells (MSCs) have gained relevance within this context not only due to their ability to promote healing through engraftment and further differentiation, but also because of their paracrine effects. Recently, it has been suggested that the secretion of exosomes may be a dominant mechanism by which MSCs exert their healing function, thus granting them a potential new role as active players in cell-free-based therapies. This chapter focuses on the recent advances on MSC-based therapies for the treatment of cutaneous wounds, namely on their mechanisms of action and the strategies adopted to improve their therapeutic efficacy.

13.1 Introduction

Chronic wounds affect 2% of the general population in the Western world, having tremendous social and economic impacts [1, 2]. They are mostly associated to medical conditions such as diabetes, vascular and autoimmune diseases, being responsible for a decreasing quality of life of the affected individuals, and frequently leading to major complications, namely amputations and even, early death [1]. The lack of effective therapies is one of the causes potentiating the chronic wound burden, making the search for alternative therapeutic strategies an issue of utmost importance. Supported by recently accumulated evidence, mesenchymal stem/stromal cells (MSCs) have increasingly become valid candidates to be applied in cell-based therapies due to their beneficial effects on tissue regeneration [3]. The potential roles and applications of MSCs for the treatment of cutaneous wounds will be the focus of this chapter.

The main function of the skin is to form an effective barrier against the external environment, conferring protection and participating in the defence against pathogens, in the regulation of body temperature and in the prevention of dehydration [4]. Cutaneous wound healing is, therefore, a crucial factor for the survival of organisms and the loss of integrity of this barrier, either as a result of injury or a disease, must be rapidly and efficiently treated [4].

Wound healing is a dynamic biological event mediated by a complex set of growth factors secreted by a variety of cell types, controlling events like clotting, inflammation, cellular migration, proliferation, extracellular matrix (ECM) deposition, angiogenesis, vasculogenesis and ultimately, leading to remodelling of the mature scar tissue [4]. There are three classic stages of wound repair (◘ Fig. 13.1) that, although distinct, do overlap: i) the inflammatory stage, ii) new tissue formation/proliferation stage and iii) the remodelling phase [4]. Inflammation is the first stage of wound repair, occurring immediately after tissue injury. In this process, haemostasis is achieved, initially by the formation of a platelet plug, followed by the formation of a fibrin matrix. In turn, platelets embedded in the recently formed fibrin clot recruit leucocytes and macrophages [5]. Neutrophils eliminate contaminating microorganisms and secrete pro-inflammatory cytokines that activate local fibroblasts and keratinocytes [4, 5]. On the other hand, macrophages are responsible for initiating the granulation tissue deposition by releasing platelet-derived growth factor (PDGF) and vascular endothelial growth factor (VEGF), which then leads to

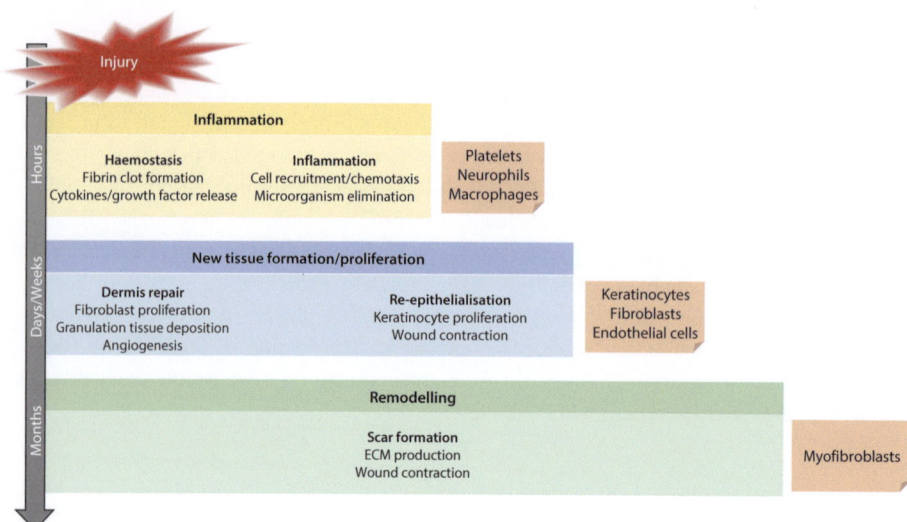

Fig. 13.1 Stages of cutaneous wound healing: inflammatory phase, phase of new tissue formation and remodelling phase

a feedback loop of continued secretion of several other growth factors that are crucial to the healing process, including transforming growth factor α and β (TGF-α, TGF-β), interleukin-1 (IL)-1 and insulin-like growth factor (IGF) [6]. Afterwards, the crosstalk between keratinocytes and fibroblasts becomes critical during the tissue formation/proliferation phase. Through paracrine signalling, keratinocytes release IL-1 thus inducing fibroblasts to secrete keratinocyte growth factor (KGF), fibroblast growth factor-7 (FGF-7), IL-6, granulocyte-macrophage colony-stimulating factor (GM-CSF) and hepatocyte growth factor (HGF). In parallel, the keratinocyte proliferation culminates in wound re-epithelialisation [5]. In the final stages, and with the contribution of bone marrow-derived mesenchymal stem cells (BM-MSCs), fibroblasts replace the fibrin matrix by granulation tissue, which is associated to recently formed capillaries [5, 7]. Ultimately, the interaction between myofibroblasts and fibroblasts leads to the production of the ECM, mainly in form of collagen. This newly formed ECM turns into a mature scar in the final remodelling stage [4–6]. Concomitantly at this stage, an almost acellular matrix enriched in type III collagen is left behind and further replaced mostly by type I collagen fibres that reinforce the repaired tissue, a process mediated by matrix metalloproteinases (MMPs) [5, 6].

Being a very complex process, numerous factors can impact the different healing phases. The factors that influence wound healing can be divided in local and systemic [8]. Local factors are the ones capable of affecting the wound characteristics, such as oxygenation/hypoxia, infection, foreign body and venous sufficiency, while systemic factors are those related to the individual's healing ability, which can be affected by physiologic or medical conditions such as age, hormones, stress, obesity, medication, nutrition, among others [8]. Diabetic patients, for instance, are unable to respond to signalling pathways of normal wound healing, thus compromising their function [9]. Chronic wounds usually fail to progress beyond the first stage of wound healing, leading to a continuous inflammatory condition characterized by excessive levels of pro-inflammatory cytokines, reactive oxygen species (ROS) and senescent cells. The maintenance of this inflammatory state also leads to excessive proteolysis promoted by MMP, that unlike in acute wounds, are not controlled by their inhibitors causing degradation of ECM, growth factors and their receptors [9, 10]. In turn, fibroblasts present reduced levels of TGF-β receptors in ulcerated tissues promoting their senescence and correlated weakened migratory capacity even in the presence of the motogenic stimulant TGF-β [11]. This event along with ECM degradation leads to a positive feedback loop that amplifies inflammation due to inflammatory cell recruitment, which again hampers the wound healing progression to the proliferative/new tissue formation stage [9]. In addition, chronic wounds contain senescent cell populations of keratinocytes, endothelial cells and macrophages, probably due to oxidative stress, DNA damage-related cell cycle arrest or abnormal metabolic changes. Chronic wounds are also subject to persistent infections and in most cases the stem cell niche is compromised or even dysfunctional [9].

The traditional and currently available therapeutic approaches for the treatment of (chronic) wounds consist in debridement [12], application of wound dressings [13], negative pressure and hyperbaric oxygen [12] or the administration of growth factors. However, current research advances in such solutions have been restricted to amelioration of patient care [14]. More curative approaches have resorted to the application of stem cells to improve the healing process, thus raising the hopes for a complete wound resolution.

13.2 Mesenchymal Stem Cells in Wound Healing: Sources and Mechanisms

MSCs have gained interest within the scientific community for they represent promising alternatives to the repair and regeneration of damaged tissues. Unlike the initial postulates, supported by strong plasticity data, the way MSCs exert their regenerative capacity is not so much by their ability for multipotent differentiation, but instead through the secretion of bioactive molecules, which renders them potent inducers of pro-healing mechanisms [15–17].

MSCs are a subset of multipotent, less committed postnatal stem cells, characterized according to The International Society for Cellular Therapy (ISCT) criteria [18], as being able to adhere to a plastic surface, of undergoing trilineage differentiation (into osteoblasts, adipocytes and chondroblasts) and expressing specific surface proteins. MSCs present several advantages over other stem cell types. As opposed to embryonic stem cells (ESCs) that require the sacrifice of the embryo for collection, MSCs' procurement is easy and obtained from non-controversial tissue sources at a relatively low cost. The use of MSCs also overcomes the safety concerns related to genome stability of ESCs and induced pluripotent stem cells (iPSCs) that have hindered their clinical applications. Moreover, cultivation of MSCs does not require the use of feeder layers or high concentrations of serum as ESCs do. Finally, although MSCs are not an infinite cell source, they have an extraordinary replicative capacity in vitro, which is of extreme relevance, due to the large amounts of cells needed for cell-based therapies [15].

Despite the fact that the perivascular niche is thought to be a common stem cell microenvironment for resident MSC-like populations [19], MSCs can be isolated from a number of neonatal tissues, including amnion, placenta [20, 21], foetal blood, liver [22] and umbilical cord blood (UCB) [23] and tissue/stroma (UC) [24, 25]; as well as from adult tissues, such as bone marrow, thymus, brain, liver, lung, kidney, aorta, muscle, spleen [19] and adipose tissue (AT) [26]. These cells display many common characteristics suggesting that all MSC populations share a similar ontogeny. However, MSCs may present variations in morphology, proliferation potential, growth rates, differentiation capacity as well as in their regenerative potential, including wound healing capacities [27]. In fact, one of the main advantages of using human neonatal MSCs is the fact that they are isolated from a tissue containing a more primitive MSC population expressing the pluripotency gene markers *Oct-4*, *Nanog* and *Sox2*, which are present in ESCs. Epigenetic modifications are also altered in neonatal tissues. Indeed, the aging of these cells is one of the hypothesis to explain the differences in MSC regenerative potential and may be associated to DNA and mitochondrial damages [28]. Several comparisons using MSCs from distinct origins but with similar number of passages have been reported as indicating differences in regenerative potential [24, 29].

The MSCs derived from bone marrow (BM-MSCs) were the first to be isolated, being, therefore, the most investigated. BM-MSCs present increased differentiation potential to osteogenic and chondrogenic lineages [29], when compared with MSCs from other sources, namely adipose tissue or umbilical cord blood [30]. In turn, AT-MSCs have been reported to better contribute to the formation of capillary networks, whereas pericytes genesis has been shown to be potentiated by UCB-MSCs [31]. UC-MSCs [24, 25] and BM-MSCs [24] have demonstrated to promote early motogenic effects on keratinocytes and fibroblasts, respectively. In contrast to UC-MSC along with UCB-MSCs, AT-MSCs and BM-MSCs have been implicated in improving vasculogenesis, whereas amniotic membrane-derived MSCs have not [31].

Therefore, the regenerative potential of MSCs may depend on their tissue of origin, which is supported by the stem cell niche theory that postulates that cell fate is the result of the impact of stem cell niche [32]. As such, the tissue source from which MSCs are isolated is an important factor that not only conditions the healing potential, but also must be taken into consideration to better define the therapeutic strategy for further clinical applications.

13.2.1 Homing of Mesenchymal Stem Cells

In stem cell science, the term "homing" refers to the stem cells' ability to migrate to a specific destination, or "niche", within a given organism. In the case of MSCs, their ability to respond to chemotactic cues and migrate towards injured sites is recognised. Therein they may differentiate or influence the wound environment by secreting regenerative paracrine factors [33, 34]. MSC homing is also characterized by MSC arrest within vessels and subsequent passage/transmigration through the endothelium [34]. In terms of clinical applications, this property allows a less invasive and systemic administration of MSCs that are then guided by cytokine gradients and other cues originating from tissue damage loci.

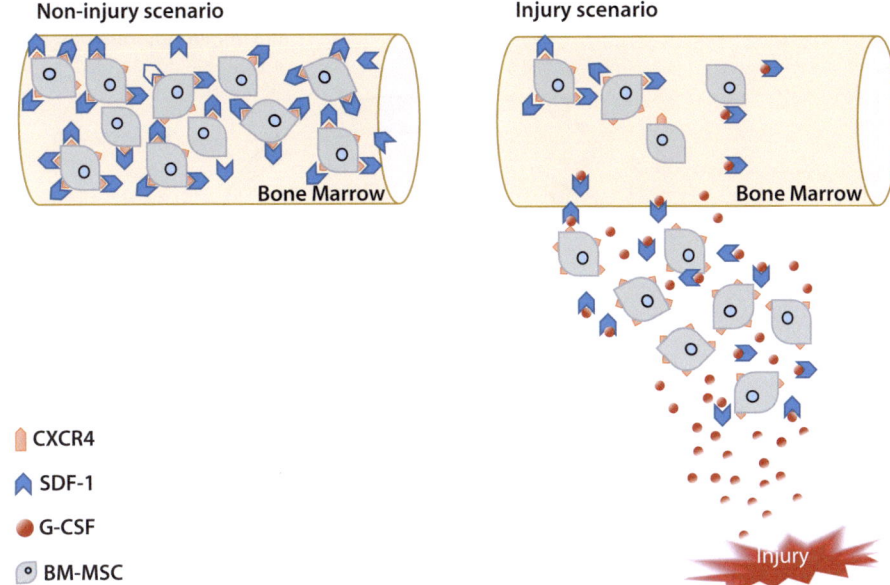

Fig. 13.2 The SDF-1/CXCR4 axis and the role of G-CSF after tissue injury. In a non-injury scenario, SDF-1 acts as a signal for the retention of BM-MSCs at the bone marrow. Conversely, in an injury scenario, cytokines, chemokines, and growth factors, like G-CSF, are released at the injured tissue, triggering the degradation of SDF-1 and consequent homing of MSC from the BM to sites of injury via the circulation

Evidence of MSC homing in homeostatic conditions has proven to be difficult to obtain, having produced some controversial results as reported in the literature [34]. However, despite the failure to isolate circulating MSCs [35], Wang et al. have shown that higher amounts of MSCs are isolated from injured mice [36], being this fact associated with increased levels of VEGF and G-CSF in peripheral blood.

The mechanism by which MSCs home to specific damaged tissues is barely understood, but it is thought to resemble the leukocyte homing cascade, due to the presence of a number of similar cell adhesion and cytokine receptors at the surface of MSCs [34] (Figs. 13.2 and 13.3). After injury, the systemic secretion of P- and E-selectins as well as VCAM-1 mobilizes leukocytes. Other chemokines such as SDF-1, IL-8, CCL2, and CXCL10 are also secreted and recognized by MSCs via specific receptors [37]. The SDF-1/CXCR4 axis has been shown to be significantly up-regulated in bone marrow and in ischemic tissues [34], being well known for their roles in hematopoietic stem and immune cell recruitment [38], as well as in MSC mobilization in experimental models of brain injury and heart myocardial infarction [39, 40]. SDF-1 acts as a signal for the retention of BM-MSCs at the bone marrow, and the G-CSF used clinically to mobilize stem cells from the BM into the peripheral circulation is thought to work by triggering the degradation of SDF-1 (Fig. 13.2) [38]. Indeed, Miranda et al. have hypothesized that G-CSF could be involved in BM-MSC recruitment to injured tissues, which is known to be important for promoting tissue regeneration [24].

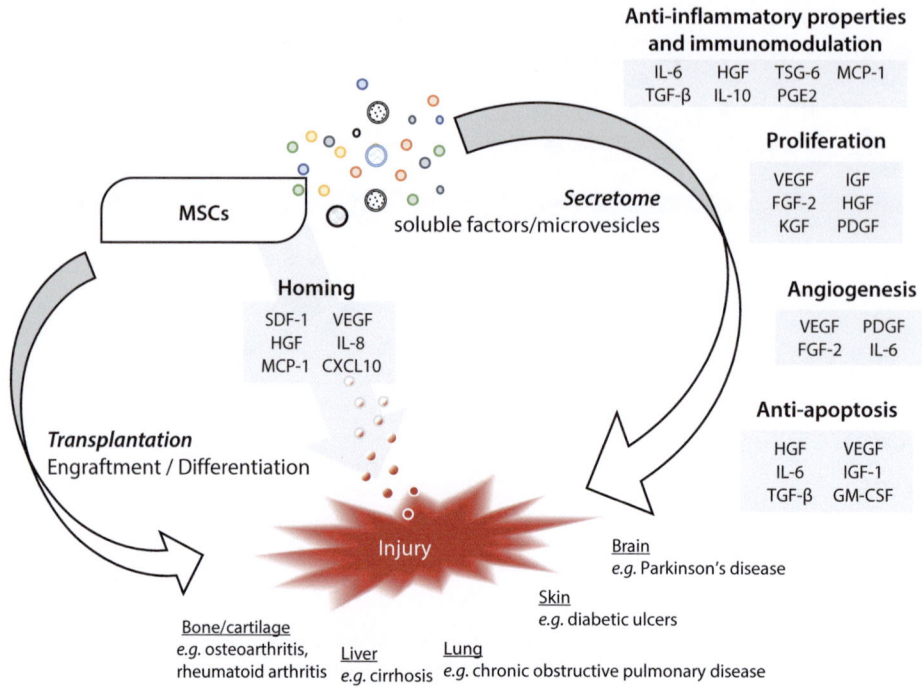

□ **Fig. 13.3** The role of mesenchymal stem cells in regenerative medicine. Mechanisms involved in MSC-based therapy with identification of relevant trophic factors

13.2.2 Mesenchymal Stem Cell Paracrine Mechanisms

Increasing amount of evidences shows that MSCs possess the ability to induce tissue regeneration without local engraftment or differentiation leading to a paradigm shift in regenerative medicine [41]. Emerging data suggest that stem cells can be considered as a reservoir of trophic factors, including growth factors and cytokines as IL-1, GM-CSF, IL-7, IL-8, IL-10, IL-11, SDF-1 or micro and nanovesicles, that are released (secretome) when needed to modulate and repair surrounding damaged tissues [42]. It has been demonstrated that trophic factors can have many roles such as regulation of inflammatory reactions, immunomodulation, anti-apoptotic and pro-angiogenic, just to mention a few (□ Fig. 13.3) [42–44]. Moreover, instead of transplantation of MSCs, the administration of their secretome can overcome many of the safety concerns regarding transplantation of viable replicating cells [45]. Understanding the cell secretome has thus attracted much attention.

MSC-mediated immune regulation is the result of the cumulative action displayed by several anti-inflammatory molecules [46], and its effects occur in a localized environment and not systemically [47]. The immunomodulatory effects of MSCs are quite relevant for their application as novel therapeutics since MSCs were shown to be able to inhibit the proliferation of CD4+ and CD8+ T cells, B cells, and natural killer (NK) cells [24, 48]. The mechanisms by which MSCs exert effects of immunomodulation are related to the suppression of pro-inflammatory behaviour of T and B cells, macrophages, dendritic cells and neutrophils [37]. These effects are mediated by the MSC secretion of growth factors such as TGF-β, HGF, IL-10, PGE2 and NO [49], IL-6 [50] and CCL2 [51], among others [37, 49]. In fact, their therapeutic role in a variety of inflammatory autoimmune diseases, such as rheumatoid arthritis [27, 44], is currently under wide pre-clinical and clinical investigation [52].

13.3 Mesenchymal Stem Cells in Skin Regeneration/Cutaneous Wound Healing

13.3.1 Endogenous MSCs

Apart from the role of stem cell niches existing in all tissues, including the skin, understanding the contribution of other endogenous stem cells to the regenerative process is a challenging quest and has even led to controversy.

Within the skin context, some authors have shown that skin-specific MSC niches, located at either the hair follicle bulge [53] or the dermal sheath [54], play differential roles in wound healing: while the former migrate to induce keratinocyte function, the latter differentiate into fibroblasts to further help on ECM deposition. After damage, epidermal repair is dependent on the migration of hair follicle bulge stem cells that through transient proliferation induce re-epithelialization [53]. The mechanism by which dermal sheath stem cells assist in wound repair is not related to migratory activity. Instead, the cells around the hair follicle assume a wound healing or myofibroblast phenotype, being involved in dermal repair [54].

On the other hand, endogenous BM-MSCs could act in the different phases of the wound healing process [24]. MSCs can induce the recruitment of endothelial cells through the secretion of VEGF and the modulation of inflammatory and immune responses through tumour necrosis factor-α (TNF-α) regulation and natural killer (NK) cell elimination. Afterwards, the proliferation of epidermal cells at the wound margins and further vasculogenesis allow the granulation tissue deposition. Finally, in later phases, BM-MSCs can affect in scar deposition by regulating interleukins and secreting prostaglandin E2 (PGE2), which further reduces collagen production [50].

13.3.2 Exogenous MSCs

Several reports support that chronic wound patients are deficient and defective in stem cells, as is also seen in in vivo healing-related disease models [55]. Hence, to overcome impaired wound healing, these patients may require a direct and active therapy through "exogenous" MSC administration whose mechanism of action can be due to cellular engraftment or through a paracrine mechanism.

13.3.2.1 Cellular Engraftment

Liu et al. have recently reported that systemic UC-MSCs, delivered intravenously, migrate to a wound site and remarkably reduce the amount of inflammatory infiltrated cells and signals while locally stimulating IL-10 and TNF-stimulated gene-6

(TSG-6). They also observed higher levels of collagen deposition as well as improved revascularization and VEGF secretion [56].

However, local application of MSCs at the wound site has also shown therapeutic benefits [57], avoiding cell loss within capillary networks. Locally applied UCB-MSCs improved the healing process by differentiating into keratinocytes as seen by the expression of keratin 19 or pan-keratin antigen [58]. On the other hand, Kong et al. have described that placenta-derived MSCs transplanted into the wound site have engrafted into the recipient's vasculature. Engraftment without apparent differentiation resulted in enhanced microvessel density and ultimately led to improved wound healing in diabetic rats, probably due to the secretion of VEGF, HGF, FGF-2, TGF-β and IGF-1 [59]. Similar findings were reported by Hong et al. who found that AT-MSCs delivered topically engrafted and promoted a higher number of CD31 positive cells at the wound site with no signs of MSC differentiation. Although not exploring in depth the signalling pathways behind such findings, the paracrine secretion of pro-healing trophic factors by AT-MSCs, implicated in endothelial cell and macrophage recruitment in vivo, has been anticipated [60]. Finally, Shin et al. have reported the beneficial effects of locally applied BM-MSCs on both, normal and impaired healing, as a result of MSC engraftment [61]. In this model, the recruitment of higher numbers of autologous CD90 and CD166 positive cells (MSCs) to the wound site was observed, as a result of the increased production of Wnt3a, VEGF and platelet-derived growth factor receptor α (PDGFR-α) [61].

13.3.2.2 Paracrine Activity

As previously mentioned, the paracrine activity of MSCs alone for the treatment of cutaneous wounds has also been directly evaluated by many authors who chose to apply MSC secretome, instead of physical cells, in their experimental setups (◘ Table 13.1). Media that have been conditioned by MSCs (CM) contain MSC-secreted growth factors, cytokines and chemokines, including IL-8, IL-6, HGF, FGF-2, TGF-β, IGF-1, TSG-6, tumour necrosis factor receptor 1 (TNFR1), VEGF, granulocyte colony-stimulating factor (G-CSF), KGF and epidermal growth factor (EGF), that play important roles in the different phases of normal wound healing [24, 25, 56, 62].

Yew et al. demonstrated that CM from BM-MSCs accelerated wound closure with increased re-epithelialization, cell infiltration, granulation formation and angiogenesis, by enhancing epithelial and endothelial migration, which was related to the observed high levels of IL-6 mediated through the activation of p38 MAPK. In turn, in vitro studies using dermal fibroblasts have revealed that the presence of CM promoted fibroblast activation, proliferation and migration, further enhancing the healing of the wounds [24, 25, 62, 63].

In addition, the application of CM from amniotic fluid (AF)-derived MSCs led to accelerated wound repair in vivo by dermal fibroblasts through the TGF-β/SMAD2 pathway [62]. During the healing process, TGF-β is involved in inflammation, angiogenesis, re-epithelialization and granulation tissue deposition, being dependent on the activation of SMAD2 which further stimulates ECM production by fibroblasts promoting healing. Along with this new tissue formation, stromal cell-derived factor 1 (SDF-1) production by exogenously administered MSCs acts as a positive feedback by stimulating the endogenous MSC migration that will also secrete cytokines; overall increasing the crosstalk with adjacent cells including kera-

Table 13.1 Summary of the studies using mesenchymal stem cell-derived conditioned media for the treatment of cutaneous wounds including study design, mechanisms and outcomes

MSC source	Experimental model	Associated factors/ signalling pathway	Outcome	Ref.
Rat bone marrow	In vitro scratch assay	MEK/Erk, MMP-2, MMP-9	Stimulation of the proliferation and migration of keratinocytes.	[65]
Human bone marrow	In vitro scratch and transwell migration assays In vivo mouse excisional wound splinting model	p38 MAPK, IL-6, IL-8, CXCL1	Induction of epithelial and endothelial cell migration resulting in accelerated wound closure with increased re-epithelialization, cell infiltration, granulation formation and angiogenesis.	[91]
Human amniotic fluid	In vitro proliferation and scratch assays In vivo mouse excisional wound splinting model	TGF-β/SMAD2	Enhancement of wound healing by dermal fibroblast proliferation and migration; accelerated wound closure.	[62]
Human umbilical cord matrix	In vitro scratch assay In vivo mouse excisional wound model	TGF-β2, HIF-1α, PAI-1	Induction of the proliferation and migration of fibroblast and wound closure. Enrichment of proliferating cells in wound site with enhanced re-epithelialization, higher cellularity in granulation tissue, and organized ECM.	[63]
Human umbilical cord matrix	In vivo mouse excisional wound splinting model	IL-10, TGF-β, VEGF, angiopoietin-1	Accelerated wound closure with increased capillary density and activation of M2 macrophages.	[92]
Human umbilical cord matrix	In vitro scratch assay In vivo mouse excisional wound splinting model	Non-identified	Enhancement of fibroblast migration with increased deposition of collagen and elastin. Improvement on wound closure with increased re-epithelialization, cellularity and vasculature, sebaceous glands and hair follicles.	[93]

tinocytes. In fact, cell-free lysates prepared from BM-MSCs have proved to be effective in inducing faster wound resolution by increasing expression of SDF-1 and CXCL-5, which are strong keratinocyte stimulators [64]. Moreover, MSC CM has also been implicated in the treatment of wounds characteristic of diabetic conditions where keratinocyte migration and proliferation are impaired. Such as shown in vitro by the treatment with CM derived from MSCs that overcame the effects of hyperglycemia by decreasing ROS overproduction, and allowing keratinocyte motility, which can be translated as an alternative therapeutic approach to ameliorate the poor wound healing conditions prompted by diabetes [65].

Finally, in a relevant comparative study, Miranda et al. have demonstrated that either CM obtained from UC-MSCs or BM-MSCs have the capacity to accelerate wound closure in vivo. However, the motogenic activity promoted by the CM from UC-MSC was significantly higher for human keratinocytes, in opposition to the effect seen with CM produced by BM-MSCs, which preferentially induced fibroblast migration. Accordingly, a comparative quantification of key factors with vital importance in the consecutive stages of wound healing revealed very different secretome profiles between the two MSCs. The relatively higher UC-MSC expression of EGF, FGF-2, and KGF strongly supported the early induction of keratinocyte migration and function necessary to trigger the later remodelling stages, where fibroblasts, triggered by IL-6, play a major role in ECM production. Concomitantly, the newly discovered UC-MSC-specific expression of G-CSF has revealed additional capacities for the CM derived from UC-MSCs to mobilize other healing-related cells (including CD34⁻/CD45⁻ precursors – MSCs). These results were noteworthy since they underpinned i) different paracrine activities between MSCs derived from different tissue sources and ii) the viability of using (complementary) CMs rather than physical cells to promote skin regeneration [27].

13.3.3 The Role of Exosomes Derived from MSCs

Recent studies have uncovered that therapeutically valuable paracrine factors secreted by MSCs are often contained in small secreted lipid vesicles (40–100 nm) termed exosomes [66–68]. Exosomes have been found to cause alterations in biological pathways by playing key roles in normal physiology as well as in different pathological conditions [66, 69]. Indeed, they are released into the extracellular space shuttling several molecules, including proteins (e.g. growth factors, cytokines and receptors, as CXCR4) and nucleic acids (mRNA, miRNA) to neighbouring cells [70]. Because they are extracellular vesicles (EVs), their cargo is protected from adverse conditions, namely differential temperatures and pH environment variations [70].

The role of MSC-derived exosomes as well as their mechanisms of action is currently under investigation, introducing another dimension to the way MSCs can be applied in regenerative medicine. Within the context of wound healing, MSC-derived exosomes have been linked to the processes of re-epithelialization [71, 72] and angiogenesis [72, 73]. ◘ Table 13.2 reviews the works where MSC-derived exosomes have been implicated in the mechanisms related to wound healing.

Zhang et al. have shown that exosomes isolated from human iPSC-derived MSCs (iMSCs) promoted fibroblast secretion of collagen and elastin, and also angiogenesis in vitro, albeit without exploring mechanisms of action [72].

Table 13.2 Summary of studies suggesting the use of mesenchymal stem cell-derived exosomes for the treatment of cutaneous wounds

MSC source	Experimental model	Associated factors/ signalling pathway	Outcome	Ref.
Human adipose tissue	In vitro scratch, proliferation and transwell migration assays In vivo mouse excisional wound model	n.i.	Stimulation of fibroblast migration and proliferation. Acceleration of wound healing with improvement on collagen deposition and maturation.	[75]
Human umbilical cord matrix	In vivo mouse excisional wound model	miR-21, miR-23a, miR-125b, miR-145, TGF-β/ SMAD2	Promotion of wound closure with reduced scar tissue formation by inducing proliferation and migration of fibroblasts although mitigating myofibroblast formation and contractibility.	[76]
Human umbilical cord matrix	In vivo rat second-degree burn wound model	Wnt/β-- Catenin, Wnt4, PI3K/ AKT	Improvement of wound healing with faster re-epithelialization. Induction of proliferation and migration of keratinocytes and dermal fibroblasts.	[71]
Human umbilical cord matrix	In vitro viability/ proliferation, transwell migration and tube formation assays In vivo rat second-degree burn wound model	Wnt/β-- Catenin, Wnt4	Induction of proliferation, migration and tube-formation capacity of endothelial cells. Improvement of angiogenesis with increased number of epidermal and dermal cells during wound healing.	[94]
Human bone marrow	In vitro scratch, viability/proliferation and tube formation assays	PI3K/AKT, MEK/Erk, IL-6/JAK/ STAT3	Stimulation of fibroblast proliferation and migration and increase in the capacity to promote tube formation of endothelial cells, through activation of genes related to cell cycle control and growth factor and cytokine expression.	[73]
Human fetal dermis	In vitro viability/ proliferation and transwell migration assays In vivo mouse excisional wound model	Notch/ jagged 1	Induction of mitogenic and motogenic activities on dermal fibroblasts. Acceleration of wound healing by promoting cell proliferation, ECM deposition and reepithelialisation.	[95]

n.i. non-identified

On the other hand, an in vitro study conducted by Shabbir et al. demonstrated that BM-MSC-derived exosomes improved the growth and migration of fibroblasts isolated from normal and chronic wounds and also induced angiogenesis by activating AKT, ERK 1/2, and STAT3 pathways. These authors found that BM-MSC exosomes carried STAT3, which is reported to promote gene expression related to growth factor production, such as HGF, IGF1, NGF, and SDF-1. Similarly, the AKT and ERK1/2 pathways were also reported to be activated in keratinocytes and dermal fibroblasts in vitro after incubation with exosomes from UC-MSCs [71] and iMSCs [74], respectively. Moreover, exosome-mediated delivery of Wnt4 signalling led to the enhancement of the healing process in a rat skin burn model by inducing β-Catenin, a dual function protein, involved in regulation and coordination of cell–cell adhesion and transcription regulation [71].

In turn, exosomes from AT-MSCs induced the expression of N-cadherin, cyclin-1, proliferating cell nuclear antigen (PCNA) and collagen I and III, associated with migration and proliferation of fibroblasts. The result was an improved collagen synthesis phenotype of these cells, which promoted a faster resolution of soft-tissue wounds [75]. Accordingly, Zhang et al. have also reported increased expression of CK19, PCNA, collagen I (compared to collagen III) in wounds treated with UC-MSC exosomes in vivo [71].

The formation of a scar tissue after skin damage, caused by myofibroblast aggregations, is particularly important since it compromises the regeneration process leading to cutaneous tissue with native elastic properties. The contribution of UC-MSCs to scar tissue deposition remains vague, but Fang et al. have found that specific exosomal miRNAs including miR-21, −23a, −125b, and − 145 were essential to the myofibroblast suppression and anti-scarring functions through inhibition of α-smooth muscle actin (α-SMA). The suppression of α-SMA, which in turn is associated to TGF-β/SMAD2 pathway, resulted in reduced scar formation and myofibroblast accumulation [76].

Uncovering the mechanisms underlying the therapeutic effects associated to the use of MSC-derived exosomes for the repair of cutaneous wound is still a challenge. Nevertheless, exosomes seem to act throughout the full wound healing process: in the initial stages, by improving re-epithelization and angiogenesis by activation of AKT and STAT3 signalling; and in the final stages, by decreasing scar tissue deposition by miRNA-mediated pathways.

13.4 Priming MSCs: Strategies for Improving Cutaneous Wound Healing

MSC response to different environments implies different outcomes of their behaviour. As such, strategies to improve MSC therapeutic efficacy, both in cell-based and cell-free-based therapies, have been attempted [77].

13.4.1 Preconditioning Strategies

Although MSCs constitutively produce several growth factors, these cells are able to reprogram their set of paracrine factors as a response to local stimuli, e.g. when administered in a wounded site.

For example, the presence of hypoxia is a typical feature of damaged tissues. Low levels of oxygen lead to hypoxia-inducible factor 1α (HIF-1α) activation, further inducing the secretion of VEGF and SDF-1. It has been shown that CM from BM-MSCs grown under hypoxic conditions further promoted cutaneous wound healing, being capable of accelerating wound closure in vivo when compared to CM from BM-MSCs cultured under normoxia conditions [78]. Similarly, the CM from hypoxic-treated AF-derived-MSCs presented significantly higher levels of VEGF and TGF-β in vitro as compared to the corresponding normoxic CM [79].

A prolonged inflammatory state is also a characteristic of the chronic wound environment that may be circumvented by MSCs. The exposure of AT-MSCs to inflammation-inducing agents, namely lipopolysaccharide (LPS) and TNF-α, has shown to improve its therapeutic efficacy by inducing skin flap survival in a diabetic rat model [80]. MSC response to such a stimulus was found to be mediated by DNA hydroxymethylation and the microRNAs miR146a, miR150 and miR155. In addition, MSC secretory capacity has been investigated within this context, where exosomes derived from LPS-primed UC-MSCs showed to enhance wound healing in diabetic rats by mediating macrophage polarization through let-7b via TLR4/NF-κB/STAT3/AKT regulatory signalling, therefore reducing inflammation within the wounded sites [81].

13.4.2 Tissue Engineering: Three-Dimensional Cultures

The multicellular 3D structure of spheroids, reached by the ability of MSCs to self-assemble in ex vivo 3D culture systems, constitutes a very efficient strategy to recreate the properties of a more physiological environment [25, 82]. Among others, 3D spheroid properties include greater cell-to-cell communication and cell-to-ECM interactions, enhanced cell viability, improved cell morphology/phenotype, stimulation of angiogenesis and anti-inflammatory modulation [25, 82]. The fact that cell proliferation and metabolism are rather different between 2D and 3D culture systems has also an important effect upon the overall cellular activity, leading to large differences in the secretion of relevant paracrine factors.

There are reports addressing the advantage of the microenvironment originated by 3D architecture for promoting wound healing. Hsu et al. have combined AT-MSC self-assembled spheroids with chitosan-hyaluronan membranes, a condition that enhanced expression of cytokine genes, namely CCL2, VEGF, FGF-1 and of migration-associated genes, such as CXCR4 and MMP-1. In a rat dorsal skin model, the wounds where AT-MSC spheroids were applied revealed faster wound closure and angiogenesis enhancement [83].

Similar observations have been described by Kwon et al. where again, CM obtained from AT-MSC 3D cultures presented higher concentrations of HGF, VEGF, FGF-2 and SDF-1, which suggests an accelerated wound closure triggered by paracrine mechanisms [84].

Regarding the potential of UC-MSC-derived CM, Santos et al. reported that dynamic spheroid 3D culturing resulted in a distinct secretome profile. Several key trophic factors involved in early and late stages of wound healing, namely HGF, TGF-β1, G-CSF, VEGF, FGF-2 and IL-6 were found to increase up to 80-fold when compared to CM of UC-MSCs cultured in 2D monolayers. In fact, in an in vivo excisional rat model, the treatment of wounds with CM from 3D cultured UC-MSCs resulted in a faster wound closure and in the fully regeneration of a more mature tissue phenotype [25].

13.4.3 Genetic Manipulation

Genetic manipulation of MSCs to express growth factors and cytokines that enhance skin regeneration has also been attempted, namely to promote angiogenesis [85], suppress inflammation [86] or to induce cell migration and function [87]. Li et al. have reported that overexpression of angiopoietin-1 in BM-MSCs-induced epidermal and dermal regeneration as well as angiogenesis in vivo, suggesting a direct contribution of angiopoietin-1 in this process [88]. Similarly, Xia et al. have demonstrated shortened healing time of radiation-wound injury when treated with BM-MSCs overexpressing VEGF and β-defensin-3. These authors showed a significant improvement in granulation tissue formation with collagen deposition and skin appendage regeneration [85]. Following the same rationale, but trying to reduce inflammatory signals, Qi et al. administered transfected BM-MSCs with TSG-6 to mouse excisional wounds, demonstrating that these cells suppressed TNF-α secretion by macrophages which led to accelerated wound healing with markedly reduced tissue fibrosis [86].

Along with the MSCs manipulated to express growth factors, chemokine receptor overexpression has also demonstrated therapeutic benefits in wounds of irradiated mice. Herein, the migration of CXCR4-overexpressing BM-MSCs was enhanced through a specific SDF-1-expression-dependent manner culminating in faster skin wound resolution [87]. Taking advantage of SDF-1 biological functions, besides inducing survival and growth of CXCR4-expressing cells [89], Nakamura et al., using a rat excisional wound model, have reported that SDF-1-engineered MSCs improved dermal fibroblast migration and new capillary vessel formation, translating into wound size decrease [90].

Nevertheless, and despite the promising results, the strategies involving the use of genetically modified MSCs for the treatment of cutaneous wounds need to consider further safety issues, e.g. those related to the viral vectors used for gene transfection, before being considered for clinical application.

13.5 Conclusion

Adapting the MSC therapeutic potential to the essential re-establishment of trophic factor, protease and metabolically competent cell equilibrium, needed to overcome impaired wound healing, constitutes an ambitious goal.

Overall, MSCs have proven to be able to influence every stage of the cutaneous wound healing process through two distinct mechanisms: provide replacement units for perished cells in tissues, or secrete trophic factors to their surroundings, thus influencing the endogenous regeneration potential [41]. The manipulation of MSCs for improving their therapeutic performance, through preconditioning, tissue engineering strategies or genetic modification, has also revealed to be beneficial. Yet, the establishment of reliable standard operational procedures for MSC-based therapies to be applied in clinical settings is still hard to achieve. The need for a better elucidation on MSC therapeutic mechanisms, to improve differentiation between different MSCs from different tissue sources, for better and consistent priming conditions, and for the best strategy of administration, are important questions whose answers will facilitate the final translation to the clinic.

> **Take-Home Message**
>
> - Impaired wound repair constitutes a severe health problem with great economic impact, due to the lack of effective therapies.
> - MSC-based therapies have gained relevance as an alternative and effective treatment for cutaneous wounds.
> - MSCs are a subset of stem cells that can be isolated from different tissue sources, which directly impacts on their therapeutic outcome.
> - The therapeutic mechanisms of MSCs involve their proliferation and differentiation after engraftment as well as their paracrine action.
> - Both endogenous and administered exogenous MSCs and/or their secretome have shown to be important players in granting MSCs therapeutic value for wound healing.
> - Exosomes derived from MSCs represent an important cell-free-based alternative therapy.
> - Strategies to improve the MSC therapeutic outcomes include preconditioning insults, culture conditions modulation or genetic manipulation of the MSCs.

References

1. Heublein H, Bader A, Giri S. Preclinical and clinical evidence for stem cell therapies as treatment for diabetic wounds. Drug Discov Today. 2015;20(6):703–17.
2. Guest JF, Vowden K, Vowden P. The health economic burden that acute and chronic wounds impose on an average clinical commissioning group/health board in the UK. J Wound Care. 2017;26(6): 292–303.
3. Domaszewska-Szostek A, Krzyżanowska M, Siemionow M. Cell-based therapies for chronic wounds tested in clinical studies. Ann Plast Surg. 2019;83(6):e96–e109.
4. Lau K, Paus R, Tiede S, Day P, Bayat A. Exploring the role of stem cells in cutaneous wound healing. Exp Dermatol. 2009;18(11):921–33.
5. Gurtner GC, Werner S, Barrandon Y, Longaker MT. Wound repair and regeneration. Nature. 2008;453(7193):314–21.
6. Singer AJ, Clark RA. Cutaneous wound healing. N Engl J Med. 1999;341(10):738–46.
7. Sasaki M, Abe R, Fujita Y, Ando S, Inokuma D, Shimizu H. Mesenchymal stem cells are recruited into wounded skin and contribute to wound repair by transdifferentiation into multiple skin cell type. J Immunol. 2008;180(4):2581–7.
8. Guo S, DiPietro LA. Factors affecting wound healing. J Dent Res. 2010;89(3):219–29.
9. Frykberg RG, Banks J. Challenges in the treatment of chronic wounds. Adv Wound Care. 2015;4(9): 560–82.
10. Nunan R, Harding KG, Martin P. Clinical challenges of chronic wounds: searching for an optimal animal model to recapitulate their complexity. Dis Model Mech. 2014;7(11):1205–13.
11. Pastar I, Stojadinovic O. Attenuation of the transforming growth Factor β–signaling pathway in chronic venous ulcers. Mol Med. 2010;16(3–4):92–101.
12. Han G, Ceilley R. Chronic wound healing: a review of current management and treatments. Adv Ther. 2017;34(3):599–610.
13. Jones V, Grey JE, Harding KG. Wound dressings. BMJ. 2006;332(7544):777–80.

14. Liu Y, Dulchavsky DS, Gao X, Kwon D, Chopp M, Dulchavsky S, et al. Wound repair by bone marrow stromal cells through growth factor production. J Surg Res. 2006;136(2):336–41.
15. Weiss ML, Troyer DL. Stem cells in the umbilical cord. Stem Cell Rev. 2006;2(2):155–62.
16. Gunawardena TNA, Rahman MT, Abdullah BJJ, Abu Kasim NH. Conditioned media derived from mesenchymal stem cell cultures: the next generation for regenerative medicine. J Tissue Eng Regen Med. 2019;13(4):569–86.
17. Harrell CR, Fellabaum C, Jovicic N, Djonov V, Arsenijevic N, Volarevic V. Molecular mechanisms responsible for therapeutic potential of mesenchymal stem cell-derived Secretome. Cell. 2019;8(5):467.
18. Dominici M, Le Blanc K, Mueller I, Slaper-Cortenbach I, Marini F, Krause DS, et al. Minimal criteria for defining multipotent mesenchymal stromal cells. The International Society for Cellular Therapy position statement. Cytotherapy. 2006;8(4):315–7.
19. da Silva Meirelles L. Mesenchymal stem cells reside in virtually all post-natal organs and tissues. J Cell Sci. 2006;119(11):2204–13.
20. In 't Anker PS, Scherjon SA, Kleijburg-van der Keur C, de Groot-Swings GMJS, Claas FHJ, Fibbe WE, et al. Isolation of mesenchymal stem cells of fetal or maternal origin from human placenta. Stem Cells 2004;22(7):1338–1345.
21. Antonucci I, Stuppia L, Kaneko Y, Yu S, Tajiri N, Bae EC, et al. Amniotic fluid as a rich source of mesenchymal stromal cells for transplantation therapy. Cell Transplant. 2011;20(6):789–96.
22. Campagnoli C. Identification of mesenchymal stem/progenitor cells in human first-trimester fetal blood, liver, and bone marrow. Blood. 2001;98(8):2396–402.
23. Erices A, Conget P, Minguell JJ. Mesenchymal progenitor cells in human umbilical cord blood. Br J Haematol. 2000;109(1):235–42.
24. Miranda JP, Filipe E, Fernandes AS, Almeida JM, Martins JP, De La Fuente A, et al. The human umbilical cord tissue-derived MSC population UCX ® promotes early Motogenic effects on keratinocytes and fibroblasts and G-CSF-mediated mobilization of BM-MSCs when transplanted In vivo. Cell Transplant. 2015;24(5):865–77.
25. Santos JM, Camões SP, Filipe E, Cipriano M, Barcia RN, Filipe M, et al. Three-dimensional spheroid cell culture of umbilical cord tissue-derived mesenchymal stromal cells leads to enhanced paracrine induction of wound healing. Stem Cell Res Ther. 2015;6:90.
26. Zuk PA. Human adipose tissue is a source of multipotent stem cells. Mol Biol Cell. 2002;13(12):4279–95.
27. Santos JM, Bárcia RN, Simões SI, Gaspar MM, Calado S, Água-Doce A, et al. The role of human umbilical cord tissue-derived mesenchymal stromal cells (UCX®) in the treatment of inflammatory arthritis. J Transl Med. 2013;11(1):18.
28. Rohban R, Pieber TR. Mesenchymal stem and progenitor cells in regeneration: tissue specificity and regenerative potential. Stem Cells Int. 2017;2017:1–16.
29. Shafiee A, Seyedjafari E, Soleimani M, Ahmadbeigi N, Dinarvand P, Ghaemi N. A comparison between osteogenic differentiation of human unrestricted somatic stem cells and mesenchymal stem cells from bone marrow and adipose tissue. Biotechnol Lett. 2011;33(6):1257–64.
30. Ardeshirylajimi A, Mossahebi-Mohammadi M, Vakilian S, Langroudi L, Seyedjafari E, Atashi A, et al. Comparison of osteogenic differentiation potential of human adult stem cells loaded on bioceramic-coated electrospun poly (L-lactide) nanofibres. Cell Prolif. 2015;48(1):47–58.
31. Rohban R, Etchart N, Pieber TR. Vasculogenesis potential of stem cells from various human tissues. BioRxiv. 2016;1–22.
32. Lin H. The stem-cell niche theory: lessons from flies. Nat Rev Genet. 2002;3(12):931–40.
33. da Silva Meirelles L, Fontes AM, Covas DT, Caplan AI. Mechanisms involved in the therapeutic properties of mesenchymal stem cells. Cytokine Growth Factor Rev. 2009;20(5–6):419–27.
34. Karp JM, Leng Teo GS. Mesenchymal stem cell homing: the devil is in the details. Cell Stem Cell. 2009;4(3):206–16.
35. He Q, Wan C, Li G. Concise review: multipotent mesenchymal stromal cells in blood. Stem Cells. 2007;25(1):69–77.
36. Wang C-H, Cherng W-J, Yang N-I, Kuo L-T, Hsu C-M, Yeh H-I, et al. Late-outgrowth endothelial cells attenuate intimal hyperplasia contributed by mesenchymal stem cells after vascular injury. Arterioscler Thromb Vasc Biol. 2007;28(1):54–60.
37. Barminko J, Gray A, Maguire T, Schloss R, Yarmush ML. Mesenchymal stem cell therapy. In: Chase LG, Vemuri MC, editors. Stem cell biology and regenerative medicine. Totowa: Humana Press; 2013. p. 15–39.

38. Bollag WB, Hill WD. CXCR4 in epidermal keratinocytes: crosstalk within the skin. J Invest Dermatol. 2013;133(11):2505–8.
39. Yu J, Li M, Qu Z, Yan D, Li D, Ruan Q. SDF-1/CXCR4-mediated migration of transplanted bone marrow stromal cells towards areas of heart myocardial infarction via activation of PI3K/Akt. J Cardiovasc Pharmacol. 2010;55(5):1.
40. Wang Y, Deng Y, Zhou G-Q. SDF-1α/CXCR4-mediated migration of systemically transplanted bone marrow stromal cells towards ischemic brain lesion in a rat model. Brain Res. 2008;1195:104–12.
41. Caplan AI, Dennis JE. Mesenchymal stem cells as trophic mediators. J Cell Biochem. 2006;98(5):1076–84.
42. El Omar R, Beroud J, Stoltz J-F, Menu P, Velot E, Decot V. Umbilical cord mesenchymal stem cells: the new gold standard for mesenchymal stem cell-based therapies? Tissue Eng Part B Rev. 2014;20(5):523–44.
43. Marfia G, Navone SE, Di Vito C, Ughi N, Tabano S, Miozzo M, et al. Mesenchymal stem cells: potential for therapy and treatment of chronic non-healing skin wounds. Organogenesis. 2015;11(4):183–206.
44. Miranda JP, Camões SP, Gaspar MM, Rodrigues JS, Carvalheiro M, Bárcia RN, et al. The Secretome derived from 3D-cultured umbilical cord tissue MSCs counteracts manifestations typifying rheumatoid arthritis. Front Immunol. 2019;10:18.
45. Lai RC, Chen TS, Lim SK. Mesenchymal stem cell exosome: a novel stem cell-based therapy for cardiovascular disease. Regen Med. 2011;6(4):481–92.
46. Ghannam S, Bouffi C, Djouad F, Jorgensen C, Noël D. Immunosuppression by mesenchymal stem cells: mechanisms and clinical applications. Stem Cell Res Ther. 2010;1(1):2.
47. Chapel A, Bertho JM, Bensidhoum M, Fouillard L, Young RG, Frick J, et al. Mesenchymal stem cells home to injured tissues when co-infused with hematopoietic cells to treat a radiation-induced multi-organ failure syndrome. J Gene Med. 2003;5(12):1028–38.
48. Bárcia RN, Santos JM, Filipe M, Teixeira M, Martins JP, Almeida J, et al. What makes umbilical cord tissue-derived mesenchymal stromal cells superior Immunomodulators when compared to bone marrow derived mesenchymal stromal cells? Stem Cells Int. 2015;2015:583984.
49. Soleymaninejadian E, Pramanik K, Samadian E. Immunomodulatory properties of mesenchymal stem cells: cytokines and factors. Am J Reprod Immunol. 2012;67(1):1–8.
50. Aggarwal S, Pittenger MF. Human mesenchymal stem cells modulate allogeneic immune cell responses. Blood. 2005;105(4):1815–22.
51. Rafei M, Hsieh J, Fortier S, Li M, Yuan S, Birman E, et al. Mesenchymal stromal cell-derived CCL2 suppresses plasma cell immunoglobulin production via STAT3 inactivation and PAX5 induction. Blood. 2008;112(13):4991–8.
52. Wang L-T, Ting C-H, Yen M-L, Liu K-J, Sytwu H-K, Wu KK, et al. Human mesenchymal stem cells (MSCs) for treatment towards immune- and inflammation-mediated diseases: review of current clinical trials. J Biomed Sci. 2016;23(1):76.
53. Levy V, Lindon C, Zheng Y, Harfe BD, Morgan B. a. Epidermal stem cells arise from the hair follicle after wounding. FASEB J. 2007;21(7):1358–66.
54. Jahoda CA, Reynolds AJ. Hair follicle dermal sheath cells: unsung participants in wound healing. Lancet. 2001;358(9291):1445–8.
55. Rodriguez-Menocal L, Salgado M, Ford D, Van Badiavas E. Stimulation of skin and wound fibroblast migration by mesenchymal stem cells derived from normal donors and chronic wound patients. Stem Cells Transl Med. 2012;1(3):221–9.
56. Liu L, Yu Y, Hou Y, Chai J, Duan H, Chu W, et al. Human umbilical cord mesenchymal stem cells transplantation promotes cutaneous wound healing of severe burned rats. PLoS One. 2014;9(2):e88348.
57. Basiouny HS, Salama NM, Mohamed Z, Maadawi E, Farag EA. Effect of bone marrow derived mesenchymal stem cells on healing of induced full-thickness skin wounds in albino rat. Int J Stem Cells. 2013;6(1):12–25.
58. Luo G, Cheng W, He W, Wang X, Tan J, Fitzgerald M, et al. Promotion of cutaneous wound healing by local application of mesenchymal stem cells derived from human umbilical cord blood. Wound Repair Regen. 2010;18(5):506–13.
59. Kong P, Xie X, Li F, Liu Y, Lu Y. Placenta mesenchymal stem cell accelerates wound healing by enhancing angiogenesis in diabetic Goto-Kakizaki (GK) rats. Biochem Biophys Res Commun. 2013;438(2):410–9.

60. Hong SJ, Jia S-X, Xie P, Xu W, Leung KP, Mustoe TA, et al. Topically delivered adipose derived stem cells show an activated-fibroblast phenotype and enhance granulation tissue formation in skin wounds. PLoS One. 2013;8(1):e55640.
61. Shin L, Peterson D. a. Human mesenchymal stem cell grafts enhance normal and impaired wound healing by recruiting existing endogenous tissue stem/progenitor cells. Stem Cells Transl Med. 2013;2(1):33–42.
62. Yoon BS, Moon J-H, Jun EK, Kim J, Maeng I, Kim JS, et al. Secretory profiles and wound healing effects of human amniotic fluid–derived mesenchymal stem cells. Stem Cells Dev. 2010;19(6):887–902.
63. Arno AI, Amini-Nik S, Blit PH, Al-Shehab M, Belo C, Herer E, et al. Human Wharton's jelly mesenchymal stem cells promote skin wound healing through paracrine signaling. Stem Cell Res Ther. 2014;5(1):28.
64. Mishra PJ. Cell-free derivatives from mesenchymal stem cells are effective in wound therapy. World J Stem Cells. 2012;4(5):35.
65. Li M, Zhao Y, Hao H, Dai H, Han Q, Tong C, et al. Mesenchymal stem cell–conditioned medium improves the proliferation and migration of keratinocytes in a diabetes-like microenvironment. Int J Low Extrem Wounds. 2015;14(1):73–86.
66. Lai RC, Arslan F, Lee MM, Sze NSK, Choo A, Chen TS, et al. Exosome secreted by MSC reduces myocardial ischemia/reperfusion injury. Stem Cell Res. 2010;4(3):214–22.
67. Shi Y, Shi H, Nomi A, Lei-lei Z, Zhang B, Qian H. Mesenchymal stem cell–derived extracellular vesicles: a new impetus of promoting angiogenesis in tissue regeneration. Cytotherapy. 2019;21(5):497–508.
68. Wu P, Zhang B, Shi H, Qian H, Xu W. MSC-exosome: a novel cell-free therapy for cutaneous regeneration. Cytotherapy. 2018;20(3):291–301.
69. Li T, Yan Y, Wang B, Qian H, Zhang X, Shen L, et al. Exosomes derived from human umbilical cord mesenchymal stem cells alleviate liver fibrosis. Stem Cells Dev. 2013;22(6):845–54.
70. Marquez-Curtis LA, Gul-Uludag H, Xu P, Chen J, Janowska-Wieczorek A. CXCR4 transfection of cord blood mesenchymal stromal cells with the use of cationic liposome enhances their migration toward stromal cell-derived factor-1. Cytotherapy. 2013;15(7):840–9.
71. Zhang B, Wang M, Gong A, Zhang X, Wu X, Zhu Y, et al. HucMSC-exosome mediated-Wnt4 signaling is required for cutaneous wound healing. Stem Cells. 2015;33(7):2158–68.
72. Zhang J, Guan J, Niu X, Hu G, Guo S, Li Q, et al. Exosomes released from human induced pluripotent stem cells-derived MSCs facilitate cutaneous wound healing by promoting collagen synthesis and angiogenesis. J Transl Med. 2015;13(1):49.
73. Shabbir A, Cox A, Rodriguez-Menocal L, Salgado M, Van Badiavas E. Mesenchymal stem cell exosomes induce proliferation and migration of normal and chronic wound fibroblasts, and enhance angiogenesis in vitro. Stem Cells Dev. 2015;24(14):1635–47.
74. Kim S, Lee SK, Kim H, Kim TM. Exosomes secreted from induced pluripotent stem cell-derived mesenchymal stem cells accelerate skin cell proliferation. Int J Mol Sci. 2018;19(10):pii: E3119.
75. Hu L, Wang J, Zhou X, Xiong Z, Zhao J, Yu R, et al. Exosomes derived from human adipose mensenchymal stem cells accelerates cutaneous wound healing via optimizing the characteristics of fibroblasts. Sci Rep. 2016;6(1):32993.
76. Fang S, Xu C, Zhang Y, Xue C, Yang C, Bi H, et al. Umbilical cord-derived mesenchymal stem cell-derived Exosomal MicroRNAs suppress Myofibroblast differentiation by inhibiting the transforming growth factor-β/SMAD2 pathway during wound healing. Stem Cells Transl Med. 2016;5(10):1425–39.
77. Noronha Nc NDC, Mizukami A, Caliári-Oliveira C, Cominal JG, Rocha JLM, Covas DT, et al. Priming approaches to improve the efficacy of mesenchymal stromal cell-based therapies. Stem Cell Res Ther. 2019;10(1):1–21.
78. Chen L, Xu Y, Zhao J, Zhang Z, Yang R, Xie J, et al. Conditioned medium from hypoxic bone marrow-derived mesenchymal stem cells enhances wound healing in mice. PLoS One. 2014;9(4):e96161.
79. Jun E, Zhang Q, Yoon B, Moon J-H, Lee G, Park G, et al. Hypoxic conditioned medium from human amniotic fluid-derived mesenchymal stem cells accelerates skin wound healing through TGF-β/SMAD2 and PI3K/Akt pathways. Int J Mol Sci. 2014;15(1):605–28.

80. Liu G-Y, Liu Y, Lu Y, Qin Y-R, Di G-H, Lei Y-H, et al. Short-term memory of danger signals or environmental stimuli in mesenchymal stem cells: implications for therapeutic potential. Cell Mol Immunol. 2016;13(3):369–78.
81. Ti D, Hao H, Tong C, Liu J, Dong L, Zheng J, et al. LPS-preconditioned mesenchymal stromal cells modify macrophage polarization for resolution of chronic inflammation via exosome-shuttled let-7b. J Transl Med. 2015;13(1):308.
82. Anton D, Burckel H, Josset E, Noel G. Three-dimensional cell culture: a breakthrough in vivo. Int J Mol Sci. 2015;16(3):5517–27.
83. Hsu S, Hsieh P-S. Self-assembled adult adipose-derived stem cell spheroids combined with biomaterials promote wound healing in a rat skin repair model. Wound Repair Regen. 2015;23(1):57–64.
84. Kwon SH, Bhang SH, Jang H-K, Rhim T, Kim B-S. Conditioned medium of adipose-derived stromal cell culture in three-dimensional bioreactors for enhanced wound healing. J Surg Res. 2015;194(1):8–17.
85. Xia Z, Zhang C, Zeng Y, Wang T, Ai G. Transplantation of BMSCs expressing hVEGF 165/hBD3 promotes wound healing in rats with combined radiation-wound injury. Int Wound J. 2014;11(3):293–303.
86. Qi Y, Jiang D, Sindrilaru A, Stegemann A, Schatz S, Treiber N, et al. TSG-6 released from intradermally injected mesenchymal stem cells accelerates wound healing and reduces tissue fibrosis in murine full-thickness skin wounds. J Invest Dermatol. 2014;134(2):526–37.
87. Yang D, Sun S, Wang Z, Zhu P, Yang Z, Zhang B. Stromal cell-derived factor-1 receptor CXCR4-overexpressing bone marrow mesenchymal stem cells accelerate wound healing by migrating into skin injury areas. Cell Reprogram. 2013;15(3):206–15.
88. Li Y, Zheng L, Xu X, Song L, Li Y, Li W, et al. Mesenchymal stem cells modified with angiopoietin-1 gene promote wound healing. Stem Cell Res Ther. 2013;4(5):113.
89. Kortesidis A, Zannettino A, Isenmann S, Shi S, Lapidot T, Gronthos S. Stromal-derived factor-1 promotes the growth, survival, and development of human bone marrow stromal stem cells. Blood. 2005;105(10):3793–801.
90. Nakamura Y, Ishikawa H, Kawai K, Tabata Y, Suzuki S. Enhanced wound healing by topical administration of mesenchymal stem cells transfected with stromal cell-derived factor-1. Biomaterials. 2013;34(37):9393–400.
91. Yew T-L, Hung Y-T, Li H-Y, Chen H-W, Chen L-L, Tsai K-S, et al. Enhancement of wound healing by human multipotent stromal cell conditioned medium: the paracrine factors and p38 MAPK activation. Cell Transplant. 2011;20(5):693–706.
92. Shohara R, Yamamoto A, Takikawa S, Iwase A, Hibi H, Kikkawa F, et al. Mesenchymal stromal cells of human umbilical cord Wharton's jelly accelerate wound healing by paracrine mechanisms. Cytotherapy. 2012;14(10):1171–81.
93. Fong C-Y, Tam K, Cheyyatraivendran S, Gan S-U, Gauthaman K, Armugam A, et al. Human Wharton's jelly stem cells and its conditioned medium enhance healing of excisional and diabetic wounds. J Cell Biochem. 2014;115(2):290–302.
94. Zhang B, Wu X, Zhang X, Sun Y, Yan Y, Shi H, et al. Human umbilical cord mesenchymal stem cell exosomes enhance angiogenesis through the Wnt4/β-catenin pathway. Stem Cells Transl Med. 2015;4(5):513–22.
95. Wang X, Jiao Y, Pan Y, Zhang L, Gong H, Qi Y, et al. Fetal dermal mesenchymal stem cell-derived exosomes accelerate cutaneous wound healing by activating notch signaling. Stem Cells Int. 2019;2019:1–11.

Correction to: Cellular Reprogramming and Aging

Correction to:

G. Rodrigues, B. A. J. Roelen (eds.), *Concepts and Applications of Stem Cell Biology,* Learning Materials in Biosciences, https://doi.org/10.1007/978-3-030-43939-2_5

This chapter was inadvertently published with incorrect affiliation of second author Bruno Bernardes de Jesus which has been corrected as below:

The affiliation of Bruno Bernardes de Jesus has been changed to "Department of Medical Sciences and Institute of Biomedicine – iBiMED, University of Aveiro, Aveiro, Portugal".

The updated online version of this chapter can be found at
https://doi.org/10.1007/978-3-030-43939-2_5

© Springer Nature Switzerland AG 2020
G. Rodrigues, B. A. J. Roelen (eds.), *Concepts and Applications of Stem Cell Biology,*
Learning Materials in Biosciences, https://doi.org/10.1007/978-3-030-43939-2_14

MIX
Papier aus verantwortungsvollen Quellen
Paper from responsible sources
FSC® C105338

If you have any concerns about our products,
you can contact us on
ProductSafety@springernature.com

In case Publisher is established outside the EU,
the EU authorized representative is:
**Springer Nature Customer Service Center GmbH
Europaplatz 3, 69115 Heidelberg, Germany**

Printed by Libri Plureos GmbH
in Hamburg, Germany